T0339962

Maritime Transport and Regional Sustainability

Maritime Transport and Regional Sustainability

Edited by

Adolf K.Y. Ng
Professor, Transport and Supply Chain Management, University of
Manitoba, Winnipeg, MB, Canada

Jason Monios
Associate Professor, Maritime Logistics, Kedge Business School,
Marseille, France

Changmin Jiang
Associate Professor, Transport and Supply Chain Management,
University of Manitoba, Winnipeg, MB, Canada

ELSEVIER

Elsevier
Radarweg 29, PO Box 211, 1000 AE Amsterdam, Netherlands
The Boulevard, Langford Lane, Kidlington, Oxford OX5 1GB, United Kingdom
50 Hampshire Street, 5th Floor, Cambridge, MA 02139, United States

Notices
Knowledge and best practice in this field are constantly changing. As new research and
experience broaden our understanding, changes in research methods, professional practices, or
medical treatment may become necessary.

Practitioners and researchers must always rely on their own experience and knowledge in
evaluating and using any information, methods, compounds, or experiments described herein.
In using such information or methods they should be mindful of their own safety and the safety
of others, including parties for whom they have a professional responsibility.

To the fullest extent of the law, neither the Publisher nor the authors, contributors, or editors,
assume any liability for any injury and/or damage to persons or property as a matter of
products liability, negligence or otherwise, or from any use or operation of any methods,
products, instructions, or ideas contained in the material herein.

Library of Congress Cataloging-in-Publication Data
A catalog record for this book is available from the Library of Congress

British Library Cataloguing-in-Publication Data
A catalogue record for this book is available from the British Library

ISBN: 978-0-12-819134-7

For information on all Elsevier publications
visit our website at https://www.elsevier.com/books-and-journals

Publisher: Joe Hayton
Acquisition Editor: Brian Romer
Editorial Project Manager: Michelle Fisher
Production Project Manager: Selvaraj Raviraj
Cover Designer: Mark Rogers

Typeset by SPi Global, India

Working together
to grow libraries in
developing countries

www.elsevier.com • www.bookaid.org

Contents

Part I
Theoretical settings

1. Setting the scene on maritime transport and regional sustainability

Adolf K.Y. Ng, Jason Monios, Changmin Jiang

2. Environmental governance in shipping and ports: Sustainability and scale challenges

Jason Monios

Part II
Adapting to climate change impacts

3. Storm resilience and sustainability at the Port of Providence, USA

Richard Burroughs, Austin Becker

4. Insights from recent economic modeling on port adaptation to climate change effects

Laingo M. Randrianarisoa, Kun Wang, Anming Zhang

5. Sustainability cruising and its supply chain

Grace W.Y. Wang

6.   How does the UK transport system respond to the
      risks posed by climate change? An analysis from the
      perspective of adaptation planning

Tianni Wang, Zhuohua Qu, Zaili Yang, Adolf K.Y. Ng

Part III
Improving environmental practice

7.   Green port initiatives for a more sustainable port-city
      interaction: The case study of Barcelona

Marta Gonzalez-Aregall, Rickard Bergqvist

Part IV
Opening up the Arctic seas

9. **Navigational risk factor analysis of Arctic shipping
    in ice-covered waters**
    Mingyang Zhang, Di Zhang, Chi Zhang, Wei Cao

Part V
Other key issues

List of figures

List of tables

Contributors

Numbers in parentheses indicate the pages on which the authors' contributions begin.

Mawuli Afenyo (179,223), Department of Supply Chain Management, Asper Business School; Transport Institute, University of Manitoba, Winnipeg, MB, Canada

S.S. Appadoo (311), Department of Supply Chain Management, University of Manitoba, Winnipeg, MB, Canada

Austin Becker (33), Department of Marine Affairs, University of Rhode Island, Kingston, RI, United States

Rickard Bergqvist (109), Department of Business Administration, School of Business, Economics and Law, University of Gothenburg, Gothenburg, Sweden

Richard Burroughs (33), Department of Marine Affairs, University of Rhode Island, Kingston, RI; School of Forrestry and Environmental Studies, Yale University, New Haven, CT, United States

Wei Cao (153), Intelligent Transportation Systems Research Center; National Engineering Research Center for Water Transport Safety, Wuhan University of Technology, Wuhan, China

Jihong Chen (267), College of Transport and Communications, Shanghai Maritime University, Shanghai, China

Lidan Du (289), College of Transport and Communications, Shanghai Maritime University, Shanghai, China

Ying-En Ge (289), College of Transport and Communications, Shanghai Maritime University, Shanghai, China

Marta Gonzalez-Aregall (109), Department of Business Administration, School of Business, Economics and Law, University of Gothenburg, Gothenburg, Sweden

Changmin Jiang (3,251), Department of Supply Chain Management, Asper School of Business, University of Manitoba, Winnipeg, MB, Canada

Faisal Khan (179), Memorial University of Newfoundland, St. John's, NL, Canada

Amit Kumar (311), School of Mathematics, Thapar Institute of Engineering & Technology, Patiala, India

Yui-Yip Lau (239), Division of Business and Hospitality Management, College of Professional and Continuing Education, The Hong Kong Polytechnic University, Kowloon, Hong Kong

Paul Tae-Woo Lee (267), Ocean College, Zhejiang University, Zhoushan, China

Xiaoyu Li (251), Department of Supply Chain Management, Asper School of Business, University of Manitoba, Winnipeg, MB, Canada

Yufeng Lin (223), Transport Institute; Department of Supply Chain Management, Asper School of Business, University of Manitoba, Winnipeg, MB, Canada

Steven Messner (195), e360 LLC, Sonoma, CA, United States

Akansha Mishra (311), School of Mathematics, Thapar Institute of Engineering & Technology, Patiala, India

Jason Monios (3,13), Kedge Business School, Domaine de Luminy, Rue Antoine Bourdelle, Marseille, France

Adolf K.Y. Ng (3,85,179,209,223,251), Department of Supply Chain Management, Asper School of Business; St. John's College; Transport Institute, University of Manitoba, Winnipeg, MB, Canada

Zhuohua Qu (85), Liverpool Business School, Liverpool John Moores University, Liverpool, United Kingdom

Laingo M. Randrianarisoa (45), Sauder School of Business, The University of British Columbia, Vancouver, BC, Canada

Naima Saeed (209), Department of Working Life and Innovation, School of Business and Law, University of Agder, Grimstad, Norway

Xiaodong Sun (239), School of Business Administration, East China Normal University, Shanghai, China

Grace W.Y. Wang (73), Maritime Economist, District of Columbia, United States

Kun Wang (45), School of International Trade and Economics, University of International Business and Economics, Beijing, China

Tianni Wang (85), Liverpool Business School; Liverpool Logistics, Offshore and Marine Research Institute, Liverpool John Moores University, Liverpool, United Kingdom; Transport College, Shanghai Maritime University, Shanghai, China

Zhongyu Wang (289), College of Transport and Communications, Shanghai Maritime University, Shanghai, China

Gordon Wilmsmeier (133), School of Management, Universidad de los Andes, Bogota, Colombia; University of Applied Sciences Bremen, Bremen, Germany

Zaili Yang (85), Liverpool Logistics, Offshore and Marine Research Institute, Liverpool John Moores University, Liverpool, United Kingdom

Anming Zhang (45), Sauder School of Business, The University of British Columbia, Vancouver, BC, Canada

Chi Zhang (153), Intelligent Transportation Systems Research Center; National Engineering Research Center for Water Transport Safety, Wuhan University of Technology, Wuhan, China

Di Zhang (153), Intelligent Transportation Systems Research Center; National Engineering Research Center for Water Transport Safety, Wuhan University of Technology, Wuhan, China

Mingyang Zhang (153), Intelligent Transportation Systems Research Center; National Engineering Research Center for Water Transport Safety, Wuhan University of Technology, Wuhan, China; Aalto University, School of Engineering, Department of Mechanical Engineering, Maritime Technology, Espoo, Finland

Yong Zhou (289), College of Transport and Communications, Shanghai Maritime University, Shanghai, China

Foreword

We live in a deeply interconnected world. Over 80% of the volume of global merchandise trade is carried by sea, from port to port, and across the closely interlinked network of global supply chains. Maritime transport is vital to global trade and, being a derived demand, develops in line with global developments, trade, and production patterns. The drivers of these developments, as well as their implications for shipping and regional sustainability, are complex and include some major global challenges, including the challenge of increasing climate variability and change and the associated need for maritime transport infrastructure, and operations to adapt to the potentially important direct and indirect impacts, which include climate-related damage, delay, and disruption with economic and trade-related repercussions.

This book with a focus on Maritime Transport and Regional Sustainability is both timely and particularly pertinent, with learned contributions covering a wide array of topics, ranging from the important issue of climate change adaptation for seaports to environmental sustainability, governance, and management in shipping, and maritime transport through the Arctic—an area which holds economic promise, but also poses significant technical and environmental challenges. It also offers perspectives on some other key issues, including maritime and other transport-related developments in China, which are of considerable interest given their potential implications, including for global transport and trade. The book makes an important contribution to the understanding of the issues considered and should be read by all with an interest in the developments that are shaping the future of maritime transport.

Regina Asariotis
Chief, Policy and Legislation Section
TLB, Division on Technology and Logistics
United Nations Conference on Trade and Development (UNCTAD)

Part I

Theoretical settings

Chapter 1

Setting the scene on maritime transport and regional sustainability

Adolf K.Y. Ng[a,b], Jason Monios[c], Changmin Jiang[a]
aDepartment of Supply Chain Management, Asper School of Business, University of Manitoba, Winnipeg, MB, Canada, bSt. John's College, University of Manitoba, Winnipeg, MB, Canada, cKedge Business School, Domaine de Luminy, Rue Antoine Bourdelle, Marseille, France

1 Introduction

This book addresses the contemporary development and connections between maritime transport and regional sustainability, including its geography, planning, economics, management, policy and regulations, governance, and organizational behaviors. It introduces how the maritime sector tackles challenges posed by climate change so as to achieve regional sustainability in selected areas around the world, including the Arctic. Through undertaking deep analysis on different types of transport from diversified angles and perspectives, including case studies from both the developed world and the emerging economies, it enables readers to understand the major issues and challenges that planners, policymakers, managers, and researchers face, and enables them to develop the ability of applying theoretical knowledge into practice. Hence, the first major audience includes all researchers and postgraduate (both research and professional) students. It will inspire them to generate original, innovative ideas for further research in sustainable transport and regional systems. The second audience consists of senior policymakers and industrial practitioners. It will equip them with comprehensive knowledge and understanding regarding contemporary developments in maritime transport in the context of regional sustainability.

A key feature of this book is that it closely knits maritime transport with its regional surroundings. Although many previous works address the importance of sustainability in (maritime) transport, they mainly focus on how maritime transport, as an "operational system," can achieve well-defined sustainability (e.g., carbon emission targets). Few works directly address how maritime transport affects the well-being and functioning of surrounding regions. This leaves

Maritime Transport and Regional Sustainability. https://doi.org/10.1016/B978-0-12-819134-7.00001-0

a major gap where researchers, policymakers, and industrial practitioners, in many cases, only treat transport as a field of operational interest rather than considering the role played by (maritime) transport in a regional system. In fact, the impacts of transport on surrounding regions have become only more important in recent years, especially with a number of strong governmental incentives to invest and commit in ports and other maritime transport facilities. For example, the *Silk Road Economic Belt and the 21st Century Maritime Silk Road* (hereinafter referred to as the "Belt & Road Initiatives," or BRI), launched by the Chinese government in 2013, aims "to promote the connectivity of Asian, European, and African continents and their adjacent seas, establish and strengthen partnerships among the countries along the BRI, set up omni-dimensional, multitiered and composite connectivity networks, and realize diversified, independent, balanced, and sustainable development in these countries." Indeed, as things stand, the Chinese government is also seriously considering the possibility of applying some of its BRI philosophies and perspectives to the Arctic area (e.g., the Canadian Arctic) through the "Polar Silk Road" initiatives and, rather peculiarly, declares China as a "Near-Arctic State" (PRC Government, 2018). Although the outcomes of such an initiative remain to be seen, it is without doubt extremely important to study in more detail how maritime transport can contribute to the sustainability of regions.

Through close collaboration between reputed scholars, researchers, and practitioners from diversified backgrounds, this book identifies the commonalities that contribute to the coherent transport–regional relationship—how operations, planning, and management impact on sustainable regional governance and establish a strong theoretical framework for this purpose. This is going to pose substantial interest to researchers, policymakers, and practitioners involved in any aspect of the transport, environmental, and planning sectors (e.g., port authorities, airport managers, planners, etc.). Readers can understand the major trends related to the topic, develop effective approaches and skills so as to address diversified challenges in transport and sustainable regional planning, and for researchers, inspire new ideas on research in transport and regional studies and planning. Hence, we believe that this book offers a unique perspective. Through collating the most updated information and research outcomes, especially original research conducted by contributors, the volume serves as an efficient channel in facilitating the transference of the latest original knowledge to researchers and practitioners, bridging the gap between academic knowledge and professional practice. It is a quality companion and a solid platform for further research, planning, and the development of appropriate policies and effective practice.

Also, this book is a major output of the *International Workshop on Climate Change Adaptation Planning for Ports, Transport Infrastructures, and the Arctic* (CCAPPTIA conference), held in Winnipeg, Manitoba, Canada, on 3–4 May 2018 and the *Yangtze River Research and Innovation Belt* (Y-RIB), held in Zhoushan, China, on 2–5 December 2018. Bringing together leading experts and stakeholders from academia, government, industry, and interest groups, CCAPPTIA and Y-RIB aim to consolidate and coordinate global research and development activities that understand the current decision-making process in climate adaptation planning, identify attributes that catalyze collaboration between ports and other transport infrastructures, including those in the Arctic and countries/regions along BRI, in sustainable planning, identify ways that can facilitate the transfer of adaptation strategies and solutions to regions under different geographical and cultural contexts, and assess strategy and policy implementation under the context of climate adaptation planning. All the editors and many contributors attended and contributed to CCAPPTIA conference and/or Y-RIB. Such initiatives have directly contributed to the establishment of the *International Forum on Climate Change Adaptation Planning for Ports, Transport Infrastructures, and the Arctic* (ccapptia.com) in 2019.

The rest of the chapter is structured as follows: Section 2 briefly describes the contents of this book, while Section 3 provides some insight for further research and collaborations.

2 Description of different chapters

The 17 chapters in this volume are divided into five major subsections, namely "theoretical settings," "adapting to climate change impacts," "improving environmental performance," "opening up the Arctic seas," and "other key issues."

2.1 Theoretical settings (Chapter 2)

In Chapter 2, Monios reviews different institutions and players in the environmental governance of shipping and ports. From the International Maritime Organization (IMO) to national policy to local planning ordinances, a variety of regulations and planning regimes interact to govern this complex global sector. Yet the existence of overlapping jurisdictions and the reluctance to invest leads to certain challenges being under-addressed, ranging from carbon emissions at sea and in ports, to noise and congestion in port regions to climate change adaptation. By revealing the scale conflicts and the relative power plays in this multifaceted governance regime, this chapter identifies current institutional weaknesses where action is needed by policymakers. The conclusion is that, while maritime policy is polycentric with a strong global component, national policies for dealing with climate change adaptation can be expected to play an increasing role in the evolution of maritime transport and trade in the coming decades.

2.2 Adapting to climate change impacts (Chapters 3–6)

In Chapter 3, Burroughs and Becker study storm resilience and sustainability at the port of Providence, Rhode Island, United States. Sustainability trajectories provide a means to consider social capacity to guide interactions between nature, in this case hurricanes, and society. In Providence, port stakeholders viewed a visualization of the impacts of historical hurricane intensity on a slightly shifted path that resulted in an estimated water level of 6.4 m above NAVD 88. Port business representatives observed projected water elevation with respect to specific business facilities. Results from surveys and interviews of 15 businesses show that meeting, training, planning, and data back-up were common proactive measures; many had back-up generators; but few had flood and wind proofed buildings, elevated properties, or electrical systems. In Providence, port businesses have taken initial actions along the trajectory but lack the port-wide coordination of effective stakeholder actions that marks a robust sustainability trajectory.

In Chapter 4, Randrianarisoa, K. Wang, and Zhang review economic modeling works on port adaptation. They focus on summarizing the modeling approaches and the major factors considered in such works. Specifically, they compare how disaster uncertainty and port market structure have been incorporated in these models, in terms of similarities and differences. The findings are then reconciled and compared. There are several findings that are robust across the modeling frameworks, thus offering useful policy and managerial implications. Meanwhile, there are some distinct findings, driven largely by the specific model setups and assumptions. Possible explanations are provided so as to better understand such variations. Finally, based on the existing modeling work, it proposes several promising avenues for future modeling research.

In Chapter 5, G. Wang studies sustainability cruising and its supply chain. The cruise industry is made up of the interconnection of various stakeholders in the maritime space. Their interdependence displays the need for cooperation in order to increase industry adaptability and flexibility. The growth of the industry has shown the need for a reliable value chain within the cruise sector. It identifies sustainable practices in order to improve the efficiency and reliability of the cruise network. Under this domain, it identifies sustainable practices and vulnerabilities from different stakeholders within the cruise industry. Sustainability of cruise practices can be observed from the perspectives of cruise ports, cruise liner companies, and the cruise supply chain in a maritime cluster. These practices are essential for the long-run sustainability and development of the cruise industry. Finally, it offers a holistic viewpoint of industry sustainability from multiple stakeholders as well as an incentive reliability framework. It is designed to offer a potential focus for policy improvement.

In Chapter 6, T. Wang, Qu, Yang, and Ng study the adaptation experience of United Kingdom's road and rail systems in managing the risks posed by climate change. In particular, they explore the current and potential issues in climate adaptation planning through in-depth investigation of four cases in the United Kingdom. Although considerable adaptation measures and actions have been implemented at both the national and regional levels, the road and rail systems still confront diverse challenges. They include insufficient scientific data, aging infrastructure, unclear planning horizon, and unspecialized climate risk management. Through combining the analysis of the relevant literature, local reports, news, and interviews with domain transport experts, it offers a broad view of adaptation planning of roads and railways in the United Kingdom and useful insight in creating an integrated inland transport adaptation system. An analysis of road and rail adaptation measures to climate change does not only benefit the stated sectors by cross-reference but also generates new solutions in terms of using one system to enhance the resilience of the other when climate risks occur.

2.3   Improving environmental performance (Chapters 7 and 8)

In Chapter 7, Gonzalez-Aregall and Bergqvist examine port-based strategies that improve environmental performance and promote sustainable solutions for combating climate change in the context of large urban areas. Focusing on Barcelona, the investigation analyzes past and present hinterland initiatives that could successfully facilitate the city's and its port's growth and resilience. The geographically sensitive location of the port of Barcelona along the Mediterranean Sea results in high volume of freight and cruise passenger traffic. It presents an interesting case in addressing the sustainability challenges associated with freight transport paths and urban areas.

In Chapter 8, Wilmsmeier discusses the relevance of climate adaption and mitigation for the Colombian port system. It discusses the identified threats and general adaptation and mitigation needs in the national port climate change action plan. Colombia's climate action plan, or *Intended Nationally Determined Contribution* (INDC), includes the goal to reduce its greenhouse gas (GHG) emissions by 20% by 2030, as compared to a projected business-as-usual scenario. The INDC document stresses that climate action is fundamentally a developmental issue. Thus, innovative and strong development in the various sectors of the economy will support efforts to reach this goal. In addition, this chapter focuses on the mitigation efforts by presenting results for current baseline measures for implementing and monitoring mitigation solutions in the port sector.

2.4   Opening up the Arctic seas (Chapters 9–13)

In Chapter 9, M. Zhang, D. Zhang, C. Zhang, and Cao analyze the navigational risk factors of Arctic shipping. Typical navigational risks in the ice-covered waters in the Arctic area, such as a ship stuck in ice and collision accidents (e.g., ice-ship and ship-icebreaker collisions), are identified and estimated, which are defined as three typical accident scenarios in Arctic ice-covered waters. In this chapter, models of stuck ships and collision accidents are developed and analyzed to identify and classify the factors influencing these risks. It first identifies navigational risk factors using the HFACS-based ship collision risk analysis model, and the test mining approach based on accident reports and scientific literature. After then, the "fault tree" approach is used to analyze the fundamental risk factors that contribute to three typical accident scenarios. Finally, it presents the findings of a qualitative analysis carried out to analyze navigational risk factors, in the context of three typical accident scenarios. The navigational risk factors in different scenarios are proposed by comparative analysis. Finally, it concludes by providing theoretical guidance for ice navigation in risk control and safety management.

In Chapter 10, Afenyo, Khan, and Ng review risk assessment techniques which are both static and dynamic in nature. It explores the challenges to implementing these techniques in the Arctic scenario, especially for shipping. A potential scenario is used to illustrate how these tools are used. The tier IV fugacity approach is employed in combination with the Monte Carlo simulation technique to model the fate and transport of oil spill. This tracks where the oil goes after the spill. To address data limitation, variability, and uncertainty issue further, the Bayesian approach through influence diagrams and the Object-oriented Bayesian Network (OOBN) are used. The results illustrate that the tools available are capable of addressing the problems of oil spill risk assessment.

In Chapter 11, Messner considers the potential consequences from air emissions associated with the increased shipping activity in the arctic with a specific focus on black carbon (BC) particle emissions. Environmental impacts from increased shipping through the Arctic will occur from BC emissions from fuel oil combustion. In addition to impacts on local health from haze, research efforts indicate that increased BC emissions from fuel oil combustion from shipping could have a noticeable local warming effect—warming from the dual effects of BC solar radiation absorption and albedo reduction from deposition on snow and ice. In this case, IMO requirements to lower sulfur content in vessel fuels by 2020 should reduce health impacts relative to local haze. However, considerable uncertainty remains on how BC emissions will be reduced and thus on how to reduce the overall impacts of increased shipping on the climate in the Arctic area. It argues that future policy changes will need to consider include improved particulate controls, cleaner fuels, and improved engine technologies in Arctic-going vessels.

In Chapter 12, Saeed and Ng analyze the potential opportunities and challenges associated with the opening of the Arctic seas for Norway. Climate change in the Arctic is opening up access to sea routes of the Arctic region and this could boost trade and, consequently, may result in a sharp increase in shipping traffic in Norway. Specifically, northern Norway would become a focal point for port- and shipping-related activities. However, these opportunities can only be fully utilized if the stakeholders involved can manage the challenges of the opening of the Arctic seas. Like opportunities, challenges also cover a range of spheres, including environmental pollution, ecological damage, and geopolitical risks associated with new resources and trade opportunities. With the help of relevant documents, case studies, and in-depth interviews, it identifies opportunities and challenges related to the opening of the Arctic seas and discusses measures that could be taken to minimize the risks.

In Chapter 13, Lin, Ng, and Afenyo discuss what Arctic shipping could mean for communities in the Arctic area. The opportunities that climate change presents for shipping in the Arctic are well-documented. However, the "landside" of Arctic shipping is often overlooked. Their analysis is supported by an in-depth case study of the town of Churchill located in northern Manitoba, Canada. It focuses on the damaged railway incident (that took place in 2017) that disrupted Churchill's vital links between the northern communities and other parts of Canada. It highlights the need for infrastructural investments in these regions so as to open up the Canadian Arctic area and get ready for the opportunities that Arctic shipping can potentially bring to northern Canadian. Finally, it generates useful information on how transport disruption influences the social sustainability of remote communities and thus affects intermodal transport development in the Arctic.

2.5 Other key issues (Chapters 14–18)

In Chapter 14, Lau and Sun investigate the responsibility of cruise tourism in China. Traditionally, the cruise market has been dominated by the North American and European regions. Since the 1990s, cruise passengers have increasingly begun to search unexpected cultures, attractive cruising destinations, exotic experience, and interesting shore excursions. In this case, China is becoming an attractive, emerging market. However, existing research mainly focuses on how cruises bring positive economic impacts to regions and port cities, while the negative impacts remain largely overlooked, including China. Understanding such a framework of responsible cruise tourism, including nonprofit organizations, governments, cruise passengers, local communities, and cruise liners, is proposed. It strives to increase the awareness of researchers, policymakers, and practitioners to the responsibility and sustainability of cruise tourism in the long term.

In Chapter 15, Jiang, Ng, and Li conduct an in-depth case study on the development of high-speed rail (HSR) in China. There have been considerable efforts by many governments to engage remote or peripheral regions within national or continental networks. Indeed, they often play hugely significant, and sometimes decisive, roles in the evolution and development of transport and regional systems. However, hitherto, the impacts of such initiatives on transport and regional systems are under-researched. The chapter first reviews the Chinese HSR's network development, where it analyzes the impacts of government initiatives on transport infrastructure and regional accessibility. Next, it investigates the Chinese HSR by projecting its future development and discusses its potential impacts on spatial distribution of employment and population, intermodal network, HSR freight, and long-distance HSR services. At the national level, they argue that government initiatives may pose positive impacts on transport and regional systems. However, at the international level, such impacts may be more limited.

In Chapter 16, Lee and Chen look at the "connect" or "be connected" strategy in the context of the BRI with a focus on the Korean peninsula. There are three ways to connect Korea to Europe, i.e., existing south-west bound maritime routes, railway by Trans-China Railway and Trans-Siberian Railway, and the "Polar Silk Road." Although China has not explicitly mentioned Korea in its BRI documents, recently, the Chinese government has been persuading the Korean government to join it. In particular, as dialogues between North and South Koreas have been progressing, despite its uncertainty, the Korean Peninsula is likely to be connected to the Eurasian train system. Having considered such circumstance, it is worthwhile to investigate the impacts of the BRI in connecting Korea to Europe. It aims to overview China's BRI and overseas port development strategy and explore some key points to make connectivity between the Korean peninsula and Europe efficient in the context of the BRI, drawing its implications for logistics providers and transport carriers.

In Chapter 17, Ge, Du, Z. Wang, and Zhou compare the competitiveness of existing and potential corridors for freight transport between the Indian Ocean region and China in the BRI context. A multiobjective programming model is proposed to allocate cargoes to each corridor in which four objectives, such as minimum transport cost, minimum energy consumption, minimum greenhouse emissions and highest safety, are balanced. The proposed model gives the optimal cargo volume allocation over the four corridors of interest, which has been shown numerically to vary as the priority of each objective varies. Subsequently, the rate of the return on investments in the three potential corridors is analyzed.

In Chapter 18, Mishra, Kumar, and Appadoo propose the "Mehar method" for solving unbalanced generalized interval-valued trapezoidal fuzzy number transportation problems. Hitherto, there is an existing method that transforms an unbalanced generalized interval-valued trapezoidal fuzzy number transportation problem (IVTrFNTP) into a balanced generalized IVTrFNTP and the methods in solving the latter problem. However, the authors show that, when applying the existing method in transforming an unbalanced generalized IVTrFNTP into a balanced generalized IVTrFNTP, the obtained dummy supply and/or dummy demand is not a generalized interval-valued trapezoidal fuzzy number (IVTrFN), and thus it is not valid. Understanding such, a new method (the "Mehar method") is proposed in this chapter so as to transform an unbalanced generalized IVTrFNTP into a balanced generalized IVTrFNTP. The validity of this new method is also discussed.

Acknowledgments

The publication of this book, the organization of the CCAPPTIA conference, and the establishment of CCAPPTIA (ccapptia.com) are made possible through the generous support from the Social Science and Humanities Research Council of Canada (SSHRC)'s Insight Grant Program (Nos. 435-2017-0735 and 435-2017-0728) and the "GENICE" project (genice.ca) funded by Genome Canada and Research Manitoba. Also, we would like to thank the publisher (Elsevier) and its "China Transportation Series" for its useful advice and guidance.

3 Looking to the future

"Maritime transport and regional sustainability" is a huge topic and, admittedly, it is not possible for one book or research project to cover it fully. Having said so, we believe that the contents of this book give a general, comprehensive coverage of the topic and offer a good platform for future research. Following on from the topics already covered in this volume, we can identify several critical issues that require further investigation:

- How should climate change adaptation and resilience be implemented?
- Should climate change adaptation and resilience be quantified? If so, how should this be done?
- There is a clear imbalance of research between different aspects of Arctic shipping, in terms of both geographical/regional (e.g., Russia vs. Canada) and sectoral (e.g., marine side vs. landside). What should be done to address such gaps?
- What are the impacts of rapid infrastructural development on regional societies and economies, such as the BRI initiatives, the Greater Bay Area in southern China, and the isolated Arctic areas? How can relevant socioeconomic factors be studied?

Each of these questions involves many issues regarding methodological approach and data availability, as well as political, economic, and regulatory issues regarding implementation of possible solutions. As the editors, we hope (and believe) that this volume will act as the "gateway" in opening the door for more valuable research on these related topics.

Reference

PRC Government, 2018. China's Arctic Policy. . White Paper Published by the State Council Information Office of the People's Republic of China (PRC).

Chapter 2

Environmental governance in shipping and ports: Sustainability and scale challenges

Jason Monios
Kedge Business School, Domaine de Luminy, Rue Antoine Bourdelle, Marseille, France

1 Introduction—Environmental challenges in shipping and ports

Despite the fanfare surrounding the Kyoto Protocol and the Paris Agreement, neither introduced legally binding emissions targets for shipping. A large body of research has developed in recent years, analyzing and quantifying emissions from the maritime sector (Shi et al., 2018), which it is hoped will eventually form a baseline for future reduction targets. These emissions can be divided broadly into greenhouse gas (GHG) emissions affecting climate change and local air pollution, primarily sulfur oxides (SOx), nitrogen oxides (NOx), and particulate matter (PM). In 2007–12 shipping accounted for 2.8% of global GHG emissions or double the level produced by air travel (IMO, 2014). Local pollutants are a more pressing issue in coastal areas due to their impact on human health. According to the World Health Organization (WHO) air pollution (in total, not just from shipping) results in four million deaths per year (WHO, 2018). Shipping contributes a significant amount to this total, especially in coastal areas. Shipping accounts for approximately 15% of NOx and 5%–8% of SOx emissions globally (Zis et al., 2016). Brandt et al. (2011) found that emissions from shipping caused about 50,000 premature deaths in Europe alone in 2000.

While emissions tends to be the main environmental issue discussed, there are numerous other environmental challenges at sea, including accidents, oil spills, and water pollution from ballast water. EMSA (2016) reports on figures for EU-flagged vessels and/or within EU waters, revealing that in one year alone there were 3296 incidents involving 3669 ships, including 36 lost ships and 115 fatalities. Sixty-two percent of these incidents were attributed to human error and 278 resulted in pollution to the water through release of bunker fuel and other residual oils and lubricants. Ballast water is another important topic that has taken decades to address. Microorganisms can be transported across the globe in ballast water and devastate local species as a result of the ballast water discharge. Various annual cost levels have been estimated, such as $14.2 billion in the United States and €1.2 billion in Europe (David and Gollasch, 2015). At the port level, environmental problems include noise, dust, waste, and water pollution (Bergqvist and Monios, 2019). In addition, one of the biggest challenges currently facing policymakers is how to deal with the growth in Arctic shipping (Fedi, 2019).

While addressing the challenge of climate change continues to gain traction, and actions are being taken by industry and policymakers (see next section), the reality is that it is not possible to decarbonize maritime transport using current technology (Bows-Larkin, 2015; Psaraftis, 2019). Small design enhancements (partly as a result of IMO policies—see next section) continue to optimize operations and thus reduce fuel usage, and the only truly successful strategy thus far has been slow steaming. A recent study by Cariou et al. (2019) found that the shipping sector has achieved CO_2 reductions of 33% since 2007, mostly a result of slow steaming and to a lesser extent a reduction in distance traveled due to changes in network design. The only realistic large-scale fuel is LNG which would only save around 20% carbon emissions, but in fact the world fleet using LNG is not much over 100 vessels, mostly LNG carriers, ferries, and service vessels (Corkhill, 2018). Despite a few high-profile cases (e.g., CMA CGM ordering nine vessels in 2017) there is as yet little evidence of widespread transition to LNG. In theory, if the entire world fleet combined the existing possibilities of ultra-slow steaming, switching to LNG, network optimization and vessel design, along with some small additional energy provision from batteries, wind, and solar, then the sector could perhaps go some way toward meeting the IMO's target of 50% reduction by 2050 (Bows-Larkin, 2015). Even studies that propose hydrogen and ammonia as serious fuels recognize that they are decades away from commercial availability (OECD, 2018). At the same time, according to the IPCC, emissions must be reduced to zero by 2050. This target is not possible without drastically reducing the amount of shipping. Are we prepared for that?

2 Environmental governance—The IMO

2.1 Role of IMO and main policies

The International Maritime Organization (IMO) is the maritime branch of the United Nations, formally established in 1948 and operational since 1958. Its role is the "responsibility for the safety and security of shipping and the prevention of marine pollution by ships" (IMO, 2019). An IMO convention is usually considered in force once it has been ratified by two-thirds of member states but it does not apply to countries that have not ratified it, and enforcement is reliant on the individual member states (according to whether the directive applies to flag or port states—see below), rather than the IMO itself, resulting in different levels of enforcement.

Two of the most famous IMO conventions are the International Convention for the Safety of Life at Sea (SOLAS), the current version agreed in 1974 and coming into force in 1980 (although there had been earlier versions, dating as far back as 1914), and the International Convention for the Prevention of Pollution from Ships (MARPOL). The latter is the convention regarding environmental management which is the focus of this chapter. The difference in dates represents an expanding focus of the IMO from safety and navigation (although still important as witnessed in recent regulations regarding VGM and the Polar Code—cf. Fedi et al., 2010; Fedi, 2019) to the environment. MARPOL was originally adopted in 1973 but did not enter into force until 1983 due to challenges with ratification, and there have been several updates since the original. Responsibility for discussing and evaluating new regulations is delegated to the Marine Environment Protection Committee (MEPC), which meets every 6–9 months and is currently up to its 73rd session. There are now six annexes that have entered into force during 1983–2005. The first five cover pollution and waste in the sea, but the sixth, which has received much attention in recent years, concerns air pollution.

MARPOL Annex VI was adopted in 1997 and came into force in 2005. Its main focus was on reducing air pollution by limiting NOx, SOx, and PM emissions. Annex VI has itself been subject to revisions over the years, and an amendment came into force in 2010 establishing emission control areas (ECAs). The ECAs are located in the North Sea, the Baltic Sea, North America, and the United States Caribbean area. ECAs are often referred to as SECAs because of their prominent sulfur limit of 0.1% as of 2015 (Cullinane and Bergqvist, 2014), but North American ECAs also include NOx restrictions. In addition, the amendment set a reduced global cap of sulfur levels from 3.5% to 0.5% by 2020. Carriers meet this restriction by either switching to low sulfur fuel while sailing through an ECA or installing scrubbers on the exhaust system. One concern raised in the industry at the time was whether sufficient low sulfur fuel could actually be produced if regulations require increased use (Notteboom et al., 2010; Cullinane and Bergqvist, 2014). What has in fact occurred is that the scrubber option has proved very popular. The resulting realization of the popularity of open-loop scrubbers which discharge the used water into the sea (as opposed to closed-loop scrubbers which retain most of the water used to clean the exhaust for later disposal) has led to concerns regarding whether this method should continue to be allowed, due to concerns that the discharged water, although treated, still contains some pollutants. Moreover, reliance on scrubbers both decreases energy efficiency leading to increased GHG emissions and delays much-needed longer term action on changing fuels (Lindstad and Eskeland, 2016). The revised Annex VI also imposed tighter NOx restrictions on ship engines, depending on when they were built by separating vessels into three categories: Tier I (2000), Tier II (2011), and Tier III (2016).

Despite the large contribution to global GHG emissions from shipping, there remain no restrictions on CO_2 emissions. One way the IMO has contributed to GHG reductions is by targeting ship efficiency, which would obviously reduce fuel use and hence emissions of all types. An amendment to MARPOL in 2011 introduced the Energy Efficiency Design Index (EEDI) and Ship Energy Efficiency Management Plan (SEEMP) which require that certain new ships must adhere to the EEDI and all ships to the SEEMP (Lister et al., 2015).

The IMO has been active in the resolution of various other environmental problems in shipping. For example, the ecosystem damage from ballast water discharge had already been known for decades before action was taken. The first action was in the original MARPOL 1973 convention to initiate studies, after which the MEPC worked to develop a suitable policy during the 1990s. It was not until 2004 that a convention was finally adopted, the International Convention for the Control and Management of Ship's Ballast Water and Sediments, which nevertheless did not achieve enough signatories to be ratified until 2016 and thus came into force in September 2017 (David and Gollasch, 2015). The fact that the convention did not apply until 13 years after its adoption and 44 years since the decision to study the problem reflects the challenges of global environmental governance.

2.2 Market-based mechanisms

The IMO has also been exploring the potential of market-based mechanisms (MBMs) instead of regulations to reduce CO_2 emissions such as emission trading schemes (Franc and Sutto, 2013) and bunker fuel levies (Kosmas and Acciaro, 2017). These have the advantage of being the same for all carriers, thus providing more certainty to the market and being less likely to distort competition. A total of 11 specific measures have been considered by the IMO, originally proposed by member states from 2008, ranging from supportive funds and incentives schemes to several variants of an emission trading scheme. The analysis was contentious, and resulted in no decision, not even a reduction in the number of measures to consider in future (Psaraftis, 2019). The MBM discussion in the IMO was effectively abandoned in 2013. Meanwhile, the EU has stated that if the IMO does not adopt MBMs by 2021, then the EU will include shipping within the EU into the existing EU emissions trading scheme.

Both cap-and-trade and carbon tax have their proponents and detractors, and detailed analysis lies beyond the scope of this chapter. A fixed tax gives certainty on the price, but how much carbon will be saved remains unknown, and it is impossible to set a price in advance to achieve a set reduction. It is possible that the price will simply be absorbed and no significant impact on behavior will result, or at least not enough to meet the proposed reduction targets. The other problem is that policymakers will never set a sufficiently high price (Bows-Larkin, 2015; Krugman, 2018) for fear of alienating industry lobbyists. According to Krugman (2018): "claims that a carbon tax high enough to make a meaningful difference would attract significant bipartisan support are a fantasy at best, a fossil-fuel-industry ploy to avoid major action at worst." Psaraftis (2019) points out that a carbon tax high enough to make significant difference would need to be in the level of three figures per ton of oil. Alternatively, cap-and-trade theoretically sets the maximum carbon emissions, and then the price will be fixed by the market. A downside of both MBMs is that they do not incentivize major structural change but rather small optimizations that merely delay the required structural transition away from fossil fuels (Lohmann, 2008).

2.3   IMO targets

The lack of GHG targets by the IMO had been criticized, and in fact had already led to the EU creating their own GHG documenting requirement in 2015 (see later section). In response, the IMO introduced an amendment to MARPOL Annex VI requiring vessels above 5000 GT to record their fuel oil consumption. Fedi (2017) points out that the IMO regulation is less comprehensive than the EU regulation, partly because it uses proxies for transport work rather than including details on the cargo carried by each ship.

In April 2018, the IMO announced a commitment of the shipping sector to reduce emissions by 50% in comparison to 2008 levels by 2050. Establishing a target for the first time is certainly a positive step, but the lack of mandatory actions and the long timeframe allows the possibility of delay and further modification of the deadline, and the fact that the United States was not in agreement may limit compliance with the target. Therefore, without clear and strong global regulations, meeting this ambitious target remains difficult. Moreover, national and international CO_2 reduction targets are based on 1990 levels not 2008 (in 1990 GHG emissions from international shipping were 70% lower than 2012—DGIPPA, 2015), and in fact the IPCC target to restrict the world to 1.5 degrees of warming is carbon neutral by 2050, not 50% reduction. The longer term goal of the IMO is to "consider decarbonization in the second half of the century," which is more realistic as regards technology, as hydrogen for instance may become commercially viable by then, but that will be too late to meet global carbon targets.

To achieve their 50% goal, the short-term measures recommended by the IMO are primarily related to increasing optimization of vessel design and operations, including slow steaming. Alternative fuels are not considered realistic until the medium term (defined as applying after 2030). Interestingly, MBMs are also only considered by the IMO as a medium-term measure. What remains curious is that, despite many academic, government, IMO, and industry articles on possible reductions, there is remarkably little discussion about the fact that without targets and policies, no reductions beyond small optimizations are likely to be forthcoming. Given the lack of mandated actions in the IMO roadmap, the weaknesses of EEDI, and the fact that the MBM discussed is closed, Psaraftis (2019) is one of the few commentators to state the blunt conclusion that significant GHG reductions from shipping are "only a wish at this point in time."

3   Environmental governance—Other organizations

3.1   The national level

Shipping is mostly governed at two levels—national, as all vessels must be registered/flagged in a particular country and international, in terms of certain rules regarding safety (e.g., IMO SOLAS convention) and environmental performance (e.g., IMO MARPOL convention). These two levels frequently interact. The United Nations Convention on the Law of the Sea (UNCLOS) stipulates three kinds of states: flag states, port states, and coastal states. The flag state is where ships are registered, port states relate to the right of the state to institute inspections on vessels in their ports, and coastal states relates to the jurisdiction of vessels sailing within the territorial waters of a country.

Flags of convenience (FOCs) are a well-known issue where many carriers changed their country of registration in order to save tax, reduce labor costs due to lower labor regulations in the flag states, and benefit from the fact that these states lacked resources (and inclination) to enforce regulatory compliance for IMO environmental and safety regulations (Roe, 2013; Alderton and Winchester, 2002). Flag determines the pay and conditions (thus saving money for operators) but the actual quality of seafarers and vessels (officially at least) must meet IMO/ILO and classification society standards, respectively. Accidents arise not because of the lack of quality of the crew but due to overwork, lower quality working conditions, and lack of time and resources to follow IMO regulations as a result of demands of the employers. Roe (2013, p. 168) decries the "territorial hypocrisy," whereby shipping lines "trade off policies at national and global level to achieve the best of both worlds (and the worst for the environment, safety, security, and competition)."

One policy response to this lack of adherence to IMO regulations has been port state control (PSC), which enables port states to inspect vessels and apply penalties. An accident and major oil spill by the *Amoco Cadiz* in 1978 led to an organization of countries to improve such inspections, resulting in the Paris Memorandum of Understanding in 1982 signed by 15 EU states (later expanded to 27 members, including Canada). Other similar MoUs followed around the world. These groups work together via a shared database, with the aim to inspect 25% of all incoming vessels. A risk profile is used to identify ships to inspect, based on past inspections and other vessel and flag information (Li and Zheng, 2008). In 1995 the EU passed directive 95/12/EC on PSC (with later amendments) which made PSC mandatory for EU states. Initially the IMO had concerns about the EU-level directive but later accepted it (Blanco-Bazán, 2004; Van Leeuwen and Kern, 2013). Van Leeuwen and Kern (2013) make an interesting point regarding how EU states may prefer, individually or collectively, to institute more stringent rules as flag states, but they know that owners will simply flag out, but as port states, they can set their own rules and ensure that ships entering their ports meet international standards. So they can take steps to ensure ships sailing in their waters are safe, even if they cannot do it for their flagged fleet.

Flagging out is of concern for countries with a large amount of maritime trade, not just for environmental and work conditions, but for reasons of national security. Ships flying the national flag may be requisitioned in times of war, and the country may want to maintain a certain level of skills in the country's citizens, which is relevant not just while sailing but in port clusters and the wider sector. Thus different policies have been attempted by various countries to retain registrations on their national flag. An interesting example of multi-level governance is the modification of EU state aid regulations in order to allow member states to operate a tonnage tax, which is effectively a subsidy, replacing corporation tax for carriers with a (lower) tax on tonnage. Each country has its own version, with various requirements on, for example, training (Roe, 2013). A different approach and indeed a long-standing policy (a reaction to the lack of national shipping capacity during WW1) is the 1920 US Jones Act, which requires that all transport between US ports be carried by US-flagged vessels, which must be crewed by US citizens, whereas the UK tonnage act requires vessels to train UK citizens; it does not require them to employ them, knowing that this would be too costly and the ships would simply flag out (Oyaro Gekara, 2010). The US Act prevents that because if they flag out then they would not be permitted to work on those routes. The Act is widely unpopular, although that is partly for the additional requirement that ships also be US built, which is costly and difficult in this era due to lack of shipbuilding capacity.

3.2 The port and city level

Ports are governed usually between the city and the national level. The main governance is usually at local level via the port authority and its interaction with the city and local planning regulations and approvals, but there is usually also a national port policy which may be quite general and "hands off" (e.g., the United Kingdom) or may be more prescriptive (Monios, 2017). Planning also has a national component, as port developments are usually large enough to require national approval. This approval may be based on various criteria, and these days there is much more attention to the environmental aspect. For instance, a port development at Southampton in the United Kingdom was refused on the basis of environmental issues.

In terms of the daily operations of ports, there are many other areas of environmental management to consider that fall within the purview of the port authority in conjunction with local and national regulations. Furthermore, it is in fact local pressure that is starting to encourage ports to deal with local environmental problems. While the focus is often on air pollution, other issues include noise, dust, waste and water pollution (Acciaro et al., 2014; Lam and Notteboom, 2014; Bergqvist and Monios, 2019). These issues occur within the port (both land and water) as well as the hinterland. Considering how much port traffic has its origins and destinations far inland, cities must question to what degree the positive impacts of the port (jobs, taxes, direct revenue if it is city-owned) outweigh the negatives (congestion, pollution, sometimes only small number of jobs provided, often needing financial support for infrastructure) (Monios et al., 2018).

Key actions taken by ports include cold ironing to reduce emissions while at berth (Winkel et al., 2016; Innes and Monios, 2018), using electricity to power handling equipment (Spengler and Wilmsmeier, 2019), requiring slow steaming or use of LNG while in the port area (Winnes et al., 2015) and incentivizing of rail and barge hinterland transport rather than road (Bergqvist et al., 2015; Gonzalez-Aregall et al., 2018). These actions are mostly voluntary rather than legislated, although to some degree they are pushed by the pressure of citizen and environmental groups. Cold ironing provides a good example of multiple levels of governance working on an issue. While the EU incentivizes adoption by all major EU ports, it is not mandated, so it is usually a decision for the port authority. Some countries, such as Sweden, offer subsidies to ports for installation costs as well as subsidizing the cost of electricity to the vessels. In the United States, the state of California has introduced its own legislation that cold ironing must provide a portion of vessel power, starting from 50% in 2014 and increasing to 80% by 2020. As well as aiming to reduce emissions for both local health and to avoid climate change, ports are already beginning to face the challenge of climate change adaptation by protecting their infrastructure and operations from sea-level rise and storm surges (Ng et al., 2016).

3.3 Supranational organizations—The EU

The European Union (EU) is not eligible to become a member of the IMO but it has observer status, and has been developing its own shipping policies since 1993. According to Van Leeuwen and Kern (2013), this was not so much because it was unsatisfied with IMO policy but due to a lack of enforcement (as discussed above with flag states and PSC). After some high-profile oil spills (*Erika* in 1999 and *Prestige* in 2002), the EU brought in three sets of laws (known as the Erika packages) in 2000–05, pushing new standards on double hulls, PSC, flag states, and also creating the European Maritime Safety Agency (EMSA) (Urrutia, 2006; Van Leeuwen and Kern, 2013; Fedi, 2019).

Roe (2007, 2013) makes some interesting points regarding the EU's desire to represent its member states in the IMO, which was resisted by both the IMO and the states themselves. On the other hand, Van Leeuwen and Kern (2013) show how the EU has been effective in pushing the IMO for more stringent environmental standards and shorter timescales for implementation, particularly on double hulls, PSC, and SOx. This was done not just by lobbying the IMO but simply by pressing ahead with its own regulations covering EU jurisdictions. In 2005 Directive 2005/33/EC defined ECAs in the English Channel, North Sea, and the Baltic Sea, which set a sulfur cap of 1.5%, in addition to which ships at anchorage or in an EU port were required to use fuel with a maximum of 0.1% sulfur. These regulations were therefore much tighter than the MARPOL Annex VI regulations at the time. Directive 2012/33/EU established additional restrictions on sulfur content of fuels in line with the revised MARPOL Annex VI and discussed the possibility of extending ECAs.

The EU also implemented directives incentivizing cold ironing and LNG in EU ports. Directive 2014/94/EU on the deployment of Alternative Fuel Infrastructures states that "Member States shall ensure that the need for shore-side electricity supply for inland waterway vessels and sea-going ships in maritime and inland ports is assessed in their national policy frameworks. Such shore-side electricity supply shall be installed as a priority in ports of the TEN-T Core Network, and in other ports, by 31 December 2025, unless there is no demand and the costs are disproportionate to the benefits, including environmental benefits" (European Commission, 2014). The directive also says that member states "shall ensure, through their national policy frameworks, that an appropriate number of refueling points for LNG are put in place at maritime ports to enable LNG inland waterway vessels or sea-going ships to circulate throughout the TEN-T Core Network by 31 December 2025 at the latest." These directives mandate a response but not necessarily direct action, thus responsibility for such action lies with individual member states.

The EU's MRV (monitoring, reporting, and verification) regulation entered into force in 2015, requiring, as of January 2018, compulsory monitoring of CO_2 emitted by vessels larger than 5,000 GT calling at EU ports. This is the first step toward potentially setting targets and then applying MBMs, but there are no limits or actions as yet. Fedi (2017) highlights the quality of this regulation compared to the IMO version, although also noting some limitations: being only regional, possibly temporary, not including other pollutants (NOx, SOx, and PM) and missing the opportunity to incentivize cold ironing.

3.4   Industry and voluntary organizations

Various national and international port organizations exist that allow ports to share best practice and work toward more sustainable operations. In Europe, the European Sea Ports Organisation (ESPO) promotes environmental management in European ports. The current top ten environmental priorities of ESPO ports are air quality, energy consumption, noise, relationship with local community, garbage/port waste, ship waste, port development, water quality, dust, and dredging operations. In the Americas, the American Association of Port Authorities (AAPA), with 150 members in North, Central, and South America, has developed a guide for environmental management, the Environmental Management Handbook (EMH).

In 2008 the International Association of Ports and Harbors (IAPH) produced the C40 World Ports Climate Declaration, establishing the World Port Climate Initiative (WPCI). This group includes 55 ports around the world that pursue various green measures such as giving discounts to vessels scoring above a certain threshold on the Environmental Ship Index (ESI). This initiative has since been expanded with the launch in 2018 of the World Ports Sustainability Programme (WPSP). This is a joint initiative by the IAPH, the AAPA, the ESPO, The Worldwide Network of Port Cities (AIVP), and the World Association for Waterborne Transport Infrastructure (PIANC). The program's aims are linked to the 17 sustainable development goals set by the United Nations, under five key themes: resilient infrastructure, climate and energy, community outreach and port-city dialog, safety and security, and governance and ethics.

Similar voluntary industry groups exist in shipping, such as the Clean Cargo Working Group established in 2003, which developed an industry standard for measuring and reporting CO_2 emissions by carriers. The analysis includes data from more than 20 carriers, accounting for approximately 85% of global TEU capacity. Within the initiative there exist several working groups on topics such as green logistics and the Eco Stars fleet recognition scheme.

While such initiatives that focus attention on more environmental practices at sea and in ports through sharing of best practice and commitments to emission reduction are welcome, these schemes are voluntary, and therefore progress on significant emission reductions remains slow. While GHG reductions at sea have been notable (mostly through slow steaming to save fuel costs, cf. Cariou et al., 2019), GHG reductions from voluntary port schemes remain low (Sköld, 2019) and it will be a long time before they are both sufficiently stringent and widely adopted. This is understandable, given the commercial nature of shipping, but that is why regulators and policymakers must be prepared, not merely to nudge and incentivize, but to take more concrete action.

Moreover, it is essential that voluntary industry initiatives are not used as greenwash. Lister et al. (2015) showed that ship owners have lobbied actively against IMO regulations such as ECAs, and Sköld (2019) discusses how a study found evidence that shipping industry organizations such as the International Chamber of Shipping (ICS), the World Shipping Council (WSC), and the Baltic and International Maritime Council (BIMCO) have actively obstructed the development of climate change policies by the IMO (InfluenceMap, 2017). A recent study found that "the five largest publicly-traded oil and gas majors (ExxonMobil, Royal Dutch Shell, Chevron, BP, and Total) have invested over $1 billion of shareholder funds in the three years following the Paris Agreement on misleading climate-related branding and lobbying" (InfluenceMap, 2019). Where profit and planet align (e.g., reducing fuel use through slow steaming), there is a win-win situation. But the high costs associated with a major transition toward low carbon shipping (e.g., transitioning away from fossil fuels entirely) will preclude any voluntary action by industry.

4 The role of scale and institutional inertia

Roe (2013, p. 168) is highly critical of what he terms the governance failure in the maritime sector, "evidenced by the inadequacies of shipping or ports policy to address the problems of environmental, security, safety and economic concerns." Other authors are not particularly critical of the IMO, although they recognize the difficulties of obtaining policy agreement among all the many countries and other organizations. Both Roe (2009) and Van Leeuwen (2015) characterize shipping governance as polycentric, having multiple centers of power and decision-making. Roe argues that the current system remains overly rigid regarding the IMO and nation states, producing gaps in which multinational carriers can evade regulation; thus a more fluid polycentric system is required, producing policy from a wider network of stakeholders. Van Leeuwen frames the situation slightly differently, suggesting that indeed the system is already polycentric, in which other organizations are active in setting policy, such as the EU driving more stringent regulations and industry organizations developing new technologies and operating practices.

By necessity, the IMO faces challenges regulating 174 countries with different strategic interests. Ratification of conventions can take years, even decades, as shown above. Then the flag states have to turn the IMO convention into national law which may take more years. Despite these challenges, the major conventions are well ratified. SOLAS has 164 states and 99.18% of tonnage, and MARPOL has 157 states and 99.15% of tonnage. The main problem is enforcement, which, as shown above, is lacking by flag states but has been taken into hand by port states via PSC. These challenges are unavoidable. The IMO will never be granted the power to unilaterally force the global industry to act in certain ways, nor would that be desirable. Another problem highlighted by Psaraftis (2019) regarding agreement on contentious policies is that developing countries voted against both EEDI and MBMs because of the "common but differentiated responsibilities and respective capabilities" principle.

At the same time, the analysis above has shown that individual countries or blocs can institute their own laws if they are so minded. Yet, one can still recognize that the power of lobbying and selfish interests do constrain the IMO more than they should. The fact that the IMO website contains statements such as the following reveals the political challenges of making clear environmental statements: "Emissions from ships exhausts into the atmosphere can *potentially be harmful* to human health and cause acid rain and *may also contribute* to global warming" (italics mine). The lack of action on CO_2 is the primary issue, but even the other actions could be considered weaker than they should be. Lister et al. (2015) note that the lowest level of sulfur for SECAs is still 100 times the allowed level of sulfur in road truck diesel. There is also a glaring geographical imbalance, in that the current SECA locations do not cover all parts of the world, particularly poorer areas such as the highly concentrated shipping lanes in Asia, and there are none in South America or Africa. One could even argue that the favoring of MBMs by stakeholders at all levels is disingenuous, knowing that they will take decades to be agreed and will never be set at suitably stringent levels to achieve major change.

Such political challenges are familiar at all levels, from national policy to the Paris Agreement, with the United States even later withdrawing from the agreement that it had signed. Nevertheless, some countries are willing to take leadership. For example, Germany, the largest economy in Europe, has recently decided to phase out coal-fired power plants by 2038. Some countries have announced bans on fossil-fuelled car sales by 2040 and some even by 2030, while several cities have decided to ban fossil-fuelled vehicles from their streets from 2030 or even as soon as 2025. Banning fossil-fuelled ships from a country's waters by a certain date would likewise be a brave decision, but a multinational bloc like the EU could take such a strong decision, ensuring that no member country benefited at the expense of another, as long as neighboring ports outside the EU were not able to capitalize on such a policy. On the other hand, the problems resulting from the US Jones Act reveal the challenges of such a localized policy.

The future for maritime governance looks likely to be less centralized, rather than more. While the IMO will remain the major venue for policymaking, increased action at all levels is expected, from cities and regions (e.g., California cold ironing law) to national (United States setting its own agenda) and international (particularly the EU vs. the IMO regarding CO_2 targets). Fedi (2017) suggests that the divergence of the EU's MRV regulation increases the likelihood that other regional schemes could emerge, thus weakening the universality principle of the IMO.

Nevertheless, policy challenges do not result only from scale and institutional complexity. Even within a single jurisdiction the political difficulties of imposing restrictions on industry and the public are enormous. This is not helped by "scientific reticence" (Hansen, 2007) which has not underlined the seriousness of climate change, but that is changing now. There is little excuse remaining for politicians and decision makers. There are fewer places to hide for lobbyists and those arguing in bad faith for solutions known to be insufficient. Such bluster has been enabled by the dominant incrementalist policy paradigm, but that too is due for change as the latest science reveals that climate change is already happening and effects once thought over a century away are now predicted in the next two decades.

5   Conclusion

There are four major environmental policy challenges facing the existing maritime governance system. These will require strategies, actions, and, most of all, courageous responses from policymakers and actors at all levels.

First, local pollution—SOx, NOx, and PM. Air pollution is becoming one of the biggest killers in the world, with over 4 million deaths worldwide annually. Strong action is needed, at minimum ECAs covering all coastlines—looking at a world map depicting ECA locations, it is striking that no ECAs cover African, Asian, or South American coastlines. On the other hand, there is now some concern that ECAs may result in increased GHG emissions as well as the problem of sea pollution through open scrubbers (Lindstad and Eskeland, 2016), thus more thinking may be needed here. A global fleet switch to LNG would solve this problem by practically eliminating SOx, NOx, and PM, even if its CO_2 reduction is insufficient to meet carbon targets. Should the IMO go further and, rather than incentivize the switch to gas, simply ban the use of fuel oil? The IMO is not likely to introduce such a strong move so it will be up to the individual coastal states to ban these pollutants in their ports and along their coasts. How many countries will take such a decision? Will the EU take the lead? Both the IMO and the EU remain fixated on MBMs despite their lack of promise.

Second, the unavoidable subject of climate change mitigation. We have seen above that the IMO's 50% target reduction by 2050 may be possible with a combination of ultra-slow steaming, global switch to LNG plus ship designs, and top ups with battery and wind. However, there is as yet nothing even on the drawing board that can go beyond that toward full decarbonization, with no commercially feasible alternative fuel that could power international shipping at its current level, so to achieve full decarbonization the only answer, at least in the short term, is to drastically reduce the distance sailed. Are we ready for that? Do we have a choice? As LNG takeup is barely progressing, even after over a decade of hype, even reaching the 50% target becomes unlikely. Moreover, these calculations do not include transport growth. If shipping grows as predicted then the savings will be neutralized by increased emissions, pushing the target even further away. Is a stronger stick needed to drive the switch to LNG rather than the ineffective carrots offered so far? Additionally, given that LNG is only around 20% less carbon intensive than fuel oil and can thus only be considered a temporary measure, should fossil fuels be banned altogether rather than making huge investments to switch to LNG for only a few decades until hydrogen becomes feasible?

Third, we must not forget the port perspective, particularly as it concerns local pollution affecting many people who do not always benefit from the port activity. Zero local emission solutions must become standard, which may include a mix of cold ironing, use of LNG or battery power and slow steaming in the port area. The question is who should make this rule—should it be left to individual port authorities? Should cities in which ports are located take the decision, or perhaps at regional or national level? The IMO level is not appropriate for this decision, but a supranational entity such as the EU could at least provide support and subsidy toward such a move. Meanwhile, ports are facing the challenge of climate change adaptation. Strong decisions will need to be taken regarding future proofing port locations for rising sea levels and storm surges, but research shows a dangerous inertia between the different scales of government (Ng et al., 2019).

Finally, even while ports and policy makers are grappling with climate change adaptation, the situation has already become graver. The new paradigm is for deep adaptation, based on the latest science which reveals that climate change effects will not be linear and incremental but sharp, sudden, and soon (Bendell, 2018). Sea-level rise of several meters is now possible by the end of the century, with dangerous weather occurring every other day. Major disruption is now predicted within the next two decades, which will affect ports directly but also shocks will come from the wider economy as a result of massive migrations of coastal populations, threats to food and energy security, and countries spending billions on adaptation, all of which will influence demand for shipping and ports. Already, in 2017, 18 m people were made homeless by weather events (Internal Displacement Monitoring Centre, 2018). Globally, 145 m people live within one meter above the current sea level (Anthoff et al., 2006) and almost 1 billion people live in low-elevation coastal zones (Neumann et al., 2015). The International Organization for Migration estimates 200 m climate refugees by 2050 (IOM, 2019). According to the World Bank (2018), climate change will push 143 million people to migrate at least within their own countries by 2050.

These challenges raise unprecedented questions regarding the viability of existing port locations, retreat from coastlines, and many international disruptions from droughts and crop failures to migrations and war that will radically reshape current production, consumption, and hence shipping practices. To a large extent, dealing with these challenges will be made at national level, through policies to decarbonize and boost renewables and plan for the retreat of millions of residents and businesses from exposed coastal locations. Thus while maritime policy is polycentric with a strong global component, national policies for dealing with climate change adaptation can be expected to play an increasing role in the evolution of maritime transport and trade in the coming decades.

References

Acciaro, M., Vanelslander, T., Sys, C., Ferrari, C., Roumboutsos, A., Giuliano, G., Lam, J.S.L., 2014. Environmental sustainability in seaports: a framework for successful innovation. Marit. Policy Manag. 41 (5), 480–500.

Alderton, T., Winchester, N., 2002. Globalisation and de-regulation in the maritime industry. Mar. Policy 26, 35–43.

Anthoff, D., Nicholls, R.J., Tol, R.S.J., Vafeidis, A.T., 2006. Global and Regional Exposure to Large Rises in Sea-Level: A Sensitivity Analysis (Working paper; No. 96). Tyndall Centre for Climate Change Research, Norwich, UK.

Bendell, J., 2018. Deep Adaptation: A Map for Navigating Climate Tragedy. IFLAS Occasional Paper 2. Available at: http://insight.cumbria.ac.uk/id/eprint/4166/. (Accessed April 16, 2019).

Bergqvist, R., Monios, J., 2019. Green ports in theory and practice. In: Bergqvist, R., Monios, J. (Eds.), Green Ports; Inland and Seaside Sustainable Transportation Strategies. Elsevier, Cambridge, MA, pp. 1–17.

Bergqvist, R., Macharis, C., Meers, D., Woxenius, J., 2015. Making hinterland transport more sustainable a multi actor multi criteria analysis. Res. Transp. Bus. Manag. 14, 80–89.

Blanco-Bazán, A., 2004. IMO—historical highlights in the life of a UN agency. J. Hist. Int. Law 6 (2), 259–283.

Bows-Larkin, A., 2015. All adrift: aviation, shipping and climate policy change. Clim. Pol. 15 (6), 681–702.

Brandt, J., Silver, J.D., Christensen, J.H., Andersen, M.S., Bønløkke, J.H., Sigsgaard, T., Geels, C., Gross, A., Ayoe, B., Hansen, A.B., Hansen, K.M., Hedegaard, G.B., Kaas, E., Frohn, L.M., 2011. CEEH Scientific Report No 3: Assessment of Health Cost Externalities of Air Pollution at the National Level Using the EVA Model System. Cent. Energy, Environ. Health Rep. Ser, Aarhus Univ., Nat. Environ. Res. Inst, Roskilde, p. 98.

Cariou, P., Parola, F., Notteboom, T., 2019. Towards low carbon global supply chains: a multi-trade analysis of CO_2 emission reductions in container shipping. Int. J. Prod. Econ. 208, 17–28.

Corkhill, M., 2018. Big Boys Join the LNG-Fuelled Fleet. LNG World Shipping. Available at: https://www.lngworldshipping.com/news/view,big-boys-join-the-lngfuelled-fleet_51714.htm. (Accessed July 1, 2019).

Cullinane, K., Bergqvist, R., 2014. Emission control areas and their impact on maritime transport. Transp. Res. D 28, 1–5.

David, M., Gollasch, S. (Eds.), 2015. Global Maritime Transport and Ballast Water Management—Issues and Solutions. Springer, London.

Directorate General for Internal Policies Policy Department A: Economic and Scientific Policy, 2015. Emission Reduction Targets for International Aviation and Shipping. Available at: http://www.europarl.europa.eu/RegData/etudes/STUD/2015/569964/IPOL_STU(2015)569964_EN.pdf. (Accessed March 9, 2019).

EMSA, 2016. Annual Overview of Marine Casualties and Incidents 2016. Available at: http://www.emsa.europa.eu/news-a-press-centre/external-news/item/2903-annual-overview-of-marine-casualties-and-incidents-2016.html. (Accessed April 4, 2019).

European Commission, 2014. Directive 2014/94/EU of the European Parliament and of the Council of 22 October 2014 on the Deployment of Alternative Fuels Infrastructure. Available at: http://eur-lex.europa.eu/legal-content/EN/TXT/?uri=celex%3A32014L0094. (Accessed April 4, 2019).

Fedi, L., 2017. The monitoring, reporting and verification of ships' carbon dioxide emissions: a european substantial policy measure towards accurate and transparent carbon dioxide quantification. Ocean Yearbook Online 31 (1), 381–417.

Fedi, L., 2019. Arctic shipping law from atomised legislations to integrated regulatory framework: the polar code (R)evolution? In: Lasserre, F., Faury, O. (Eds.), Arctic Shipping. Climate Change, Commercial Traffic and Port Development. Routledge (forthcoming).

Fedi, L., Lavissiere, A., Russell, D., Swanson, D., 2010. The facilitating role of IT systems for legal compliance: the case of port community systems and container Verified Gross Mass (VGM). Supply Chain Forum: Int. J. 20 (1), 29–42.

Franc, P., Sutto, L., 2013. Impact analysis on shipping lines and European ports of a cap-and-trade system on CO_2 emissions in maritime transport. Marit. Policy Manag. 41 (1), 61–78.

Gonzalez-Aregall, M., Bergqvist, R., Monios, J., 2018. A global review of the hinterland dimension of green port strategies. Transp. Res. D 59, 23–34.

Hansen, J.E., 2007. Scientific reticence and sea level rise. Environ. Res. Lett. 2, 024002.

IMO, 2014. Third IMO GHG Study 2014. International Maritime Organization (IMO), London, UK.

IMO, 2019. International Maritime Organisation. Available at: http://www.imo.org/. (Accessed April 4, 2019).

InfluenceMap, 2017. Corporate Capture of the International Maritime Organization How the Shipping Sector Lobbies to Stay Out of the Paris Agreement. Available at: https://influencemap.org/report/Corporate-capture-of-the-IMO-902bf81c05a0591c551f965020623fda. (Accessed April 4, 2019).

InfluenceMap, 2019. Big Oil's Real Agenda on Climate Change. Available at: https://influencemap.org/report/How-Big-Oil-Continues-to-Oppose-the-Paris-Agreement-38212275958aa-21196dae3b76220bddc. (Accessed April 2, 2019).

Innes, A., Monios, J., 2018. Identifying the unique challenges of installing cold ironing at small and medium ports—the case of Aberdeen. Transp. Res. D: Transp. Environ. 62, 298–313.

Internal Displacement Monitoring Centre, 2018. Global Report on Internal Displacement 2018. Available at: http://www.internal-displacement.org/global-report/grid2018/. (Accessed March 8, 2019).

International Organization for Migration, 2019. A Complex Nexus. Available at: https://www.iom.int/complex-nexus#estimates. (Accessed March 8, 2019).

Kosmas, V., Acciaro, M., 2017. Bunker levy schemes for greenhouse gas (GHG) emission reduction in international shipping. Transp. Res. D: Transp. Environ. 57, 195–206.

Krugman, P., 2018. The Depravity of Climate-Change Denial; Risking civilization for profit, ideology and ego. In: The New York Times. Available at: https://www.nytimes.com/2018/11/26/opinion/climate-change-denial-republican.html?fbclid=IwAR2WUqnyktb-tgxx2-NYt4Kny4NSnHj-FuJG-CzsJKlMLEbFLuVkkxGq2HbI. (Accessed November 27, 2018).

Lam, J.S.L., Notteboom, T., 2014. The greening of ports: a comparison of port management tools used by leading ports in Asia and Europe. Transp. Rev. 34 (2), 169–189.

Li, K.X., Zheng, H., 2008. Enforcement of law by the port state control (PSC). Marit. Policy Manag. 35 (1), 61–71.

Lindstad, H.E., Eskeland, G.S., 2016. Environmental regulations in shipping: policies leaning towards globalization of scrubbers deserve scrutiny. Transp. Res. D: Transp. Environ. 47, 67–76.

Lister, J., Taudal Poulsen, R., Ponte, S., 2015. Orchestrating transnational environmental governance in maritime shipping. Glob. Environ. Chang. 34, 185–195.

Lohmann, L., 2008. Carbon trading, climate justice and the production of ignorance: ten examples. Development 51 (3), 359–365.

Monios, J., 2017. Port governance in the UK: planning without policy. Res. Transp. Bus. Manag. 22, 78–88.

Monios, J., Bergqvist, R., Woxenius, J., 2018. Port-centric cities: the role of freight distribution in defining the port city relationship. J. Transp. Geogr. 66, 53–64.

Neumann, B., Vafeidis, A.T., Zimmermann, J., Nicholls, R.J., 2015. Future coastal population growth and exposure to sea-level rise and coastal flooding—a global assessment. PLoS One 10 (3), e0118571.

Ng, A.K.Y., Becker, A., Cahoon, S., Chen, S.L., Earl, P., Yang, Z. (Eds.), 2016. Climate Change and Adaptation Planning for Ports. Routledge, Abingdon.

Ng, A.K.Y., Monios, J., Zhang, H., 2019. Climate adaptation management and institutional erosion: insights from a major Canadian port. J. Environ. Plan. Manag. 62 (4), 586–610.

Notteboom, T., Delhaye, E., Vanherle, K., 2010. Analysis of the Consequences of Low Sulphur Fuel Requirements, Report Commissioned by European Community Shipowners' Associations (ECSA), pp.1–83.

OECD, 2018. Decarbonising Maritime Transport: Pathways to Zero-Carbon Shipping by 2035. OECD, Paris.

Oyaro Gekara, V., 2010. The stamp of neoliberalism on the UK tonnage tax and the implications for British seafaring. Mar. Policy 34, 487–494.

Psaraftis, H.N., 2019. Decarbonization of maritime transport: to be or not to be? Maritime Econ. Logist. 21 (3), 353–371.

Roe, M., 2007. Shipping, policy & multi-level governance. Maritime Econ. Logist. 9 (1), 84–103.

Roe, M., 2009. Multi-level and polycentric governance: effective policymaking for shipping. Marit. Policy Manag. 36 (1), 39–56.

Roe, M., 2013. Maritime Governance and Policy-Making. Springer, London.

Shi, W., Xiao, Y., Chen, Z., McLaughlin, H., Li, K.X., 2018. Evolution of green shipping research: themes and methods. Marit. Policy Manag. 45 (7), 863–876.

Sköld, S., 2019. Green port dues—indices and incentive schemes for shipping. In: Bergqvist, R., Monios, J. (Eds.), Green Ports; Inland and Seaside Sustainable Transportation Strategies. Elsevier, Cambridge, MA, pp. 173–192.

Spengler, T., Wilmsmeier, G., 2019. Sustainable performance and benchmarking in container terminals—the energy dimension. In: Bergqvist, R., Monios, J. (Eds.), Green Ports; Inland and Seaside Sustainable Transportation Strategies. Elsevier, Cambridge, MA, pp. 125–154.

Urrutia, B., 2006. The EU regulatory action in the shipping sector: a historical perspective. Maritime Econ. Logist. 8 (2), 202–221.

Van Leeuwen, J., 2015. The regionalization of maritime governance: towards a polycentric governance system for sustainable shipping in the European Union. Ocean Coast. Manag. 117, 23–31.

Van Leeuwen, J., Kern, K., 2013. The external dimension of European Union Marine Governance: institutional interplay between the EU and the International Maritime Organization. Global Environ. Polit. 13 (1), 69–87.

WHO, 2018. 9 Out of 10 People Worldwide Breathe Polluted Air, But More Countries Are Taking Action. Available at: https://www.who.int/news-room/detail/02-05-2018-9-out-of-10-people-worldwide-breathe-polluted-air-but-more-countries-are-taking-action. (Accessed March 15, 2019).

Winkel, R., Weddige, U., Johnsen, D., Hoen, V., Papaefthimiou, S., 2016. Shore side electricity in Europe: potential and environmental benefits. Energy Policy 88, 584–593.

Winnes, H., Styhre, L., Fridell, E., 2015. Reducing GHG emissions from ships in port areas. Res. Transp. Bus. Manag. 17, 73–82.

World Bank, 2018. Groundswell: Preparing for Internal Climate Migration. Available at: https://openknowledge.worldbank.org/handle/10986/29461. (Accessed February 6, 2019). (Accessed 22 February 2019).

Zis, T., Angeloudis, P., Bell, M.G.H., Psaraftis, H.N., 2016. Payback period for emissions abatement alternatives. Transp. Res. Rec.: J. Transp. Res. Board 2549, 37–44.

Part II

Adapting to climate change impacts

Chapter 3

Storm resilience and sustainability at the Port of Providence, USA

Richard Burroughs[a,b], Austin Becker[a]

[a]*Department of Marine Affairs, University of Rhode Island, Kingston, RI, United States,* [b]*School of Forrestry and Environmental Studies, Yale University, New Haven, CT, United States*

1 Introduction

Sustainable seaports must be resilient seaports. Resilient systems withstand shocks like storms. In its most basic form, resilience is the ability of a linked human and natural system to absorb disturbance, while at the same time retaining its basic structure and function (Walker and Salt, 2006). Further, it is the ability of a system and its component parts to anticipate, absorb, accommodate, or recover from the effects of a potentially hazardous event in a timely and efficient manner, including through ensuring preservation, restoration, or improvement of its essential basic structures and functions (IPCC, 2013). Resilience to hurricane events can be achieved not only through physical improvements to property but also through planning, policy adaptation, and cultivating public support (Becker and Caldwell, 2015; Raub and Cotti-Rausch, 2019).

Seaports, while shaped by the natural land and seascape, are ultimately human systems. Sustainability, the state of meeting current needs without compromising the ability of individuals to meet their needs in the future (Brundtland et al., 1987), demands that resilience to hurricanes, and ultimately climate change, should be a part of port planning. Ports and port cities grew synergistically, with the port facilitating commerce and the city developing in lock step. Hall's (2007) focus on cities favors the goal of improving social and economic conditions during urbanization while maintaining environmental quality. In this study, we examined aspects of this relationship for the Port of Providence, Rhode Island, United States with a focus on the role of stakeholders in responding to the likely impacts of hurricanes in ways that produce greater resilience and sustainability. Stakeholders responded both to surveys and to 3-D visualizations of a hurricane striking the port (for more information, see www.portof-providenceresilience.org).

Maritime Transport and Regional Sustainability. https://doi.org/10.1016/B978-0-12-819134-7.00003-4

Coastal flooding arises from a combination of sea level rise, storm surge, and rain. The recent climate assessment (IPCC, 2018) of a +1.5°C to 2°C world has medium confidence for increased extreme precipitation events eastern North America where Providence is located (IPCC, 2018; Nicholls et al., 2018). The IPCC (2013) estimates a global mean sea level rise of 0.72 m (2.4 feet) over the 1986–2005 average by 2100 if there is no mitigation from carbon release (RCP8.5) in the future. Additionally, Rhode Island state government has adopted a NOAA (2017) high estimate of 9.6 feet by 2100 (2.9 m) for some state planning purposes (CRMC, 2018). While these sea level projections vary greatly, both enforce the need for port preparedness. Hurricane futures are also difficult to predict, with one group finding that greenhouse gas-induced warming does not increase the frequency of either tropical storm or overall hurricane numbers in the Atlantic (GFDL, 2019) and others finding that overall intensity of storms is likely to increase (Kutson et al., 2010). Nonetheless the storm surges associated with historical hurricanes in Providence, RI have been particularly severe. For example in the hurricane of 1938 the water level reached 5 m above normal high tide (NWS, 2019). Anticipating a sea level rise of 0.72–2.9 m, a medium confidence for increased extreme precipitation, and historical hurricane patterns will produce significant flooding at the port in the future.

With that threat in mind many dimensions of sustainability can be affected. They include nature, life support, and human community as well as individual, social, and economic needs NRC, 1999). In explaining sustainability science Kates et al. (2001, p. 641) focus attention on interactions between society and nature as well as "society's capacity to guide those interactions along more sustainable trajectories." Key to guiding an urban port in this manner will be the perceptions and actions of primary stakeholders. Kates (2012) further elaborates this theme by noting the integrative demands across social, natural, and engineering domains as well as the environmental, health, and economic development communities. Ultimately sustainability successes require knowledge to be transformed into actions (Kates et al., 2012), an approach that we consider in this chapter.

Hiranandani (2014) examined port sustainability in the context of development and environment. A multipage table in his article summarizes focal issues across four major global ports. They include pollution (air, water, and ballast water), waste disposal (dredging, solid, and hazardous wastes), and land/resource use. Hiranandani (2014) lays out principal areas where port activities affect the environment. However, storms and sea level rise also produce direct physical impacts on port functioning (Becker et al., 2014). And as Schipper et al. (2017, Table 2) find in a global study that most ports have sustainability as a part of master plans ports, but they find limited to no awareness of flooding risk in the documents.

We apply the sustainability paradigm to the social dimensions of ports. Following Burroughs (2012) we propose that in a positive sustainability trajectory, individuals in a geographic region make decisions and implement programs that assure continuity and improvement of one or more of the dimensions of sustainability as noted in the NRC report above. A sustainability trajectory sequence includes identification of a target and societal values related to it, engaging science to create feasible solutions, selecting and implementing authoritative means to 7 meet targets, and assessing results. Stakeholders are instrumental in successful trajectories. In the case of ports, stakeholders include a wide range of individuals, organizations, and agencies, such as shippers, tenants, government agencies, neighborhood groups, insurers, and, of course, the port operator itself (Becker et al., 2014). In practice, sustainability decision-making consists of iterative analysis, as society grapples with both new information from sustainability science and new values, which collectively create new targets. For the case of the Port of Providence described in this chapter, we further limit the scope to hurricanes. Storms disrupt sustainability by causing direct, indirect, and intangible impacts on the port stakeholders. Preparing for and recovering from storms is a primary challenge for ports. A positive trajectory has the potential to limit damage, to restore a system such as cargo flow in a port after a storm, or create new but affirmative values as a new situation arises for the port. In either case, the target includes meeting individual needs, building human communities, and accumulating social capital as well as respecting environmental limits. A sustainable future for a port rests on taking resilience-enhancing actions.

2 Planning for inundation

Inundation, whether temporary due to a storm event or permanent due to sea level rise, forces port sustainability considerations. Because ports must be located in areas subject to rising sea levels and storm surge, port resilience should be a focus of future planning.

Although resilience planning progress has been made, particularly with respect to changes in residential land use and building codes (Melillo 2014), few actions have yet been taken to protect the complex system of ports and shipping that facilitate the nation's maritime-based freight economy (Becker et al., 2012; Ng et al. 2016). Indeed, while port operators themselves acknowledge the important role that climate change will play in future operations (Becker et al., 2012, 2014), there are still few examples of plans, let alone implementation actions.

Complicating the response is the fact that ports consist of complex and interdependent public/private decision-making governance structures (Notteboom and Winkelmans, 2002, 2003), thus making general or universal recommendations difficult. Natural hazards associated with climate change threaten the system as a whole, as well as the infrastructure that individual organizations depend upon. Individual organizations and agencies often do not have the proper incentives or understanding of the system's interconnectedness to justify investment in long-term resilience (Becker and Caldwell, 2015). Despite the availability of impacts assessment tools and established methods for stakeholder engagement, overcoming barriers to resilience investments for complex systems such as ports remains a significant challenge. Conflicting timescales, institutional uncertainties, and lack of resources make the process more difficult (Ekstrom and Moser, 2014; Eisenack et al., 2014; Tompkins and Eakin, 2012).

3 The Port of Providence, Rhode Island

To inform planning and develop deeper knowledge of storm issues for port stakeholders, we undertook a year-long study that culminated in a workshop with representatives of the private and public sectors of the port (Becker et al., 2017). Together, the businesses that make up the port of Providence supply Connecticut, Massachusetts, and Rhode Island with petroleum products and handle bulk and break-bulk imports and exports. Many businesses depend on the port's functionality, including trucking companies, Providence and Worcester Rail, dredging firms, tug companies, marine pilots, and cargo handling services. Hospitals and educational institutions with power plants, manufacturing companies, gas users, electricity generation plants, and aviation fuel consumers among others benefit from fossil fuel trade at the port. The hinterland for fuel distribution extends well into Massachusetts.

The study area for this project includes ProvPort, the main port terminal, and number of other waterfront businesses and industries, which together take up nearly 93 ha of waterfront in Providence and East Providence (Becker et al., 2010). ProvPort itself is about 42 ha of land that are owned by the City of Providence and operated by a nonprofit organization with five board members. ProvPort contracts the services of Waterson Terminals LLC to operate and maintain the port. In 2015 the public and private terminals at the port handled a total of 8,043,000 short tons (U.S. Army Corps of Engineers, 2016).

The port is located at the northern end of Narragansett Bay, an ecologically sensitive estuary that provides breeding grounds for marine life in the region. The length and orientation of Rhode Island's Narragansett Bay, and its proximity to the Atlantic hurricane zone, make it susceptible to extreme storm surges from the southerly winds that are generated when a hurricane passes to the west of the Bay. The most recent major storm, Hurricane Carol in 1954, produced 4.4 m of storm surge in Providence. Most of the port lands in the study area are 1–3 m above mean high water. A 7.6 m hurricane barrier north of the port protects the downtown City area, but the port is located seaward of this barrier.

To further define the perceived impacts of a storm with stakeholders, we selected a 111–129 mph hurricane traveling at 40 mph and approaching Rhode Island from the south at high tide which would be equivalent to the 1938 hurricane but shifted 80 miles east (Becker et al., 2017). Modeling such a storm using Sea, Lake, and Overland Surges from Hurricanes model (SLOSH) produced a maximum total water level of ~6.4 m above NAVD 88. With this hypothetical storm 198 ha or 86% of the study area would be covered by one foot of water or more. Three-dimensional visualizations of the storm surge inundation were used to inform stakeholders about the impacts on various waterfront parcels.

Vulnerability is defined as the degree to which physical, biological, and socio-economic systems are susceptible to, and unable to cope with, adverse impacts. Clearly this port is vulnerable. To understand the degree of vulnerability as perceived by port businesses and the resilience-building actions they had taken, we interviewed many of them, conducted a workshop, and undertook a survey.

4 Survey

We conducted online surveys of port businesses to determine what actions had been taken at the firm level to create a more resilient and hence sustainable port. We were able to involve totally 17 private firms in our work. They included seven handling petroleum, five recycled metal, and four salt. In 2015 these categories of cargoes accounted for 90% of the cargo volume for the port (U.S. Army Corps of Engineers, 2016).

Eleven businesses reported owning their property and six reported their operations as independently operated. Seven businesses stated they have 1–19 employees, five businesses stated 20–99 employees, and two businesses reported over 100. Based on stakeholder responses total employment of the businesses surveyed is ~600 to as many as ~2000 workers.

Nine businesses have more than 100 unique customers (individual purchasers), while 12 stated 100 or more businesses rely on their services. This suggests a sizeable supply chain effect if port businesses were impacted, with port products reaching many customers and businesses throughout the hinterland.

Businesses require access to land and sea corridors to be effective participants in the supply chain (Fig. 1). Nine of our respondents to the initial survey depend on access to the shipping channel and of those six require the channel to be maintained as deep draft or 35 (10.7 m) to 40 feet (12.2 m). Seven require

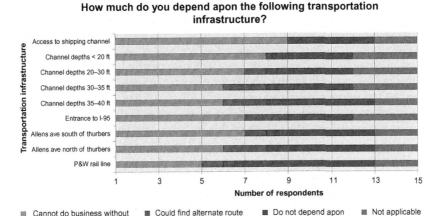

FIG. 1   This figure shows that 9 out of 15 businesses state that they could not do business without access to the 40-foot-deep shipping channel.

access to route I-95 and five the railroad line to move cargo into and out of the port area. Annual vessel calls per business range from 15 to 250 per year. At least one representative stated if the 40 ft channel were lowered (to 30 or 20 ft) business could be facilitated with smaller vessels, but at a higher cost to the business.

Storm preparedness can be measured by individual firm investments in physical reconfiguration at the business. Fig. 2 shows that many firms have backed up computer systems, installed emergency generators, and taken wind/flood proofing actions on site. Less common firms invest in raising electrical systems or moving to less flood-prone areas.

Planning by individual firms (Fig. 2) includes identifying offsite locations for equipment and cargoes, and developing hazardous material as well as business recovery plans. Only two firms have created structure stability analyses. Structural stability of piers must be maintained if cargo is to be handled after a storm. In addition firms have completed meetings, inundation maps, and prestorm contracting. In prestorm contracting waterfront businesses can identify debris removal and other needed activities in advance of the event.

Subsequent to the survey we completed a workshop to gain further insights on stakeholders that included businesses as well as government (Becker et al., 2017; Becker, 2017, see also www.portofprovidence resilience.org).

5 Discussion

All coastal ports are *vulnerable* to storm impacts because land/sea cargo transport almost always occurs at or just above sea level where storm surge, tide, and wave action can damage infrastructure. Since almost all cargo enters the port of Providence by ship or barge and leaves by truck, the viability of these land and sea arteries determines port function.

Unlike most other ports in the United States, the Providence does not have a central port authority that is responsible for operations and planning. Instead, the port includes a variety of private businesses, each responsible for its own sustainability efforts. Because the mandate for public port authorities typically includes prioritization of the local, state, or regional economy and the "public good" more generally, they may be more likely to invest in long range resilience planning (see, e.g., MassPort). The Port of Providence has grown considerably over the last century, with the expansion of Allens Avenue, the construction of the Interstate 95 highway, and landfill along the edges of the harbor to accommodate new maritime uses. Much of this infrastructure lies within the floodplain, but has not been tested by a major storm event since 1954.

When considering *resilient* infrastructure most generally, Husdon et al. (2012) focus on anticipation of the event like a hurricane that builds in ability to resist, absorb, and adapt while recovering rapidly. Actions of individual firms can build resilient ports, and we have assessed the extent to which individual firms had taken actions (Fig. 2). Since in 25% of the companies in a national

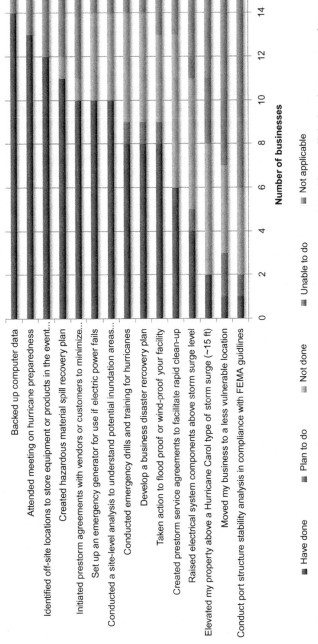

FIG. 2 Most business have backed up computer data, attended a meeting on hurricane preparedness, and identified an off-site location to store equipment or products; however, business has in general not created prestorm service agreements to facilitate rapid clean up and raised electrical systems above storm surge levels (~15 ft).

survey suffering an information technology outage of 2–6 days went bankrupt immediately (Husdon et al., 2012); this aspect of port businesses could be particularly important. Our early survey found that computer back-ups were nearly ubiquitous among the Providence respondents. This addresses an important part of information technology in our setting.

A *sustainable* port maintains system perturbations within tolerable limits such that long-term operations are assured. In the spirit of the Bruntland commission and focused on human systems, actions taken now would control storm impacts such that continued operations in the future would be feasible. An easily recognized objective is continued cargo flow, but allied to that are employment, community economic, and environmental health as well as continued effective linkage to global manufacturers and the port hinterland. Storms alone do not control sustainability as many other factors related to trade also intervene. Pressures include changes in values about waterfront use, altered lands due to changing sea level, shifts in cargo demands or technologies, and urban gentrification. For example the shift away from coal has reduced the volume of a formerly common cargo handled in Providence.

Resilient and sustainable in the context of ports implies for O'Keeffe et al. (2016) that scientific information is coupled with stakeholder involvement since effective adaptation measures will best be informed by tacit knowledge of port workers. However as they note, when ports engage in adaptation, it is often as a result of legislative or policy directives, which suggests that port managers are not convinced that bottom-up approaches should be encouraged. Our work was designed to alter this mindset by using visualization of storm impacts to bring inundation realities to individuals most directly connected to port operations and record their impressions as an entry point to informed responses and broad adoption of needed measures.

Visualizations can test both knowledge obtained and response to it (Lieske et al., 2014; Rickard et al., 2017), but we are most interested in results that cause collective action to reduce risk. One ultimate test of the effectiveness of a visualization is the extent to which individuals or groups are effectively informed and choose to constructively act to address the threat (Portman, 2014). Thus, testing individual responses to inundation of an urban area beyond the port can be instructive (Lindner et al., 2019). A combination of the ability to recognize individual flooded locations and for many direct experience with storms resulted in demographic groups varying between 46% and 62% higher likelihood of evacuation for a category 4 storm. Our sample, while directed toward port businesses, suggests high willingness to act: intention to implement individual preparedness strategies ranged from one to as many as 14 adopters out of a total of 15 businesses. While the buy-in for businesses is promising, much more remains to be done by individual firms and, most importantly, at the port-wide collaborative level. Without a robust organizational home for collective action, progress is expected to be slow until the next crisis.

As noted earlier, Providence does not have a port authority. This raises the question of the extent to which responding to storms is best considered as a collective action or a property owner/lessee individual action. Earlier work in Providence identified over 120 potential resilience strategies that could be implemented by various stakeholders of the port (Becker and Caldwell 2015). In fact, as Fig. 2 demonstrates, over one dozen individual actions have been taken by many Providence waterfront businesses. These individual actions could be taken by businesses in many US ports and to that extent the Providence case is broadly instructive.

However, protecting the port through the construction of a new storm barrier was found to be the best way to accomplish participants' resilience goals (Becker et al., 2017, Fig. 15). A new hurricane barrier seaward of the port could protect all port businesses. With protection clearly favored, implementation, in the eyes of the participants, rested on public-private collaboration or government initiatives (Becker et al., 2017, Fig. 18). Importantly, meeting individual goals requires collective action, especially when the costs are in the hundreds of millions of dollars. Thus while the need is identified for the Port of Providence, the organizational structure to accomplish it remains uncertain.

At local government level, both the Providence Harbor Management Committee and the Providence Emergency Management Agency could facilitate more resilience discussions. The former was recently created by the city to draft a new harbor management plan. The latter has recently completed a City of Providence Hazard Mitigation Plan update, which explicitly assesses hurricane readiness. Federal government through the US Army Corps of Engineers built the current Fox Point hurricane barrier and will be involved should another structure serving the community be created.

Missing from the Providence waterfront is a port authority, a common feature of large ports such as the Port of Seattle or Port of San Diego, CA. A port authority could become a strong advocate for a new more seaward barrier, if existing entities do not do so. The storm we have proposed and the responses to it make clear that a combination of collective and individual actions is appropriate. Port authorities can perform that function and other organizations can also.

At the base of this discussion is a consideration of the mix of collective (across firms, property owners, public entities) or individual actions to advance sustainability trajectories. To the extent that resilience preparedness is a function of individual action the Providence case applies to multiple settings around the United States where private entities are best positioned to make investments that matter. However, when collective action supported by government, such as the construction of a hurricane barrier is called for, a different organizational structure may be more appropriate.

6   Conclusions

This study represents a practical application of the terms vulnerability, resilience, and sustainability in port operations. While some aspects of maritime industry can be physically relocated out of the flood plain (e.g., tank farms can be located on elevated areas and product piped from the shorefront berth facility), we recognize that retreating from the shorefront is not a realistic option for most port activities. The inevitability of future maritime commerce in such vulnerable areas brings sustainability questions into clear focus here.

In our setting, impacts of hurricanes on port operations force firms and ports to respond before, during, and after the storm to assure sustainability. The context for urban port sustainability extends beyond business interruptions to include environmental impacts, gentrification of port area due to changing values, changes in cargo handling technology, and shifting port hinterlands among other influences, which are subjects for future research.

Sustainability trajectories (Kates et al., 2001; Burroughs 2012) require the assessment of multiple stakeholder actions in aggregate to determine whether a system is becoming more or less resilient over time. Applied to ports, trajectories enable one to assess the importance of actions taken in terms of the overall sustainability. In Providence individual firms have taken steps to become more sustainable in the face of hurricanes and need to take many additional actions. Missing, so far, is an institutional setting to assess and, where warranted, advance protection measures that could be taken for the port system. Further research will profitably engage both the appropriate designs for protective action and the collective will to make it happen.

References

Becker, A., 2017. Using boundary objects to stimulate transformational thinking: storm resilience for the Port of Providence, Rhode Island (USA). Sustain. Sci. 12, 477–501.

Becker, A., R. Burroughs, E. Kretsch, D. McIntosh, and J. Haymaker, 2017. Stakeholder vulnerability and resilience strategy assessment of maritime infrastructure: Pilot Project for Providence, RI. University of Rhode Island Transportation Center, Report #: FHWA-RIDOT-RTD-17-3, 55 pp. (plus appendices).

Becker, A., Caldwell, M., 2015. Stakeholder perceptions of seaport resilience strategies: a case study of Gulfport (Mississippi) and Providence (Rhode Island). Coast. Manag. 43 (1), 1–34. https://doi.org/10.1080/08920753.2014.983422.

Becker, A., Inoue, S., Fischer, M., Schwegler, B., 2012. Climate change impacts on international seaports: knowledge, perceptions, and planning efforts among port administrators. Clim. Change 110 (1–2), 5–29. https://doi.org/10.1007/s10584-011-0043-7.

Becker, A., Matson, P., Fischer, M., Mastrandrea, M., 2014. Towards seaport resilience for climate change adaptation: stakeholder perceptions of hurricane impacts in Gulfport (MS) and Providence (RI). Prog. Plan. 99, 1–49. https://doi.org/10.1016/j.progress.2013.11.002.

Becker, A., Wilson, A., Bannon, R., McCann, J., Robadue, D., Kennedy, S., 2010. Rhode Island Ports & Commercial Harbors A GIS-Based Inventory of Current Uses and Infrastructure. http://www.crc.uri.edu/projects_page/rhode-island-ports-harbors-inventory/?section=publications.

Brundtland, G., Khalid, M., Agnelli, S., Al-Athel, S., Chidzero, B., Fadika, L., Hauff, V., Lang, I., Shijun, M., de Botero, M.M., et al., 1987. Our Common Future. Oxford University Press, New York (World Commission on Environment and Development, Brundtland Report).

Burroughs, R., 2012. Sustainability trajectories for urban waters. In: Weinstein, M.C., Turner, R.E. (Eds.), Sustainability Science: The Emerging Paradigm and the Urban Environment. Springer, New York, pp. 329–349.

CRMC, 2018. http://www.beachsamp.org/wp-content/uploads/2018/07/BeachSAMP_Ch2_Trends_061218_CRMCApproval.pdf, last visited April 14, 2019.

Eisenack, K., Moser, S.C., Hoffmann, E., Klein, R.J.T., Oberlack, C., Pechan, A., Rotter, M., Termeer, C.J.A.M., 2014. Explaining and overcoming barriers to climate change adaptation. Nat. Clim. Chang. 4 (10), 867–872.

Ekstrom, J.A., Moser, S.C., 2014. Identifying and overcoming barriers in urban climate adaptation: case study findings from the San Francisco Bay Area, California, USA. Urban Clim. 9, 54–74. https://doi.org/10.1016/j.uclim.2014.06.002.

Geophysical Fluid Dynamics Laboratory (GFDL), 2019. https://www.gfdl.noaa.gov/global-warming-and-hurricanes/#global-warming-and-atlantic-hurricanes Last visited May 10, 2019.

Hall, P.V., 2007. Seaports, urban sustainability, and paradigm shift. J. Urban Technol. 14 (2), 87–101. https://doi.org/10.1080/10630730701531757.

Hiranandani, V., 2014. Sustainable development in seaports: a multi-case study. World Maritime Univ. J. Maritime Affairs 13, 127–172. https://doi.org/10.1007/s13437-013-0040-y.

Husdon, S., Tufton, E., Cormie, D., Inglis, S., 2012. Engineering resilient infrastructure. Proc. Inst. Civ. Eng. Civ. Eng. 165 (CE6), 5–12. https://doi.org/10.1680/cien.11.00065.

Intergovernmental Panel on Climate Change (IPCC), 2013. Climate Change 2013. The Physical Science Basis. Working Group I Contribution to the Fifth Assessment Report of the Intergovernmental Panel on Climate Change, http://www.climatechange2013.org/images/uploads/WGI_AR5_SPM_brochure.pdf.

IPCC, 2018. Summary for policymakers. In: Global Warming of 1.5°C. An IPCC Special Report on the Impacts of Global Warming of 1.5°C Above Pre-Industrial Levels and Related Greenhouse Gas Emission Pathways, in the Context of Strengthening the Global Response to the Threat of Climate Change, Sustainable Development and Efforts to Eradicate Poverty. World Meteorological Organization, Geneva, Switzerland. 32 pp.

Kates, R.W., 2012. From the unity of nature to sustainability science: ideas and practice. In: Weinstein, M., Turner, R.E. (Eds.), Sustainability Science: The Emerging Paradigm and the Urban Environment. Springer, New York, pp. 3–20.

Kates, R., Clark, W., Corell, C., Hall, J., Jaeger, C., Lowe, I., McCarthy, J., Schellnhuber, H., Bolin, B., Dickson, N., Faucheux, S., Gallopin, G., Grubler, A., Huntley, B., Jager, J., Jodha, N., Kasperson, R., Mabogunje, A., Matson, P., Mooney, H., Moore, B., O'Riordan, T., Svedin, U., 2001. Environment and development: sustainability science. Science 292 (5517), 641–642.

Kates, R.W., Travis, W.R., Wilbanks, T.J., 2012. Transformational adaptation when incremental adaptations to climate change are insufficient. Proc. Natl. Acad. Sci. U. S. A. 109 (19), 7156–7161.

Kutson, T., McBride, J.L., Chan, J., Emanuel, K., Holland, G., Landsea, C., Held, I., Kossin, J.P., Srivastava, A.K., Sugi, M., 2010. Tropical cyclones and climate change. Nat. Geosci. 3, 157–163.

Lieske, D., Wade, T., Roness, L., 2014. Climate change awareness and strategies for communicating the risk of coastal flooding: a Canadian Maritime case example. Estuar. Coast. Shelf Sci. 140, 83–94.

Lindner, B., Alscheimer, F., Johnson, J., 2019. Assessing improvement in the public's understanding of hurricane storm tides through interactive visualization models. J. Coast. Res. 35 (1), 130–142.

Melillo, J., Terese, M., Richmond, T.C., Yohe, G.W., 2014. Climate Change Impacts in the United States: The Third National Climate Assessment. U.S. Government Printing Office, Washington, DC.

National Oceanographic and Atmospheric Agency (NOAA), 2017. Global and Regional Sea Level Rise Scenarios for the United States. NOAA Technical Report, NOS CO-OPS 083.

National Weather Service (NWS), 2019. https://www.weather.gov/okx/1938HurricaneHome. Last visited March 10, 2019.

National Research Council (NRC), 1999. Our Common Journey: A Transition Toward Sustainability. National Academy Press, Washington, DC.

Ng, A., Becker, A., Cahoon, S., Chen, S.-L., Earl, P., Yang, Z., 2016. Climate Change and Adaptation Planning for Ports. Routledge, New York.

Nicholls, R.J., et al., 2018. Stabilization of global temperature at 1.5°C and 2.0°C: implications for coastal areas. Trans. R. Soc. A376, 20160448.

Notteboom, T., Winkelmans, W., 2002. Stakeholders relations management in ports: dealing with the interplay of forces among stakeholders in a changing competitive environment. In: Paper Presented at the IAME 2002, International Association of Maritime Economists Annual Conference, Panama City, Panama.

Notteboom, T., Winkelmens, W., 2003. Dealing with stakeholders in the planning process. In: Dullaert, W., Jourquin, B.A.M., Polak, J.B. (Eds.), Across the Border: Building Upon a Quarter Century of Transport Research in the Benelux. De Boeck, Antwerpen, pp. 249–265.

O'Keeffe, J., Cummins, V., Devoy, R., Lyons, D., Gault, J., 2016. Stakeholder awareness of climate adaptation in the commercial seaport sector: a case study from Ireland. Mar. Policy. https://doi.org/10.1016/j.marpol.2016.04.044i.

Portman, M., 2014. Visualization for planning and management of oceans and coasts. Ocean Coast. Manag. 98, 176–185.

Raub, K., Cotti-Rausch, B., 2019. Helping communities adapt and plan for coastal hazards: coastal zone management program recommendations for National Tool Developers. Coast. Manag. 47 (3), 1–16.

Rickard, L., Schuldt, J., Eosco, G., Scherer, C., Daziano, R., 2017. The proof is in the picture: the influence on imagery and experience in perceptions of hurricane messaging. Weather Clim. Soc. 9, 471–485.

Schipper, C., Vreugdenhill, H., de Jong, M., 2017. A sustainability assessment of ports and port city plans: comparing ambitions with achievements. Transp. Res. D 57, 84–111.

Tompkins, E., Eakin, H., 2012. Managing private and public adaptation to climate change. Glob. Environ. Chang. 22 (1), 3–11. https://doi.org/10.1016/j.gloenvcha.2011.09.010.

U.S. Army Corps of Engineers, 2016. Waterborne Commerce of the United States: Calendar Year 2015 Part 1 Waterways and Harbors Atlantic Coast. Institute for Water Resources IWR-WCUS-15-1.

Walker, B., Salt, D., 2006. Resilience Thinking: Sustaining Ecosystems and People in a Changing World. Island Press, Washington, DC.

Further reading

Parris, A., Bromirsji, P., Burkett, V., Cayan, D., Culver, M., Hall, J., Horton, R., Knuuti, K., Moss, R., Obeysekera, J., Sallenger, A.H., Weiss, J., 2012. Global Sealevel Rise Scenarios for the US National Climate Assessment. NOAA Technical Report, National Oceanic and Atmospheric Administration.

Chapter 4

Insights from recent economic modeling on port adaptation to climate change effects

Laingo M. Randrianarisoa[a], Kun Wang[b], Anming Zhang[a]

[a]*Sauder School of Business, The University of British Columbia, Vancouver, BC, Canada,* [b]*School of International Trade and Economics, University of International Business and Economics, Beijing, China*

1 Introduction

Scientific studies suggest that climate change may have led, and will likely lead, to an increase in both the occurrence and the strength of weather-related natural disasters (e.g., Keohane and Victor, 2010; Min et al., 2011; IPCC, 2013). Seaports are, especially, highly vulnerable to disasters, such as hurricanes, strong wind, and heavy rainfall (Wang and Zhang, 2018). Just a very recent and prominent example is typhoon Mangkhut hitting South China in September 2018. It completely shut down the Port of Shenzhen and the Port of Hong Kong (the world's third and fifth largest container ports in 2018, respectively) for more than three days, and it took several more days for the terminal operators to resume normal operations. As a result, more than 200 containerships were considerably delayed for loading and unloading, and the economic losses were huge along the supply chain.[a]

As seaports are critical nodes in global supply chains, any major loss or degradation of port services would have significant "knock-on" effects on national/global supply chain performance and overall economy (OECD, 2016; Jiang et al., 2017). Understandably, seaports around the world are now increasingly aware of the climate change-related threats and seriously consider the associated adaptation investments (Becker et al., 2013; Ng et al., 2018a,b; Yang et al., 2018). For the last several years, port adaptation has also attracted increasing

---

a. Hong Kong, Shenzhen Port shut down for three days from September 14 to 16 when Mangkhut passed by (see the article in http://www.sofreight.com/news_27402.html). There are more than 200 containerships were adversely affected and delayed in loading and unloading (see the article in https://baijiahao.baidu.com/s?id=1611936695711743447&wfr=spider&for=pc).

Maritime Transport and Regional Sustainability. https://doi.org/10.1016/B978-0-12-819134-7.00004-6
45

attention from the academic field. There have been empirical (and case-based) studies to evaluate a port's risk of the climate change-related disasters (e.g., Yang et al., 2018), existing and planned adaptation measures (e.g., Ng et al., 2018a; He et al., 2019), and the suggested coordination among stakeholders (e.g., Becker et al., 2013; Messner et al., 2015; Ng et al., 2018a,b). These empirical studies have contributed to a better understanding of the basic issues and challenges faced by ports in adapting to climate change-related disasters in practice.

Compared to the empirical studies, economic modeling of port adaptation is relatively rare but is fast developing. These theoretical studies are exemplified by the recent work of Xiao et al. (2015), Wang and Zhang (2018), Liu et al. (2018), Randrianarisoa and Zhang (2019), and Wang et al. (2019). They developed analytical frameworks by applying industrial organization and game theoretical models, which provide a good tool to analyze complicated strategic interactions among different stakeholders and factors present in port adaptation investment decisions. The factors considered include disaster's uncertainty, interport and intraport market structures, and shipping market demand characteristics.

This chapter provides a review of existing modeling work on port adaptation to climate change-related disasters. First, we summarize the major issues and factors considered in the existing economic models and describe the basic modeling approaches, accordingly. Second, we reconcile and discuss the findings of the current analytical work. On one hand, we attempt to sort out what the main results are, regardless of the different model assumptions and specifications. These consistent findings are meaningful for general managerial and policy implications. On the other hand, we examine the seemingly "inconsistent" findings among the studies, especially when models differ in setups or assumptions. This review aims to help the readers better understand the complicated nature of the port adaptation decision-making along with the potential drivers of the decisions. Finally, we propose some avenues for future theoretical research on port adaptation investment.

The rest of this chapter is organized as follows: Section 2 summarizes modeling frameworks of the existing theoretical work, and the factors considered. Section 3 discusses and compares the major analytical conclusions of these papers. Future research avenues are discussed in Section 4. Section 5 concludes this chapter.

2 Theoretical framework

As mentioned above, economic modeling of port adaptation is emerging. In this section we review several papers on port adaptation, namely, Xiao et al. (2015), Wang and Zhang (2018), Liu et al. (2018), Randrianarisoa and Zhang (2019), and Wang et al. (2019). In Section 2.1, we highlight the basic elements and issues incorporated in each of their economic models, since these studies focus on several different aspects of port adaptation investment decisions. This part is summarized in Table 1. In Section 2.2, we focus on the recent and representative analytical models of Wang and Zhang (2018) and Randrianarisoa and Zhang (2019).

TABLE 1 Summary of the main issues and factors considered in existing economic models

(a) The climate change-related disaster uncertainty

Paper	Disaster occurrence uncertainty	Investment efficiency	Investment period
Xiao et al. (2015)	Uniform distribution	Deterministic	Two-period (information accumulation)
Wang and Zhang (2018)	Knightian uncertainty	Deterministic	Single period
Liu et al. (2018)	Exponential function	Deterministic	Single period
Randrianarisoa and Zhang (2019)	Bernoulli trial/Knightian uncertainty	Stochastic	Two-period (information accumulation)
Wang et al. (2019)	Knightian uncertainty	Deterministic	Single period

(b) The port structure and decision dimensions

Paper	Port system	PA, TOC adaptation	PA, TOC pricing	PA interport cooperation	PA intraport cooperation	TOC market structure	PA ownership
Xiao et al. (2015)	Single port	Yes, yes	No, No	NA	Yes	No	Private and public
Wang and Zhang (2018)	Two-port	Yes, yes	Yes, Yes	Yes	Yes	No	Private and public
Liu et al. (2018)	Two-port	Yes, NA	No, NA	Yes	NA	NA	Private
Randrianarisoa and Zhang (2019)	Two-port	Yes, no	Yes, Yes	Yes	No	No	Private and public
Wang et al. (2019)	Two-port	Yes, no	Yes, Yes	Yes	No	Yes	Private

Note: PA stands for "port authority" and TOC stands for "terminal operator company."

2.1 Basic economic issues in existing modeling work

Port adaptation investments are affected by two broad set of factors, namely (i) the nature of the uncertainties regarding the climate change-related disaster per se, which includes, among others, storm surge, hurricane, flooding, increased precipitations, high wind, tidal surge, typhoon surge, cyclone, and earthquake, and (ii) the port market structure. Table 1(a) and (b) summarize the existing theoretical studies focusing on various specific factors regarding the disaster uncertainties and port market structure. First, disaster uncertainties may evolve dynamically such that ports have to decide when to invest, i.e., now or later. While early investment is beneficial for shippers and presents several economic advantages for the ports, there is an option value to wait for later adaptation investment, especially if better information can be accumulated. Second, as shown in Table 1(b), the existing theoretical models have been conducted for a single port or two-port system. Most of them have examined the pricing and adaptation investment decisions of landlord ports, the specific type consisting of a private or public port authority (PA), and private terminal operator companies (TOCs). Existing research has also analyzed cooperation between PA and TOCs within a single port, and cooperation across two ports, when making adaptation investment. Some frameworks have incorporated the TOC's intraport competition and interport relations. Next, in Sections 2.1.1 and 2.1.2, we review detailed modeling elements.

2.1.1 Uncertainties affecting port adaptation

Although the potential damage is severe, the uncertainties associated with the climate change-related disasters are still very high. Xiao et al. (2015), as the first analytical work on this topic, modeled disaster uncertainty in a two-period dynamic setting, assuming the disaster to be a Bernoulli trial. By definition, a Bernoulli trial is a random experiment with two possible outcomes: success and failure. In our context, success represents the occurrence of a disaster and failure indicates that the disaster does not happen. The disaster occurrence probability is uniformly distributed at both periods, but the probability in the second period is more accurate (i.e., more narrowly bounded uniform distribution) due to information learning. It implies that there is an option value in investing later with better information accumulation on the disaster occurrence probability.

Similar to Xiao et al. (2015), Randrianarisoa and Zhang (2019) adopted a two-period dynamic model, allowing port to invest in adaptation earlier or later. However, the uncertainty is assumed to be on the efficiency of port adaptation, instead of the disaster occurrence probability. It is noted that efficiency of port adaptation measures the return on investment in disaster prevention infrastructures—in terms of damage reduction—relative to the investment

---

b. The unit of measure can be percentage (%) or ratio. For instance, ports facing frequent inundation have three options in terms of adaptation: update storm defenses, elevate to compensate for projected sea levels, or relocate entirely. If the ports decide to invest in the second option, the efficiency of adaptation would be the return on the investment in elevation of some parts or the entire port facilities.

costs.[b] Over time, thanks to improved technology and better knowledge and planning of the adaptation projects, port adaptation is likely to be more efficient. Specifically, Randrianarisoa and Zhang (2019) assumed that the adaptation efficiency at later period stochastically dominates that in the earlier period. There is, thus, an option value to postpone port adaptation with higher expected efficiency of port adaptation investment. However, like Xiao et al. (2015), waiting may not always be optimal, especially when the disaster occurrence probability is high, exposing the port to no protection in early period.

Unlike Xiao et al. (2015), the work of Wang and Zhang (2018) and Wang et al. (2019) excluded dynamic choice of the ports in the timing to invest in adaptation. Instead, they focused on the degree of disaster uncertainty and its effect on the port adaptation. Specifically, Wang and Zhang (2018) first introduced the Knightian uncertainty (Knight, 1921) into the port adaptation modeling. This Knightian uncertainty assumes that the probability of a Bernoulli trial (disaster occurrence) is a random variable, with an expectation and variance. Knightian uncertainty has been well applied in economic decision and investment literature (Camerer and Weber, 1992; Nishimura and Ozaki, 2007; Gao and Driouchi, 2013), but it is the first time to be introduced to model climate change-related disaster uncertainty faced by ports. This concept is ideal for the port adaptation case, as it captures the existing great ambiguity on climate change-related disaster occurrence. Though scientists and port stakeholders try very hard to estimate the disaster occurrence probability, the prediction is always inaccurate and falls within a wide confidence interval, mainly due to our limited scientific knowledge. The Knightian uncertainty also captures the information accumulation on disaster occurrence probability by assuming the variance decreases over time. In Wang and Zhang (2018), to simplify the analysis, such variance totally disappears in later period, which is an extreme case of the information updating on disaster occurrence probability. But unlike Xiao et al. (2015) and Randrianarisoa and Zhang (2019), the port adaptation can only be made before information gaining. This is the major variation among these studies, causing seemingly contradictory conclusions (we will discuss in later section in detail).

Randrianarisoa and Zhang (2019) also considered Knightian uncertainty for the disaster occurrence probability and investigated how it would affect the model setup. Though they could not derive any explicit analytical results due to analytical tractability, they provided a framework for implementing simulation exercises and empirical analyses. Last, Liu et al. (2018) have considered neither dynamic period nor information gaining. They assumed a given disaster occurrence probability, with ports making adaptation investment in a single period.

To summarize, existing modeling has accounted for the major elements of the uncertainties on port adaptation decisions, such as option value in timing choice decision due to information accumulation, uncertainty of disaster occurrence probability and of efficiency of adaptation investment.

2.1.2 Port market structure

The port market structure could be complex depending on the port type, PA ownership, downstream TOC market, and interport and intraport competing or cooperative relationship. This leads to significant complexity when trying to capture the impact of port market structure on port adaptation investment decisions. Since it is extremely difficult to incorporate all these port market structural elements within one single model, existing studies chose to focus on the subset of these specific characteristics, as exhibited in Table 1(b). We discuss these major elements as follows:

- *Single or two-port system*

The first economic modeling work on port adaptation by Xiao et al. (2015) considered a single port. As their study is focused on the optimal timing of adaptation and intraport structure, the interport competition and cooperation are excluded. However, in real world, it is common to observe cluster of ports in one region competing or cooperating with each other. For example, in Pearl River Delta, Port of Hong Kong competes with Port of Shenzhen to be the gateway of South China; Hamburg-Le Havre (HLH) port ranges have several competing ports to be the gateways to West and North Europe; Port Authority of New York and New Jersey controls Port of Newark, Port of Perth Amboy and Port of New York, Georgia Port Authority controls Port of Savannah and Port of Brunswick on the East coast of the United States. These ports are likely to be exposed to a common disaster threat, leading to strategic interactions in port adaptation decisions.

Later theoretical studies extended to a two-port system, enabling to analyze interport competition and cooperation on the regional port adaptation investments. For example, Wang and Zhang (2018) explored the optimal adaptation levels when two ports cooperate and compete with each other. Wang et al. (2019) further recognized the joint venture (JV) of TOCs across two ports. With a dynamic investment setting, Randrianarisoa and Zhang (2019) modeled the asymmetric timing of port adaptation between two ports, i.e., one port invests early while the other late. Such equilibrium is compared to the cases of both investing in early or late period. While these studies assume that two ports' services are substitutable, Liu et al. (2018) considered the case of two ports with complementary services.

- *Intraport vertical structure*

Ports can be categorized into four types (Liu, 1992): service port, tool port, landlord port, and private port. A service port is characterized by a PA that is responsible for the provision of all port facilities. A tool port consists of a public PA that provides infrastructures and superstructures, while the provision of services is licensed to private operators. For a landlord port, the domain of the PA (public or private) is restricted to the provision of infrastructures, while investment in superstructures and port operations is the responsibility of licensed private companies. Finally, the provision of all the facilities and services of a private port is left to one single private entity. Xiao et al. (2015) and Wang and Zhang (2018) considered that the adaptation investments are made by the port authorities and port tenants, while Randrianarisoa and Zhang (2019) and Wang et al. (2019) focused on the case where the port authorities decide on the adaptation investments. In practice, the decisions on adaptation vary across ports. In some cases, the PAs fully support the costs of adaptation investments. This has been the case of Port of Boston's "Infrastructure Disaster Resiliency" project for 2016–20, which is entirely financed by Massport. In some other cases, collaboration between a wide range of public and private stakeholders is required. Becker et al. (2012) list the potential actors that can be involved in decision-making at different stages of adaptation planning. These actors, among others, are regulators, terminal operator companies, shippers, insurers, scientists, engineers, planners, and financers. The private/public stakeholders can also represent the federal, regional, and local government, city, port authorities, and nonprofit environmental and local organizations. For example, the adaptation project "Pier S Shoreline Protection: Seawall Retrofit" for Port of Long Beach to begin in 2020 is funded by the Port of Long Beach, Vopak (a private company that operates the chemical off load and storage facilities at the site), Nielson Beaumont Marine (the owner of the small boat marina), and the barge operators. Equivalently, the "Upgrade of Pavement Subgrade at Howland Hood Marine Terminal (part of the 2017 Sandy Program)" project at Port of New York New Jersey (NYNJ) was funded by insurance, federal public assistance, and PA of NYNJ.

With the exception of Liu et al. (2018) to study private ports, most theoretical papers have addressed the case of landlord ports, as the majority of ports around the world belong to this type (Cheon et al., 2010). A typical landlord port consists of an upstream PA (public or private) and downstream private TOCs. For example, PSA International, Hutchison Port Holding, APM terminals, DP World, and China Merchant Holding are the major TOC corporations operating worldwide. First, PA and TOCs have a vertical relationship in which the TOCs sign concession contracts with the PA to get access to the port basic infrastructures. TOCs, as tenants, own the superstructures of the port to handle the daily port operations and charge service fees to shipping companies/shippers (Trujillo and Nombela, 2000; De Monie, 2005; Notteboom, 2006). Second, PA is primarily responsible in investing in port adaptation, as it owns the basic port infrastructures and lands. Several adaptation measurements such as building breakwaters, storm barriers, and flood-control gates are not specific to particular terminal or berth (Becker et al., 2012); therefore adaptation investment has to been done by PA. TOCs might also be able to make adaptation investment specifically on their owned berth, terminals, and facilities, such as elevating terminal, upgrading the drainage system, and redesigning and retrofitting of the terminal facilities (Becker et al., 2012).

Thus, in principle, for a landlord port in the market, both PA and TOCs have two decision variables: pricing and adaptation investment. Since adaptation investment normally takes long time to plan and complete, while the port price is easy to adjust in a short term, the port pricing decisions could be conditional on port adaptation investment. This seems intuitive because a well-adapted port is able to charge a premium as their users are better protected against climate change-related disasters threat. In this way, the port pricing decisions are endogenized and linked with the port adaptation decisions.

Xiao et al. (2015) concentrated on the adaptation investment made by PA and TOCs by assuming exogenous port prices. Thus, the port strategic pricing behaviors and its impact on ex ante port adaptation investment cannot be analyzed. Recognizing this limitation, Wang and Zhang (2018) incorporated the pricing decisions of both PA and TOCs conditional on port adaptation investment in their setting. A vertical structure is imposed to model that the PA first decides on the concession fee to be charged to the downstream TOCs, and TOCs in turn decide the service charge to shipping companies/shippers. Wang and Zhang (2018) also allowed TOCs to invest in adaptation. Randrianarisoa and Zhang (2019) and Wang et al. (2019) are basically in the spirit of Wang and Zhang (2018) to endogenize port pricing decisions conditional on the port adaptation. However, these two models rule out the adaptation decisions by TOCs in order to guarantee model tractability.

● *Public or private port authorities*

For landlord port, one essential issue in existing theoretical model is PA's ownership. Public PA is assumed to maximize social welfare, and the private PA for its own profit. Normally, social welfare includes the profits of PA and TOCs, along with the shipper's surplus (Xiao et al., 2015; Wang and Zhang, 2018). As the adaptation investment decisions have to be made ex ante, the social welfare, profits, and shippers' surplus all refer to their expected values. As an important extension, Randrianarisoa and Zhang (2019) further considered the positive spill-over effect of port adaptation on the nearby communities and regional economy. Thus, the social welfare is extended to a larger social scope.

Section 2.1 summarizes the major issues and factors included in the existing economic models on port adaptation. In the next subsection, we review the representative modeling framework in detail. It helps explain more clearly how the economic issues are actually modeled. Moreover, with a basic economic model framework, we discuss possible feasible extensions to accommodate different economic issues as future research avenues.

2.2   Economic modeling

In this section, we introduce and discuss the economic model developed by Wang and Zhang (2018) upon which the models of Randrianarisoa and Zhang (2019) and Wang et al. (2019) are based on. While Xiao et al. (2015) is the first economic modeling work on port adaptation, Wang and Zhang (2018), Randrianarisoa and Zhang (2019), and Wang et al. (2019) have more rich factors being modeled. The model of Liu et al. (2018) is less sophisticated, without considering port pricing, intraport vertical structure, and investment timing issues. Thus, the extensions of the framework in Liu et al. (2018) are quite limited.

Wang and Zhang's (2018) model is divided into two parts, namely the operation stage with port pricing and the adaptation investment stage. Randrianarisoa and Zhang (2019) basically followed Wang and Zhang (2018) in the port pricing stage set-up conditional on port adaptation, while they extended the port adaptation investment stage to a two-period dynamic setup. By contrast, Wang et al. (2019) made extension only on the operation stage by considering more complicated structure in TOC market structure. Table 2 summarizes the notations and parameter definitions in the model of Wang and Zhang (2018).

Wang and Zhang (2018) considered a two-port region subject to a common disaster threat, as shown in Fig. 1. They examine the impacts of interport competition and cooperation between PAs and intraport cooperation between the PAs and TOCs on port adaptation. A multistage game is used to model both the adaptation investment stage and the operation stage conditional on the adaptation investments. The timeline of the model is given in Fig. 2. Wang and Zhang (2018) assumed the disaster to be a Bernoulli trial at the operation stage with occurrence probability x. The probability of disaster occurrence x is assumed to be ambiguous at the adaptation investment stage, which is a Knightian uncertainty.

TABLE 2 Notational glossary of the model parameters in Wang and Zhang (2018)

Parameter	Definition
V	Utility to shipper of using the port service
D	Disaster damage level to the shipper, and we assume $D < V$
η	Adaptation efficiency to reduce damage
t	Unit distance transport cost for the shipper to move cargo to the port
I_i^a	Adaptation investment made by port authority at port i
I_i^t	Adaptation investment made by TOC at port i
x	Random variable denoting probability of the disaster occurrence
Ω	Expectation of x at the adaptation investment stage
Σ	Variance of x at the adaptation investment stage
Ψ	Second moment of x, which is equal to $\Omega^2 + \Sigma$
p_i	Service fee charged by TOC to shippers at port i
ϕ_i	Concession fee charged by port authority to TOC port i
Q_i	Demand for service at port i at the operation stage
Π_i	Profit of TOC at port i at operation stage
π_i	Profit of port authority at port i at operation stage

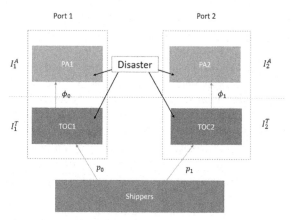

FIG. 1 The market structure of the two-port system in Wang and Zhang (2018).

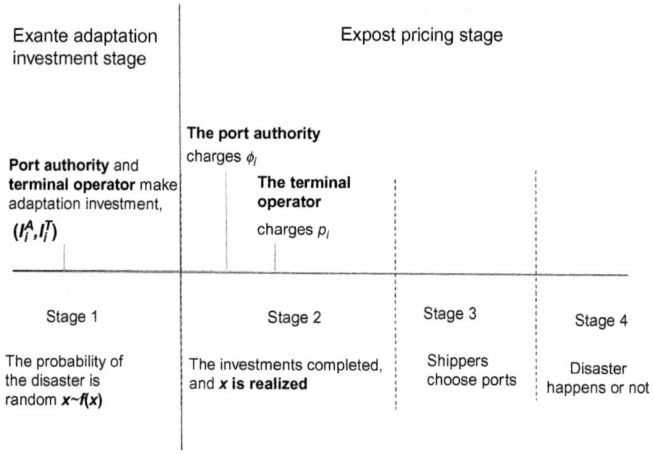

Exante adaptation investment stage | Expost pricing stage

Port authority and terminal operator make adaptation investment, (I_i^A, I_i^T)

The port authority charges ϕ_i

The terminal operator charges p_i

Stage 1 — The probability of the disaster is random $x \sim f(x)$

Stage 2 — The investments completed, and x is realized

Stage 3 — Shippers choose ports

Stage 4 — Disaster happens or not

FIG. 2 The timeline of the decisions of different parties in Wang and Zhang (2018).

Knightian uncertainty implies that the disaster occurrence probability x can be a random variable at the adaptation investment stage, with a probability density function (pdf) $f(x)$, expectation Ω, and variance Σ. But this probability only becomes realized later at the operation stage when the ports decide the price and the shippers choose a port. This improvement in information reflects a relevant setting in which a better knowledge of climate change and related disasters is accumulated during the lengthy period of adaptation investment.

2.2.1 Port demand and pricing decisions conditional on port adaptation

At the operation stage, Wang and Zhang (2018) adopted an infinite linear city model to derive shippers' demand conditional on port service charges p_i and port adaptation investments $\{I_i^a, I_i^t\}$ in response to disaster occurrence probabil-

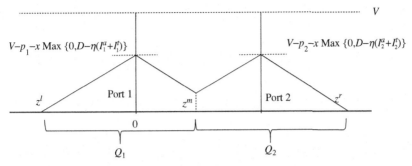

FIG. 3 Shipper's utility at each port after completion of adaptation investments in Wang and Zhang (2018).

ity x, where I_i^a is the adaptation by PA and I_i^t by TOC. This demand function has also been inherited by Randrianarisoa and Zhang (2019) and Wang et al. (2019). The infinite linear city model is demonstrated in Fig. 3.

The value to a shipper of using the port service, denoted by V, is exogenously given. Shippers who are the owners of cargo to be shipped to the destination choose which port to use before observing if the disaster occurs or not. If the disaster occurs, shippers will incur damage $D - \eta(I_i^a + I_i^t)$, where D is the damage level without any port adaptation, and $\eta(I_i^a + I_i^t)$ is the reduction of damage owing to port adaptation investments. η measures the adaptation efficiency to mitigate damage when the disaster occurs. The disaster damage to shippers can include the cargo damage and inventory delay cost. If the disaster does not occur, the shippers do not incur any cargo damage. With disaster occurrence probability x, the expected damage incurred by the shippers is $x \text{Max}\{0, D - \eta(I_i^a + I_i^t)\}$.

Shippers are assumed to be uniformly distributed along the linear city with density 1. Each shipper incurs a cost per unit distance, denoted by t, to transport cargo from its location to the port. This transport cost can also capture any horizontal differentiation (service homogeneity) of two ports' services perceived by the shippers. Shippers choose which port to use, and directly pay the service price to TOC. TOC, in turn, pays a concession fee to PA in exchange for the use of the port lands and basic infrastructures. The port charging thus takes place in a vertical structure: PA chooses its concession fee, ϕ_i, on TOC first, and then TOC chooses service charge, p_i, on shippers.

For a shipper located at point z in the two ports' common hinterland, the utility of using port 1 is $V - p_1 - zt - x \text{Max}\{0, D - \eta(I_1^a + I_1^t)\}$, and the utility of using port 2 is $V - p_2 - (1 - zt) - x \text{Max}\{0, D - \eta(I_2^a + I_2^t)\}$. For a shipper located at point z in port 1's captive hinterland, the utility is $V - p_1 - |z|t - x \text{Max}\{0, D - \eta(I_1^a + I_1^t)\}$, and for a shipper located at point z in port 2's own hinterland, the utility is $V - p_2 - (z - 1)t - x \text{Max}\{0, D - \eta(I_2^a + I_2^t)\}$. We can therefore derive the locations of the marginal shipper (i) who is indifferent between using port 1's service and not using the port service at all, denoted by z^l; (ii) the one who is indifferent between using port 2's service and not using the port services, denoted by z^r; and (iii) the one who is indifferent between using port 1 and port 2's service, denoted by z^m. The locations are given by

$$\left|z^l\right| = \frac{V - p_1 - x\,\mathrm{Max}\left\{0, D - \eta\left(I_1^a + I_1^t\right)\right\}}{t}; \quad (1a)$$

$$z^r = 1 + \frac{V - p_2 - x\,\mathrm{Max}\left\{0, D - \eta\left(I_2^a + I_2^t\right)\right\}}{t}; \quad (1b)$$

$$z^m = \frac{1}{2} + \frac{p_2 - p_1 - x\,\mathrm{Max}\left\{0, D - \eta\left(I_1^a + I_1^t\right)\right\} + x\,\mathrm{Max}\left\{0, D - \eta\left(I_2^a + I_2^t\right)\right\}}{2t}. \quad (1c)$$

The demand at the operation stage is $Q_1(p) = |z^l| + z^m$ for port 1 and $Q_2(p) = (1 - z^m) + (z^r - 1)$ for port 2. That is

$$Q_i(p) = \frac{1}{2} + \frac{2V + p_j - 3p_i + x\,\mathrm{Max}\left\{0, D - \eta\left(I_j^a + I_j^t\right)\right\} - 3x\,\mathrm{Max}\left\{0, D - \eta\left(I_i^a + I_i^t\right)\right\}}{2t}, \quad (2)$$

where $i = 1, 2$. With the above shipper demand function, private TOCs maximize profits Π_i conditional on the port adaptation and the concession fee charged by PA. That is

$$\mathrm{Max}\,\Pi_{i\,p_i} = \left(p_i - \phi_i\right)Q_i \quad (3)$$

where

$$p_i\left(\phi_i, \phi_j\right) = 0.2[(2V + t) + 2.57\phi_i + 0.42\phi_j - 2.43x\,\mathrm{Max}\{0, D - \eta(I_i^a + I_i^t)\} \\ + 0.42\,\mathrm{Max}\{0, D - \eta(I_j^a + I_j^t)\}]. \quad (4)$$

The port authorities in turn maximize their profits, π_i, if they are privately owned, and regional social welfare if they are public. The social welfare is the sum of consumer surplus, CS, which represents the shippers' benefits arising from the utilization of the port facilities, TOC profits, π_i, and PA profits, Π_i. It is noted that this social welfare does not account for any positive externality of shipping activities on general economy, while considering it will not change all the analytical conclusions qualitatively. The PA profits and regional social welfare are specified as follows:

$$\pi_i = \phi_i Q_i\left(p_i\left(\phi_i, \phi_j\right), p_j\left(\phi_i, \phi_j\right)\right) \quad (5)$$

$$CS = \int_0^{|z^l|}[V - p_1 - x\,\mathrm{Max}\{0, D - \eta(I_1^a + I_1^t)\} - zt]dz \\ + \int_0^{z^m}[V - p_1 - x\,\mathrm{Max}\{0, D - \eta(I_1^a + I_1^t)\} - zt]dz \\ + \int_{z^m}^1[V - p_2 - x\,\mathrm{Max}\{0, D - \eta(I_2^a + I_2^t)\} - (1 - z)t]dz \\ + \int_1^{z^r}[V - p_2 - x\,\mathrm{Max}\{0, D - \eta(I_2^a + I_2^t)\} - (z - 1)t]\,dz. \quad (6)$$

One notable feature of this infinite linear city shipper demand is the parameter t, capturing the intensity of interport competition (or service heterogeneity). As can be seen later, we are able to shed light on its impact on port adaptation investment.

Based on the above setup, Randrianarisoa and Zhang (2019), and Wang et al. (2019) made some changes to incorporate new elements.

- *Random adaptation efficiency by* Randrianarisoa and Zhang (2019)

Unlike Wang and Zhang (2018), who modeled disaster occurrence probability as a random variable (Knightian uncertainty), Randrianarisoa and Zhang (2019) considered a random adaptation efficiency parameter η in the shipper demand function. Specifically, they modeled η as a random variable that is distributed over the positive support $[0, \eta_{max}]$, with probability density function, $g(\eta)$ and cumulative density function $G(\eta)$. Furthermore, they built up a two-period dynamic model which assumes information updating on adaptation efficiency over time. This can be achieved with the technology and knowledge development, along with better planning and cooperation among different stakeholders of the port over time. To capture this efficiency improvement mathematically, the distribution of η in the later period is assumed to first-degree stochastically dominate that in the earlier period. In other words, the investment efficiency remains a random variable with same distribution functions over the two periods, but its observed values in the second period are higher. It implies that on average, efficiency of investment in the second period is higher than that in the first period.[c]

- *Terminal operator market structure by* Wang et al. (2019)

Most of the modeling work assumed a single TOC within one port. This greatly simplifies the model setup and derivation (Xiao et al., 2015; Wang and Zhang, 2018; Randrianarisoa and Zhang, 2019). However, one prominent feature of port market structure is that multiple TOCs may be present at one port. These TOCs may be operated by several independent companies. For example, PSA International, Hutchison Port Holding (HIT), APM terminals, DP World, and China Merchants Holding are the major TOC companies in the world. Not only do they actively compete with each other at many container ports (i.e., intraport competition of TOCs) but also they may compete across nearby ports for shippers in the common hinterland (interport competition of TOCs). In addition, the same TOC may be present at two nearby ports at the same time, thus forming an interport JV. Such market structure of intra- and interport TOC competition and JV can be well exemplified by Hong Kong Port and Shenzhen Port. Hong Kong Port has nine major container terminals, where Modern Terminal Limited (MTL) operates in four of them, HIT in four, HIT and COSCO jointly in one, DP World in one, and ACT in one. These TOCs compete with each other within Hong Kong Port. Meanwhile, MTL also invests in Shekou and

---

c. In their simulation exercise, Randrianarisoa and Zhang (2019) assumed that in the first period, investment efficiency follows log normal distribution with a mean of 0.2% and standard deviation of 0.1%. In the second period, efficiency follows the same distribution but with a higher mean of 1.2%.

Dachaiwan terminals, the major terminals of Shenzhen Port, and HIT operates another major terminal, Yantian terminal, in Shenzhen Port. Thus, these TOCs can coordinate their operations in Hong Kong Port and Shenzhen Port through the common ownership (joint venture).

Wang et al. (2019) extended the existing economic models, especially Wang and Zhang (2018), to formally examine the impact of the TOC market structure on port adaptation. The two-port structure is now revised compared to that of Wang and Zhang (2018) in Fig. 1, as shown in Fig. 4. Their study assumed a number N of TOCs at each port.

The profit of one TOC, denoted by $\pi_{r,i}$, where subscript $i \in \{1,2\}$ stands for the port, and $r \in \{1,2,....,N\}$ stands for the TOCs at one port, is given by

$$\pi_{r,i} = \left(p_i(Q_1,Q_2 | I_1,I_2) - \phi_i\right)q_{r,i}, \text{where } i = 1,2. \tag{7}$$

Wang et al. (2019) assumed quantity competition among TOCs (i.e., Cournot competition). Under a Cournot competition, the TOCs simultaneously decide on the amount of output they produce in a specific period so as to maximize their own profits. Then, given the amount of output, they set the service prices to be charged to the shippers. As observed in real business world, TOCs across the ports could be independent and compete with each other, or the same TOC may operate in two ports at the same time. If TOCs across ports compete independently, the first-order condition (FOC) for the TOC by choosing optimal quantity $q_{r,i}$ is

$$V - \phi_i - x\max\{0,D - I_i\} - \frac{t}{4}\left(3Q_i + Q_j - 2\right) - \frac{3t}{4}q_{r,i} = 0. \tag{8}$$

If the TOCs have joint venture across the ports, they maximize a joint profit, $\pi_{r,i} + \pi_{r,j}$, given by

$$\pi_{r,i} + \pi_{r,j} = (p_i(Q_1,Q_2|I_1,I_2) - \phi)q_{r,i} + (p_j(Q_1,Q_2|I_1,I_2) - \phi_j)q_{r,j} \tag{9}$$

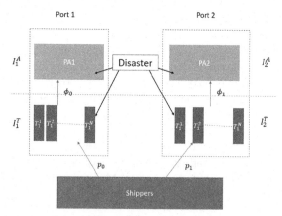

FIG. 4 Market structure of the two-port and multiple-operator system in Wang et al. (2019).

The FOCs for the TOC by simultaneously choosing optimal $q_{r,i}$ and $q_{r,j}$ are

$$V - \phi_i - x\max\{0, D - I_i\} - \frac{t}{4}(3Q_i + Q_j - 2) - \frac{3t}{4}q_{r,i} - \frac{t}{4}q_{r,j} = 0, \quad (10)$$

$$V - \phi_j - x\max\{0, D - I_j\} - \frac{t}{4}(3Q_j + Q_i - 2) - \frac{3t}{4}q_{r,j} - \frac{t}{4}q_{r,i} = 0. \quad (11)$$

Since the rest of the model derivations are analogous to Wang and Zhang (2018), we only present the crucial equations in this section. Moreover, with the current setting, we are able to generate new insights on how the TOC market structure affects port adaptation investment decisions.

2.2.2 Port adaptation investment

Existing economic models vary significantly in their treatment of disaster uncertainties and timing of adaptation investment. We have summarized this in Section 2.1. Wang and Zhang (2018) and Wang et al. (2019) assumed Knightian uncertainty of disaster occurrence probability, and this probability can be updated over time. But the adaptation investment is made at one single period. Randrianarisoa and Zhang's (2019) analysis, however, is based on a two-period dynamic setup and is focused on the timing of adaptation (earlier vs later period), with an information accumulation on the adaptation efficiency. They also considered how the assumed Knightian uncertainty for the disaster occurrence probability affects the modeling framework, thereby the main insights from the model. As these two frameworks [Wang and Zhang (2018) vs Randrianarisoa and Zhang (2019)] represent two different aspects of disaster or adaptation uncertainties, in this subsection, we review the detailed modeling approaches for port adaptation investment in both studies.

- *Knightian uncertainty in port disaster occurrence probability*

Wang and Zhang (2018) modeled a Knightian uncertain disaster occurrence probability such that PA and TOC of the two ports have to decide adaptation ex ante without any information updating on the disaster occurrence probability. Thus, the PAs maximize the expected profits or expected social welfare, depending on their ownership, and the TOCs, as private entity, maximize expected profits. The expected profits for private PAs at the investment stage are $E[\pi_i] = [\int \pi_i f(x) dx] - 0.5\omega I_i^{a2}$, and the expected profits for the terminal operators are $E[\Pi_i] = [\int \Pi_i f(x) dx] - 0.50\omega I_i^{t2}$. The expected social welfare for public PAs is $E[SW_i] = [\int SW_i f(x) dx] - 0.5\omega I_i^{a2}$. Constraint $\eta(I_i^a + I_i^t) \leq D$ must be imposed since ports cannot adapt beyond the maximum disaster damage level D. Therefore, we have: $\text{Max}_{I_i^a} E[\pi_i]$, st. $\eta(I_i^a + I_i^t) \leq D$ and $\text{Max}_{I_i^t} E[\Pi_i]$, st. $\eta(I_i^a + I_i^t) \leq D$. Here, we assume an increasing marginal adaptation investment cost with a quadratic function, $0.5\omega I_i^{a2}$ and $0.50\omega I_i^{t2}$, where superscripts a and t stand for PA and TOC, respectively (see Table 2 for reference on the parameter definition).

Wang and Zhang (2018) also modeled different interport and intraport competition and cooperation cases for private and public PAs. Specifically, the two ports can compete with each other in port adaptation such that two PAs maximize their own expected profits or regional social welfare as follows: $\mathrm{Max}_{I_i^a} E[\pi_i]$, st. $\eta(I_i^{\,a}+I_i^{\,t}) \leq D$ or $\mathrm{Max}_{I_i^a} E[SW]$, st. $\eta(I_i^{\,a}+I_i^{\,t}) \leq D$. Alternatively, the two ports can cooperate (the monopoly) so as to maximize a joint expected profit or total social welfare. That is: $\mathrm{Max}_{I_i^a,I_j^a} E[\pi_i+\pi_j]$, st. $\eta(I_i^{\,a}+I_i^{\,t}) \leq D$ or $\mathrm{Max}_{I_i^a,I_j^a} E[SW_i+SW_j]$, st. $\eta(I_i^{\,a}+I_i^{\,t}) \leq D$. They also considered intraport coordination in port adaptation by PA and TOC within the same port. In this case, the maximization problem is: $\mathrm{Max}_{I_i^a,I_i^t} E[\pi_i+\Pi_i]$, s. t. $\eta(I_i^{\,a}+I_i^{\,t}) \leq D$ or $\mathrm{Max}_{I_i^a,I_i^t} E[SW_i]$, s. t. $\eta(I_i^{\,a}+I_i^{\,t}) \leq D$.

- *Option value of adaptation timing*

Randrianarisoa and Zhang (2019) adopted a two-period dynamic setting, incorporating the PA's decision on adaptation timing (i.e., earlier vs later). An option value exists such that investing later is associated with better adaptation efficiency thanks to information accumulation and better cooperation among stakeholders. But this delay in adaptation would leave the port exposed to no protection in the first period, bringing high disaster damage risk for shippers. PA's objective function is highlighted as $\mathrm{Max}_{I,q^i,q^{ii}} W_j = W_j^i + k W_j^{ii}$, where $W_j^T = \pi_j^T + \alpha(\Pi_j^T + CS_j^T)$ denotes total welfare of port $j=1$, 2 at period T, π_j^T the PA profits, Π_j^T the TOC profits, and CS_j^T consumer surplus. The two periods are represented with the superscripts i and ii, respectively.

The weighted objective function W_j, introduced by the authors, combines PA's own profit with a share of TOC's profit and shipper's surplus. The share is captured by the parameter α. Therefore, when α is larger (smaller) and close to 1 (to zero), PA resembles more to a public entity (private entity) which maximizes social welfare (profits). It is noted that the two PAs must decide which period to install the adaptation facilities, at either period i or period ii, together with the specific investment levels, captured by the decision variable vector I. The model also allows the two ports to choose the timing of adaptation in an asymmetric manner (one adapts early, and the other late). Unlike earlier studies that assume price competition between PAs, Randrianarisoa and Zhang (2019) considered Cournot competition such that the PAs decide on quantities in each period, which are captured by the decision variable vectors q^i and q^{ii}. The decision variables and timings in their model are summarized in Table 3.

A more detailed expression of the objective function W_j is given by

$$W_j = \left(f_j^i q_j^i - C_j^i\right) + k\left(f_j^{ii} q_j^{ii} - C_j^{ii}\right) \\ + \alpha\left[\left(p_j^i - f_j^i\right)q_j^i + k\left(p_j^{ii} - f_j^{ii}\right)q_j^{ii} + \left(CS_j^i + kCS_j^{ii}\right)\right] \tag{12}$$

where f_j^T is the concession fee charged by PA j to its TOC at period T, p_j^i is the price charged by TOC to shippers, and q_j^T is the quantity produced by port j at period T. The disaster damage cost for the port is C_j^i and C_j^{ii} in the first and second periods, respectively, and their values depend on the port adaptation

TABLE 3 The decision variables and adaptation timing in Randrianarisoa and Zhang (2019)

	First period (T = i)	Second period (T = ii)
A and B invest early	$(I_A, I_B), (q_A^i q_B^i)$	(q_A^{ii}, q_B^{ii})
A and B invest late	(q_A^i, q_B^i)	$(I_A, I_B), (q_A^{ii}, q_B^{ii})$
A invests early and B waits j	$(I_A, q_A^i q_B^i)$	$(I_B, q_A^{ii}, q_B^{ii})$
A waits and B invests early j	$(I_B, q_A^i q_B^i)$	$(I_A, q_A^{ii}, q_B^{ii})$

investment level and random adaptation efficiency s η and its CDF $G(\eta)$. It is assumed that the distribution of η in the second period has a first-degree stochastic dominance over that in the first period, i.e., G_2 first-degree stochastically dominates G_1.

3 Discussions on existing theoretical findings

This section reconciles and compares the main analytical findings in our reviewed economic modeling work. On one hand, we show some robust conclusions less dependent on various model assumptions and specifications, thereby more general to be applied for business and policy implications. On the other hand, we highlight some seemingly contradictory findings, and scrutinize the underlying assumptions or mechanisms leading to such diversion.

Specifically, we can categorize the existing analytical findings into three major aspects: timing of port adaptation investment, effect of the disaster uncertainty, and effect of the port market structure. For the port market structure, we investigate several detailed elements, such as interport cooperation and competition on port adaptation, intraport vertical structure, PA ownership, and TOC market structure. We first summarize the major findings in Table 4 for easy reference.

3.1 Timing of port adaptation investment

The optimal timing of adaptation investment is modeled in Xiao et al. (2015) and Randrianarisoa and Zhang (2019) with a two-period real options model. While Randrianarisoa and Zhang (2019) assumed an information accumulation on the adaptation efficiency over periods, Xiao et al. (2015), instead, considered the information updating on the disaster occurrence probability. Both papers found that, when the average disaster occurrence probability is high, it is optimal to adapt in early period. This is because leaving the port exposed to high disaster risk without protection in the early period could be costly. However, when the disaster occurrence probability is low, waiting is a better option so as to take advantage of the information gain on disaster occurrence probability or adaptation efficiency. This is summarized in the following Proposition 1.

TABLE 4 The summary of major theoretical findings on port adaptation

	Port adaptation		Timing
Higher disaster occurrence probability	+		Early
	Xiao et al. (2015), Wang and Zhang (2018), Randrianarisoa and Zhang (2019), and Wang et al. (2019)		Xiao et al. (2015) and Randrianarisoa and Zhang (2019)
Higher disaster uncertainty	+	−	Late
	Xiao et al. (2015) and Randrianarisoa and Zhang (2019)	Wang and Zhang (2018) and Wang et al. (2019)	Xiao et al. (2015) and Randrianarisoa and Zhang (2019)
Interport competition	+		Early
	Wang and Zhang (2018), Randrianarisoa and Zhang (2019), and Wang et al. (2019)		Randrianarisoa and Zhang (2019)
Intraport coordination (between PA and TOC)	+		NA
	Xiao et al. (2015) and Wang and Zhang (2018)		
Public port ownership	+		Early
	Xiao et al. (2015), Wang and Zhang (2018), and Randrianarisoa and Zhang (2019)		Xiao et al. (2015) and Randrianarisoa and Zhang (2019)
TOCs competition	+		NA
	Wang et al. (2019)		

Note: the "+" sign indicates increase in port adaptation, and "−" sign indicates decrease.

Proposition 1 *When the disaster occurrence probability is high, it is optimal for decision makers to invest in early period, despite the option value associated with the information accumulation on the disaster occurrence probability or adaptation efficiency. On the contrast, when the disaster occurrence probability is low or intermediate, it is optimal to wait to achieve the option value by utilizing the information updating on the disaster uncertainty.*

Randrianarisoa and Zhang (2019) further incorporated two-port competition into their analysis, and found that interport competition encourages ports to adapt early not late. This is stated as Proposition 2. The rationale is that when port has competitive pressure, they are more willing to make the adaptation earlier to gain competitive advantage. Randrianarisoa and Zhang (2019) also showed that the impacts of disaster risk and adaptation on investment timing are as important as that of competition and information accumulation. For instance, when the risk of being hit by a disaster is very high, the port that invests early would attract more shippers than the one investing late. This is because shippers value resilient infrastructures, and by using the most resilient port, they would reduce their expected disaster-related damage costs. They also showed that greater disaster occurrence probability requires large adaptation investment. The same reasoning applies when the shippers change ports. Specifically, the most resilient port will always receive higher demand, given all other variables. For instance, if port A is highly vulnerable to hurricane but its rival, say port B, is at a lower risk, without adaptation by port A, the shippers will choose port B, *ceteris paribus.*

Proposition 2 *When competition is intensified, it is optimal for ports to invest earlier than later. Immediate investments are less preferred when competition is weak, even lesser in the presence of information accumulation.*

3.2 Uncertainty of disaster and adaptation efficiency

Our reviewed analytical work diverts in the approaches to model uncertainties related to adaptation. For example, Xiao et al. (2015) assumed a uniformly distributed disaster occurrence probability in two periods, with the one in the second period having a narrowed range to reflect the information updating. Randrianarisoa and Zhang (2019) assumed a constant disaster occurrence probability, while the adaptation efficiency follows a more general distribution in both periods. For analytical tractability reasons, they proceeded with simulations to derive the main insights from the model, and assumed log normal distributions for the investment efficiency. The efficiency in the later period is updated based on the early one, implying that the distribution in the second period has a first-degree stochastic dominance over that in the first period. Both studies concluded that the ports would adapt less in the presence of information accumulation either on the disaster occurrence probability or on adaptation efficiency.

Wang and Zhang (2018) modeled the disaster occurrence probability with the Knightian uncertainty, and it does not restrict to any specific form of distribution. That said, the disaster occurrence probability is a general random variable. Then, the authors investigated how port adaptation changes with such Knightian uncertainty (i.e., expectation and variance of disaster occurrence probability). They found that the port adaptation is increased with a higher expectation but a lower variance of the disaster occurrence probability.

It is noted that, under Wang and Zhang (2018), port adapts more when disaster occurrence probability has low variance (less Knightian uncertainty). But Xiao et al. (2015) and Randrianarisoa and Zhang (2019) concluded that the information accumulation (less uncertainty at later period) could reduce port's incentive to adapt. The above results are not contradictory, in the sense that Wang and Zhang (2018) referred to ex ante adaptation decision before any information updating, while Xiao et al. (2015) and Randrianarisoa and Zhang (2019) are for the ex post adaptation decision made post the information accumulation. This seems intuitive as the lower ex ante ambiguity in disaster occurrence imposes lower risk to decision maker, thus increasing the expected investment return (Camerer and Weber, 1992; Nishimura and Ozaki, 2007; Gao and Driouchi, 2013), while a lower ex post uncertainty in disaster occurrence makes the port to have better information to make the optimal level adaptation, not necessarily to overinvest. These results are stated in Proposition 3.

Proposition 3 *When adaptation is invested ex ante (before information accumulation on disaster uncertainty), less uncertainty in disaster uncertainty encourages port adaptation investment. But when adaptation is made ex post, the presence of the information accumulation on the disaster uncertainty reduces port adaptation.*

3.3 Port market structure

Port's optimal adaptation investment is affected not only by the disaster uncertainty but also by the port market structure. In this subsection, we review the analytical results related to the port market structure and its impact on port's adaptation investment.

- *Interport cooperation and competition*

Wang and Zhang (2018), Randrianarisoa and Zhang (2019), and Wang et al. (2019) considered interport competition, and analyzed its effect on port adaptation investment. All studies found that interport competition would increase the port adaptation. Wang and Zhang (2018) defined it as the "competition effect" on port adaptation. Three studies also concluded that such competition effect can be strengthened with more intensity of the interport competition (service homogeneity). This intensity is captured by the road toll or transport cost parameter t in the infinite linear model of shipper demand. This finding is stated in Proposition 4.1.

Proposition 4.1 *Interport competition increases port adaptation (the "competition effect"). More intense interport competition (less service heterogeneity) strengthens such competition effect on adaptation.*

The intuition of this proposition is that adaptation investment can be regarded as a competitive tool for ports to attract shipper demand. Adaptation across ports is a strategic substitute such that ports have incentives to invest more in adaptation to compete with each other. In addition, as discussed earlier, in a dynamic setup (Randrianarisoa and Zhang, 2019), it is also found that the interport competition makes it more likely for two ports to adapt to early period than waiting. With Knightian uncertainty on the disaster occurrence probability (Wang and Zhang, 2018), the competition effect is further strengthened by a higher expected value and variance of the disaster occurrence probability.

- *Intraport vertical structure*

As discussed earlier, a landlord port consists of PA (upstream) and TOC (downstream). In Xiao et al. (2015) and Wang and Zhang (2018), both PA and TOC can make adaptation investment. The two papers found there is a free-riding in adaptation efforts between the PA and TOC at the same port. Specifically, the aggregate adaptation investment is higher if the two parties are able to coordinate. This free riding happens because the port adaptation at the same port benefits both PA and TOC (a positive externality to each other), but the investment cost is private. This thus discourages individual incentive to invest in adaptation, leading to a suboptimal adaptation level. Therefore, it is suggested that a vertical coordination among different shareholders within the same port should be promoted to overcome the free-riding problem. Besides, in practice, the decision-making on adaptation may involve multiple levels of governments as well as public and private stakeholders. For example, the U.S Army Corps of Engineers discussed sea-level rise (SLR) and storm surge impacts on the 2010 multimillion $ project: "Savannah Harbor Expansion Project" for the Port of Savannah. The participants include US Army Corps of Engineers (USACE), Environmental organizations (Georgia Conservancy), Georgia Port Authority, and State of Georgia. Proposition 4.2 summarizes this result.

Proposition 4.2 *PA and TOC within the same port free-ride each other in making port adaptation. Coordination between the two entities would then stimulate the aggregate adaptation investments.*

- *PA ownership (public vs private)*

Wang and Zhang (2018) directly benchmarked port adaptation levels between public and private PAs, and found that public PA invests more port adaptation. Randrianarisoa and Zhang (2019) further showed that public PA is also more likely to invest early. However, higher or early port adaptation by public port does not necessarily lead to a higher expected social welfare. Xiao et al. (2015) suggested that there are risks of overinvestment (i.e., the marginal benefits of investments are zero ex post if there is no disaster) such that a regulatory intervention is not always optimal when the regulator does not have a good understanding of disaster probability distribution. Similarly, Wang and Zhang (2018) found that, with intraport coordination between PA and TOC in adaptation efforts, the public PA could overinvest over socially optimal level when trying to correct the lower adaptation incentive of the private TOC.

However, the current modeling work has not well accounted for the external benefit of port adaptation for the neighboring regions. The local communities and regional economic activities near the port areas can also be protected by the port adaptation investment. Randrianarisoa and Zhang (2019) is the only work attempting to incorporate such external benefit by adding an extra positive term in the social welfare expression. Intuitively, the socially optimal port adaptation level should be higher and installed earlier when the social welfare is enlarged to a broader scope. Meanwhile, the concern of overinvestment by public port could be partially alleviated as well. We summarize the above discussions in Proposition 4.3.

Proposition 4.3 *Public PA invests more adaptation than the private. However, there may be overinvestment, thus not necessarily resulting in the socially optimal adaptation (the first best outcome). When accounting for the positive externality of port adaptation on neighboring communities or regional economy, the socially optimal port adaptation level is higher, and the overinvestment concern associated with the public PA is partly alleviated.*

- *TOC market structure*

Among our reviewed analytical modeling work, Wang et al. (2019) is the only paper to consider the specific market structure of TOCs. As shown in Fig. 4, they assumed that there are N TOCs at each of the two competing ports. TOCs conduct Cournot (quantity) competition within and across the ports. Across-port TOCs could also form JV due to the common ownership. Wang et al. (2019) showed that the TOC market structure considerably affects port adaptation investment. Specifically, they found that port adaptation increases with the number of TOCs present at each of the two ports. Interport competition among TOCs leads to higher port adaptation than that under cross-port JV (i.e., the competition effect of terminal operator). They also followed Wang and Zhang (2018) in adopting the Knightian uncertainty assumption on disaster occurrence probability, and showed that a more competitive TOC market would make PA more aggressive to invest in port adaptation. This strengthens the positive effect of the expected disaster occurrence probability on adaptation, while weakening the negative effect of its variance.

Proposition 4.4 *PA increases adaptation investment with a larger number of TOCs at each port. Under the assumption of Knightian uncertainty for disaster occurrence probability, a larger number of TOCs at each port strengthens the effect of expected disaster occurrence probability on port adaptation investment at the port adaptation stage, while weakening the effect of its variance. Interport competition among the independent TOCs induces higher port adaptation than that of TOCs' joint venture (i.e., competition effect of TOCs).*

The intuition of Proposition 4.4 is as follows: When TOC market is more competitive, the port throughput would be enlarged due to lower port service charge, ceteris paribus. This then increases the marginal benefit of PA's adaptation investment, as the same level of adaptation can protect more cargos at the port. As a result, PA has stronger incentive to make the port adaptation investment.

4 Avenues for future research

The economic modeling of port adaptation investment is still at the developing stage. The existing analytical frameworks have already offered a couple of valuable insights into this topic, and laid solid foundation for future extension. This section thus discusses three main avenues with significant potential for future explorations.

4.1 Asymmetry in disaster uncertainty and other port features

Current analytical studies on the two-port region assume that the ports are subject to a common disaster threat, i.e., the same disaster occurrence probability, maximum damage level, and adaptation efficiency. Although this assumption greatly simplifies the model derivations and guarantees the existence of closed-form analytical solutions, it deviates from the reality. Two ports can be very close, such as Shenzhen and Hong Kong, yet their geographic conditions and landscape can be very different. As a result, the same disaster event can bring very asymmetric damages on two ports. When ports are further separated, they could be subject to quite different threats of climate change-related disasters.

Port asymmetries could also come from other sources, such as adaptation investment, PA ownership, and downstream TOC market structure. The equilibrium analyses under these asymmetries could resemble more the real-world cases. The existing models can be extended to accommodate several port asymmetries. For example, Wang and Zhang (2018) can be extended to account for different Knightian uncertainties in disaster occurrence probabilities for two ports, respectively. The maximum disaster damage parameter D can also be made specific to each port as D_i. Two ports can also have different adaptation investment cost parameter ω_i. In Wang et al. (2019), the two ports can have asymmetric number of TOCs, and different subset of across-port JVs.

However, these treatments could complicate the model derivation greatly, making it difficult to derive closed-form solutions for clear-cut economic insights. It is quite likely to rely on simulation to conduct such analyses.

4.2 Vertical concession contract between PA and TOCs

After reviewing existing economic modeling work, it is noted that the vertical strategic relationship between PA and TOCs has not been well explored, especially in the presence of multiple TOCs at one port. In addition to Wang et al. (2019), there should be more strategic interactions to be examined.

For example, it is possible for PA to form exclusive contract with a subset of TOCs to jointly finance the port adaptation. To reward TOCs' participation, PA can consider to offer favorable concession terms (i.e., lower concession fee or revenue sharing) to them. Such discussions have already been provided in several airport economics studies. For example, Fu and Zhang (2010) and Zhang et al. (2010) modeled the revenue sharing between airport and airlines. To maximize the profits or social welfare, airport can strategically determine a subset of airlines to form the revenue-sharing contract.

In addition to the noncooperative game theoretic approach widely used, future studies can also implement alternative approaches, such as Nash-bargaining (Yang et al., 2015) and cooperative game theory (Wan et al., 2016) which have been applied in transport economic analysis. These approaches could be more appropriate, as there is an increasing trend for PA to involve more stakeholders to cooperate in the adaptation planning and investment (Becker et al., 2013). More rich implications on the vertical relationship between PA and TOCs, and the effects on port adaptation can be generated.

4.3 Positive externality of port adaptation on regional economy

Existing economic modeling framework has not well captured the externality of port adaptation on local community and regional economy. Randrianarisoa and Zhang (2019) is the only one trying to account for this aspect. But their treatment is still preliminary, just adding a positive term (proportionally to the adaptation investment level) on the social welfare function. There should be more detailed issues to be examined. For example, to achieve higher positive externality of the port adaptation on the broader region, it may require more sophisticated type of adaptation with higher investment cost. Then, it also comes to a question on whether the port should finance the project or government subsidy is called for.

In addition, we question if the current economic models are enough to tackle this complex externality issue, as current studies are basically partial equilibrium models, focusing only on the shipping sector. Thus, a general equilibrium model that well incorporates the different economic sectors and stakeholders should be developed to provide a more comprehensive framework. Specifically, future research should be able to endogenize the positive externality of the port adaptation and analyze a system-wide economy equilibrium, while deriving the overall economic and social welfare effect of the port adaptation on the whole region.

5 Conclusion

In recent years, port adaptation has attracted increasing attention from the academic field. This chapter comprehensively reviews the existing economic modeling of port adaptation, represented by Xiao et al. (2015), Wang and Zhang (2018), Liu et al. (2018), Randrianarisoa and Zhang (2019), and Wang et al. (2019). We compare how the disaster uncertainty and the port market structure have been incorporated in these models in terms of the commonality and differences. The analytical findings of the studies are then reconciled and compared. Last, the future research avenues are identified.

We found that the existing theoretical studies have applied game theory and real options approach to model the timing of port adaptation, disaster uncertainties, port market structure, and their effects on port adaptation. Several robust findings have been reached despite some distinct modeling specifications and assumptions. Among others, port adapts earlier and at higher level when the port is highly vulnerable to climate-related disasters and in presence of interport competition. In general, the likelihood of occurrence for extreme events (within the lifetime of the infrastructure) can be classified into three categories, including low ($x \leq 1\%$), moderate ($1\% < x \leq 10\%$), high ($10\% < x \leq 20\%$), and virtually certain or already occurring ($>20\%$). Some world ports fall into the "high risk" category while others are not, depending on the geographical location of the ports. For instance, Japanese ports are highly vulnerable to Earthquake, the Port of Vancouver in Canada is threatened by flood from both the ocean and the river (Fraser river) side, high tides, and storm surges, and the Port of Los Angeles in the United States has low risk impacts from sea-level rise and flooding. However, ports have more incentives to wait when they can accumulate better knowledge of the disaster occurrence probability and of adaptation efficiency in the next period. Moreover, there is an intraport free riding in port adaptation between PA and TOCs at the same port. Public ports are likely to overinvest in adaptation, not necessarily leading to socially optimal outcomes compared to private ports. Meanwhile, there are some seemingly inconsistent findings, mainly driven by variation in specification and modeling factors. For example, some studies suggest ports to adapt more with less ex ante disaster uncertainty but adapt less with less ex post disaster uncertainty (information accumulation).

Based on existing models, we proposed several avenues for the future studies. First, the asymmetry in the disaster risk, the port disaster uncertainty, the adaptation efficiency, and the investment cost may be further considered in order to better reflect real-world situations. Second, the vertical relationship between PA and TOCs at the same port can be better explored. More sophisticated concession contracts involving adaptation investment, joint financing, and revenue sharing can be examined. Meanwhile, other than the conventional noncooperative game theoretical approaches, Nash-bargaining and the cooperative game approach may also be introduced to analyze more complicated vertical interactions between PA and TOCs in adaptation investment. Third, the discretely made investment decisions may be extended to the decisions made in a continuous fashion (e.g., Balliauw et al., 2019). Fourth, it would also be interesting to test the validity of the assumptions of the current models with actual data on disasters, such as hurricane, earthquake, or the typhoons striking ports like Hong Kong. Finally, the economic modeling of the positive externality of port adaptation on regional economy can be improved. To achieve a more comprehensive and objective analysis, a general equilibrium model that incorporates different economic sectors and stakeholders can be established, enabling to derive the overall economic and social welfare effects on the whole region.

Acknowledgments

We thank the participants of Workshop on Climate Change and Adaptation Planning for Ports, Transportation Infrastructures, and the Arctic at University of Manitoba, and seminar at University of International Business and Economics in Beijing for their helpful comments. Financial support from the Social Sciences and Humanities Research Council of Canada (SSHRC) and the Social Science Foundation of Ministry of Education of China (19YJC790136) is gratefully acknowledged.

References

Balliauw, M., Kort, P., Zhang, A., 2019. Capacity investment decisions of two competing ports under uncertainty: a strategic real options approach. Transp. Res. B Methodol. 122, 249–264.

Becker, A., Inoue, S., Fischer, M., Schwegler, B., 2012. Climate change impacts on international seaports: knowledge, perceptions, and planning efforts among port administrators. Clim. Chang. 110 (1), 5–29.

Becker, A.H., Acciaro, M., Asariotis, R., Cabrera, E., Cretegny, L., Crist, P., Esteban, M., Mather, A., Messner, S., Naruse, S., Ng, A., Rahmstorf, S., Savonis, M., Song, D., Stenek, V., Velegrakis, A.F., 2013. A note on climate change adaptation for seaports: a challenge for global ports, a challenge for global society. Clim. Chang. 120 (4), 683–695.

Camerer, C., Weber, M., 1992. Recent developments in modeling preferences: uncertainty and ambiguity. J. Risk Uncertain. 5 (4), 325–370.

Cheon, S., Dowall, D.E., Song, D.W., 2010. Evaluating impacts of institutional reforms on port efficiency changes: ownership, corporate structure, and total factor productivity changes of world container ports. Transp. Res. E: Logist. Transp. Rev. 46 (4), 546–561.

De Monie, G., 2005. Public-Private Partnership. Lessons at Specialization Course in Public Private Partnerships in Ports' Structure, Pricing, Funding and Performance Measurement. 10–14 October, ITTMA, Antwerp.

Fu, X., Zhang, A., 2010. Effects of airport concession revenue sharing on airline competition and social welfare. J. Transp. Econ. Policy 44 (2), 119–138.

Gao, Y., Driouchi, T., 2013. Incorporating Knightian uncertainty into real options analysis: using multiple-priors in the case of rail transit investment. Transp. Res. B Methodol. 55, 23–40.

He, Y., Ng, A.K., Zhang, A., Xu, S., Lin, Y., 2019. Climate change adaptation by ports: the attitude and perception of Chinese port organizations. Working Paper, University of Manitoba, Winnipeg.

IPCC (Intergovernmental Panel on Climate Change), 2013. Climate change. In: The Physical Science Basis (Summary for Policymakers). Working Group, I Contribution to the IPCC Fifth Assessment Report, Geneva, p. 2013.

Jiang, C., Wan, Y., Zhang, A., 2017. Internalization of port congestion: strategic effect behind shipping line delays and implications for terminal charges and investment. Marit. Policy Manag. 44 (1), 112–130.

Keohane, R. O., Victor, D. G. (2010). The regime complex for climate change. Discussion Paper, The Harvard Project on International Climate Agreements, Cambridge, MA, 2010–33.

Knight, F.H., 1921. Risk, Uncertainty and Profit. Houghton Mifflin, Boston.

Liu, Z., 1992. Ownership and Productive Efficiency: With Reference to British Ports (Ph.D. thesis). Queen Mary and Westfield College, University of London, London.

Liu, N., Gong, Z., Xiao, X., 2018. Disaster prevention and strategic investment for multiple ports in a region: cooperation or not. Marit. Policy Manag. 45 (5), 585–603.

Messner, S., Becker, A., Ng, A.K., 2015. Port adaptation for climate change: the roles of stakeholders and the planning process. In: Climate Change and Adaptation Planning for Ports. Routledge, London, pp. 41–55.

Min, S., Zwiers, F., Zhang, X., Hegerl, G., 2011. Human contribution to more-intense precipitation extremes. Nature 470, 378–381.

Ng, A.K., Monios, J., Zhang, H., 2018a. Climate adaptation management and institutional erosion: insights from a major Canadian port. J. Environ. Plan. Manag. 1–25.

Ng, A.K., Wang, T., Yang, Z., Li, K.X., Jiang, C., 2018b. How is business adapting to climate change impacts appropriately? Insight from the commercial port sector. J. Bus. Ethics 150 (4), 1029–1047.

Nishimura, K.G., Ozaki, H., 2007. Irreversible investment and Knightian uncertainty. J. Econ. Theory 136 (1), 668–694.

Notteboom, T., 2006. Concession agreements as port governance tools. Res. Transp. Econ. 17, 437–455.

OECD, 2016. Adapting transport to climate change and extreme weather: implications for infrastructure owners and network managers. Discussion Report of International Transport Forum, OECD, Paris.

Randrianarisoa, L.M., Zhang, A., 2019. Adaptation to climate change effects and competition between ports: invest now or later? Transp. Res. B Methodol. 123, 279–322.

Trujillo, L., Nombela, G., 2000. Seaports. In: Estache, A., De Rus, G. (Eds.), Privatization and Regulation of Transport Infrastructure Guidelines for Policymakers and Regulators. The World Bank, Washington, DC, pp. 113–169.

Wan, Y., Basso, L.J., Zhang, A., 2016. Strategic investments in accessibility under port competition and inter-regional coordination. Transp. Res. B Methodol. 93, 102–125.

Wang, K., Zhang, A., 2018. Climate change, natural disasters and adaptation investments: inter-and intra-port competition and cooperation. Transp. Res. B Methodol. 117, 158–189.

Wang, K., Yang, H., Zhang, A., 2019. Seaport adaptation to climate change-related disasters: impact of multiple terminal operators. Working Paper, University of International Business and Economics, Beijing.

Xiao, Y.B., Fu, X., Ng, A.K., Zhang, A., 2015. Port investments on coastal and marine disasters prevention: economic modeling and implications. Transp. Res. B Methodol. 78, 202–221.

Yang, H., Zhang, A., Fu, X., 2015. Determinants of airport–airline vertical arrangements: analytical results and empirical evidence. J. Transp. Econ. Policy 49 (3), 438–453.

Yang, Z., Ng, A.K., Lee, P.T.W., Wang, T., Qu, Z., Rodrigues, V.S., Pettit, S., Harris, I., Zhang, D., Lau, Y.Y., 2018. Risk and cost evaluation of port adaptation measures to climate change impacts. Transp. Res. Part D: Transp. Environ. 61, 444–458.

Zhang, A., Fu, X., Yang, H.G., 2010. Revenue sharing with multiple airlines and airports. Transp. Res. B Methodol. 44 (8–9), 944–959.

Chapter 5

Sustainability cruising and its supply chain

Grace W.Y. Wang[1]
Maritime Economist, District of Columbia, United States

1 Introduction

The importance of the cruise market has been addressed in tourism literature. However, when it comes to resource distribution of the maritime sectors, total quantity in terms of twenty-foot-equivalent units (TEU) and volumes are used to show the important economic values created by the given industry. It is not the case for the cruise industry where quality of service is the key to maintain competitiveness and to fulfill the high expectation of cruise passengers who look for the ultimate relaxing and luxurious vacation experience at sea (Wang et al., 2019b). In this research, we offer a holistic viewpoint to capture the uniqueness and the complexity of the cruise industry—from the structure of the maritime cruise cluster and the environmental concerns raised by the stakeholders, to the sustainability of the cruise supply chain.

First, we discuss cooperation and vertical integration in cruise maritime clusters. The cruise industry contains maritime clusters, and the interaction of the businesses within the clusters is what drives the performance of cruises. To better understand cruise activity and business, we need to define cruise clusters, their objectives, and the extent of possible integration. The benefits of vertical integration in cruise ports and its supporting industries can be further quantified.

Second, from the public standpoint, cruise economic impact is an important part of regional economic growth. This section presents change to the Regional Economic Impact System (RECONS) established by the United States Army Corps of Engineers including both direct growth impacts and indirect environmental impacts. This section contributes to a responsible economic impact model supporting both economic growth and true environmental offsets.

---

[1] DISCLAIMER: The author is the Director of Economic Studies at the Federal Maritime Commission (FMC) whose mission is to ensure a competitive and reliable international ocean transportation supply chain. The author's personal email is gw77539@gmail.com. This paper is the result of the author's independent research and does not represent the views of the FMC or the United States government.

Maritime Transport and Regional Sustainability. https://doi.org/10.1016/B978-0-12-819134-7.00005-8

Third, switching to firm level, if public's interests can be aligned perfectly with the interest of the private sector, there may be a possibility to find a balance between efficient operations for profit-driven cruise companies and to address long-term sustainability concerns for stakeholders. Cruise companies' environmental efficiency is an important part of their corporate social responsibility. This is essential for long-term sustainability for cruise companies. The measurement of the relative efficiency of cruise companies is used to compare their efficiency in environmental practices and technological investment.

The fourth and the final section is about the efficiency and effectiveness of the supporting industry in the cruise maritime cluster. The oligopolistic nature of cruise lines creates a dominant position within the cruise supply chain. Using game theory and backward induction, optimal incentive mechanisms are identified in order to align the motivations of the supplier with the cruise line. The need for a reliable supply chain regardless of circumstances has created the need for the alignment of players' self-interests to reduce self-serving bias, moral hazard, and adverse selection (Wang and Pallis, 2014).

In the following sections, we provide an in-depth discussion for each topic mentioned above to address how sustainability issues are dealt with by players in the cruise maritime cluster.

2 Understand coordination and vertical integration in cruise ports

The cruise sector has interdependent players who provide services necessary for effective cruise operations. The maritime clusters' interactions stimulate the overall economic impact of cruise activities providing both direct and indirect impacts in jobs supported and income created. The interplay of cruise members is essential for the economic growth provided by this economic engine to support local business, tourism, and port-city development. The vertical integration and cooperation of cruise cluster players is essential for economic growth. The indirect and induced economic impact comes from markets other than cruise ports within core maritime industries. The cruise operation regional economic impact is a result of the interaction of the core cruise maritime industries stimulating growth in multiple areas simultaneously. The integration of services through cooperation can create flexibility within the cruise port value chain.

The high level of concentration of market and geographic areas can lead to delays and congestion of cruise areas. This section aims to provide an understanding of the composition of maritime clusters in cruise ports and the extent of coordination and cooperation in maritime industries. This serves as a foundation for the development of cruise integrated regulation regarding challenges and issues in cruise coordination, vertical integration, cooperation, and to align interests of cruise lines, cruise terminal operators, and port authorities. Business segments within the cruise industry fall into six categories: cruise operations, port operations, ground transportation (airport shuttles), tourism, healthcare, and finance. The interactions of these categories comprise the maritime clusters and

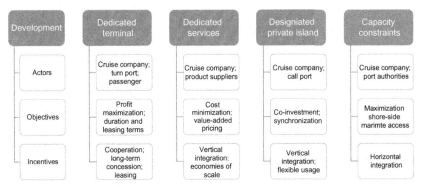

FIG. 1 Cruise cluster development and possible integration.

the extent of vertical integration (Fig. 1). Currently many cruises utilize dedicated cruise supply providers as a cost-effective segment of operations. The integration can be through integrated tourism and resort experiences. Combining travel packages is another form of integration for cruise lines to provide convenience and improve passenger experience (Wang et al., 2015; Wang and Zeng, 2017).

2.1 Cruise maritime cluster development—objectives and incentives

Maritime cluster development has a close relationship with supply chain management. Details of the interconnected maritime cluster and activities associated with it can be seen in the following examples (Fig. 2). Dedicated terminals are sometimes necessary to accommodate busy cruise traffic. Dedicated terminal also can be used to improve the efficient operations while in port. The differing interests of the terminal provider and cruise can be bridged because efficient

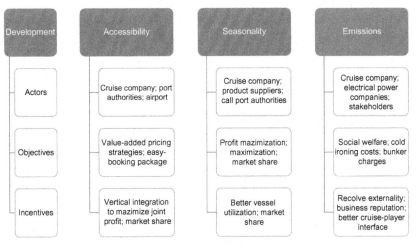

FIG. 2 Cruise cluster development and possible integration (continue).

cruise operation is in the best interests of both parties. Dedicated services are important for cruise lines because of the short time horizon in which activities in port need to be accomplished, such as services from berthing, stevedoring, and bunkering to passenger-related operations. Many cruise line companies will provide exclusive beach experiences, and some will utilize private islands to further differentiate their packages and stay competitive in the industry. Lease agreements can cause imperfect competition and price differentiation. Capacity constraints are important for the cruise sector. Shore excursions can be offered through cooperation between waterfront multipurpose terminals, cruise lines, and port authorities in order to promote tourism. Capacity constraints have to do with the volatility of cruise demand due to its nature as a luxury activity and its seasonal nature. Accessibility is another important consideration for cruise line companies because if cruise homeports are not easily accessible for cruise passengers it will increase costs to the consumer, and decrease the quantity demanded for the more elastic consumers who are very sensitive to price change. The possibility of transit packages being included for consumer convenience is an important consideration involving travel agencies, airlines, and land transportation. Air emissions and environmental issues are important for cruises; due to the different impacts on different stakeholders, cruises rely on environmental tourism, so the damage to that environment can decrease demand for cruises. The cruise terminals are focused on the social welfare of the hinterland and have their own concerns. The differing focuses between cost-efficiency and externality are essential for modeling effective cruise emission reduction.

3 Responsible cruising and its economic impacts

The United States budget for port construction and maintenance is usually used for cargo port maintenance, transport cost reduction, and total tonnage increase, while ignoring factors such as the number of passengers served. The growth and development of the cruise sector in the United States has not been correlated with an influx of federal funding. The growing industry's infrastructure may be disregarded for cargo terminal maintenance and expansion. It is important to develop a method of incorporating cruise ports into regional economic impact studies. The Regional ECONomic System (RECONS) is the system used by the US Army Corps of Engineers to estimate growth, jobs, and other economic activities (Wang et al., 2018). This section aims to reevaluate how cruise impact analysis incorporates corporate social responsibility (CSR) in regard to environmental and social welfare. Value-added activities by cruise port operations need to be included into the economic evaluation. Thus, economic growth will be balanced with environmental impacts in order to present the true long-term economic impact of cruise activity on a region.

3.1 Social, cultural, and environmental externality

Responsible cruising is essential for economic growth. The direct relationship between cruise revenue and cruise responsibility should lead to benefits for local economies, but the local economies only retain around half of the economic benefit through tourism expenditures. In the long run, cruise operations' continued growth depends on the reliability and capacity of the maritime space. Risk of the "tragedy of the commons" within cruise call locations leads to an increasing awareness of the need for responsible cruise practices in order to have market success. Implementation of clear standards of environmental and social responsibility may be one of the needed steps. The measurement of cruise companies' corporate social responsibility and externality to the local stakeholders in the efficiency study is important to benchmark how companies truly impact the economy, local communities, and the environment.

To understand potential externalities, categorization of qualitative and quantitative aspects of corporate social responsibility and sustainability within the cruise industry is needed. This categorization is used to identify how information can be inputted into the dynamic RECONS input-output model. For example, quantitative information such as carbon footprint, cold ironing capacity, energy consumption, waste reduction, exhaust gas cleaning technology, and water efficiency can be tied directly to the operation expenses and cost savings. Qualitative information such as crew trainings, health, safety, and security issues must rely on subject experts to identify the implicit values in order to quantify the impacts associated with the activities and investments. Externalities usually refer to the negative impacts of cruises to the local economies. The positive impacts of cruise economic activity on a region must be balanced with environmental offsets from the externalities of cruising. These offsets include measuring the waste management of cruise lines and identifying this impact on local communities. The input of emission activity economic offset is another important externality that can negatively impact the quality of life of local communities. The impacts are not just environmental; cruise tourism can have a negative sociocultural impact on call port communities. This is a significant issue for these communities because the host communities are not receiving equivalent benefit from on-shore activities. The excursion deals offered by cruise lines sometimes cause the tour guide to receive less than 50% of the total profit. Cruise lines also have cultural impacts through peoples' pollution where the influx of passengers can overwhelm the populations of small cities and this can significantly change the way of life for these communities and be larger than a community can handle.

3.2 Challenges to quantify environmental externality

Standard measurements of job creation, economic growth, and impacts do not fully show the importance of cruise ports as a regional economic engine. Adjusted RECONS model (Fig. 3) including both environmental offsets and cruise specific inputs gives a clearer picture. The number of passengers and crew is used as an input for RECONS to estimate passenger and crew impact based on passenger and crew estimated expenditure. Cruise port economic impact is vastly different than standard cargo port economic impact because the cruise "cargo" has on-shore expenditure creating multiplier effects to

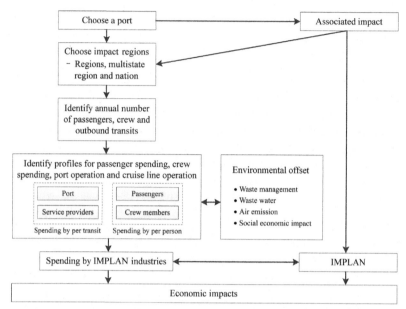

FIG. 3   RECONS configuration. *(From Wang, G., Chang, W.-H., Yue, C., 2018. Responsible cruising and its economic impact. In: International Association of Maritime Economists Conference, Kenya; Wang, G., Xiao, Y., Li, K., 2019a. Measuring environmental efficiency of cruise company considering corporate social responsibility. Mar. Policy. Accepted; Wang, G., Chang, W.-H., Yue, C., Qi, G., Li, K., 2019b. Sustainable Cruise Shipping With Environmental and Economic Considerations. Submitted.)*

local economies. The inclusion of passenger and crew member spending shows the cruise economic impact in a region due to the differences between a cruise port and a cargo port. The multiplier also captures the percentage of the economic impact retained by the regional economy. Other than passenger and crew spending, RECONS also captures the value-added activities associated with daily cruise operations by cruise port and cruise lines.

Quantifying these externalities is difficult. While taxation, "cost of damages," and "cost of control" are used to quantify the externalities, these can be positive economic generators because it creates spending and thus multiplier effects to the community. Discussion of externalities using taxation mechanism creates positive monetary flow such as revenue for the government that may not truly reflect the negative impact to the local stakeholders. The externalities and implicit costs are often neglected in impact studies. Conflicts of interest along the value chain also are an important consideration. The social costs involved must be considered in order to effectively measure the true impact. Effective taxation is another consideration due to the widespread adoption of flags of convenience by cruise lines. The lack of standardized taxation for cruise ships also needs to be considered for a more responsible economic impact model.

The balance of environmental offsets with economic growth is important to the new cruise economic impact model. While responsible cruising should minimize the negative externalities, the focus for cruise lines remains on profit maximization. Maximization of profit should lead to responsible cruise tourism if the interests of the cruise lines are aligned well with the public's interest. Responsible cruising is essential for long-term sustainability in the cruise sector due to dependence on both the environment and local communities to provide the cruise experience.

4 Marine environmental efficiency and CSR

This section aims to help cruise shipping companies identify issues that present risks to the maritime space's internal and external stakeholders. The use of environmentally friendly technology is an important part of corporate social responsibility for cruise lines in mitigating negative externalities to stakeholders and to the public. The growth of the installation of exhaust gas cleaning systems and advanced wastewater purification systems onboard cruise ships shows an increase in compliance with International Convention for the Prevention of Pollution from Ships (MARPOL) regulations. To quantify environmental efficiency, Carnival Corporation, with its subsidiaries, Costa Crociere and Holland, was selected because of its large market presence in the Caribbean (Wang et al., 2019a). This application is aimed to identify environmentally friendly strategies and the extent to which these can be effectively implemented in order to reflect corporate social responsibility in the cruise sector. Further, the goal is to provide policy advice aligning public interests with cruise stakeholders.

4.1 Quantify environmental efficiency with undesirable output

By utilizing statistical tools such as the superslack-based measure (SBM) model and the Malmquist productivity index, undesirable environmental externalities, such as air emissions, wastewater, and other pollution, as well as other undesirable outputs are taken into consideration when we measure environmental efficiency. Wang et al. (2019a) utilized empirical data from Carnival cruise lines and its subsidiaries to compare CSR based on air emissions, gray water discharge, bilge water discharge, and solid waste across the companies. A Data envelopment analysis (DEA) method was used to measure the efficiency of decision-making units through a linier combination of inputs and outputs. In order to allow for the changes of efficiency over time, the DEA method was combined with super-SBM and the Malmquist index. These methods are used to find the relative efficiency in different areas and the Malmquist productivity index was used to expand upon the findings of the super-SBM based on inputs of employers, energy consumption, and water consumption with revenues as the desirable outputs.

This methodology captures not only the desirable outputs but also the undesirable outputs such as air emissions, wastewater, solid waste, and energy consumption as discussed in the previous section as externalities in cruise operations. Analysis at the firm level shows that certain companies are well equipped with the needed technology to tackle environmental concerns and others may have facility and infrastructure advantages. Similarly, the model can identify areas of inefficiency. Some firms have disadvantages in wastewater, while others may have weaknesses in air emissions. For example, Costa Crociere was the best in air emissions environmental efficiency when compared to the other Carnival companies. Its efficiency has also been shown overtime to increase between 2010 and 2015. Carnival was the most efficient in wastewater treatment due to significant investment into technology and effective ways of managing wastewater. In both air emissions and wastewater areas, companies studied showed significant investment in technology to enhance overall efficiency. While Carnival was not as environmentally efficient as its subsidiaries it has made significant progress in emissions reduction technology and energy conservation. While the cruise industry is generally more environmentally aware due to its nature of capitalizing on the environment, it still has a long way to go. It provides a method to measure environmental and social responsibility efficiency for sustainable cruising.

5 Incentive mechanism in cruise supply chain

The reliability and sustainability of cruise operations depend on the interactions among players within the value-added supply chain. The negotiation of cruise port-cruise line contracts is important for establishing long-term interactions between the players. Cruise ports' motivations are the long-term port-city development and economic growth, whereas the cruise line's motivation is profit maximization. This causes many ports to offer incentives to receive long-term contracts, which ensures long-term economic growth (Wang et al., 2014a,b). The vertical integration and cooperation within the supply chain is important to maintain a successful and reliable cruise supply chain. The utilization of incentives to align the motivations of members of the maritime cluster which form a cohesive and stable unit.

5.1 Contract theory to overcome moral hazard and vulnerability

The cruise supply chain is made up of multiple players, the cruise line, the cruise ports (home and call), the cruise suppliers, the travel agency, and excursion providers. The interdependence of the cruise supply chain with different self-interests can create conflicts of interest leading to a less efficient and sustainable supply chain. The asymmetric information between cruise players in the cruise supply chain can lead to adverse selection (before the contract is signed) or moral hazard (after the contract is signed). These supply chain

FIG. 4 Incentive mechanism design for supplier's flexibility. *(From Qu, Wang, and Zeng, 2019. Modelling the procurement process and production disruption of a multilayer cruise supply chain. Maritime Pol. Manag. submitted; Wang G., Qu, C., 2019. Modelling Incentive Strategy of a Multi-Layer Cruise Supply Chain. Transportation Research Board, United States.)*

players all have varying self-interests and motives. The need for reliability and flexibility for the cruise supply chain is essential for all parties to profit from the consumer's experience. In order to align the motivations of the supply chain stakeholders, a game theory approach with the cruise line, the cruise port, and the cruise supplier was developed to apply incentive mechanisms to enhance customer experience (Wang and Qu, 2019). This forms a more cohesive and reliable network. The Stackelberg leader-follower game was applied to model sequential decisions in the supply chain and to determine incentives that can promote optimal outcomes. Backward induction was used in order to find the right incentive method in the leader-follower framework.

Fig. 4 shows the incentive approach for the cruise line relationship with the cruise supplier. The cruise line must deal with asymmetric information because only the supplier knows if the supplier is flexible. The flexibility in the supply chain is essential for dealing with market disruptions and maintaining reliability. The cruise line wants a highly flexible supplier in order to accommodate changing onboard consumption and/or an unforeseen emergency resupply. The flexibility of service providers is incentivized through cash rebates in exchange for high flexibility. Using punitive measures will discourage flexibility causing inefficiency in the supply chain. Through a theoretical framework for the cruise supply chain including suppliers, cruise lines, and cruise port/terminals, the behavior and decision-making process of all players can be explicitly quantified. It finds that in order to maintain reliability of the supplier, the cruise line needs to incentivize the supplier. The optimal method of ensuring a highly flexible supplier is through nonpunitive measures reinforcing flexibility in the supply chain. The high market power of cruise lines due to their oligopolistic nature within the cruise supply chain places them in the dominant position in the supply chain. The differing motivations of players are essential for identifying the correct incentive methods of aligning the player's self-interest.

6 Conclusions

Supply chains for cruise ships enable passenger to enjoy the diverse experience of cruising. As a cruise ship travels to each port of call, its supply chain changes accordingly. This complexity adds vulnerability that can test the sustainability and reliability of the chain. Given the fixed itinerary and short turnaround window, supply chain operations for the cruise industry are unique. Reliability in the supply chain is essential in order to adjust after an event (hurricane, port strike, etc.) that the cruise supply chain has reliable alternatives for and alternate routes where the supply can be adjusted. Alternatives need to be found to make the supply chains more resilient for future events. In order to improve on the reliability of the chains due to the economic significance of cruise lines to the nation and in particular the ports they visit, we investigate incentive strategies for cruise supply chains affected by natural disasters and forecasting error of consumptions. Assessment of the vulnerability of cruise emergency supply chains and resilience to disruption is essential for supply chain analysis. The utilization of incentives is used to improve potential outcomes of an event and provide resilience to event vulnerability. The development of plans of action for ship resupply, rescheduling, and rerouting is essential for overall reliability.

This game theoretical approach can be applied to different cruise regions around the world and the game has different results due to differing market setups. For example, the cruise sector in China has experienced significant growth, but the market dynamic is vastly different from that of the Mediterranean and Caribbean counterparts. The Chinese cruise market operates differently because the dominant position shifts to the cruise port due to the lack of multipurpose terminals willing to host cruise activities. This gives the cruise ports the negotiating power over the cruise lines; this is contrary to the Caribbean and Meditation markets where the oligopolistic nature of the cruise lines grants them significant negotiation power and market share. The reversal of the dominant position places the port in the dominant position of negation. This changes the overall dynamics of the value chain. Cultural and regional dynamics significantly influence the cruise supply chain's dynamics, and players. The use of incentive mechanisms needs to change toward the cruise market being served.

References

Wang, G., Pallis, A.A., 2014. Incentive approaches to overcome moral hazard in port concession agreements. Transp. Res. E: Logist. Transp. Rev. 67, 162–174.

Wang, G., Qu, C., 2019. Modelling Incentive Strategy of a Multi-Layer Cruise Supply Chain. Transportation Research Board, United States.

Wang, G., Zeng, Q., 2017. A vertical integration pricing model on a cruise supply chain. In: International Association of Maritime Economists Conference, Japan, pp. 1–16.

Wang, G., Pallis, A.A., Notteboom, T.E., 2014a. Incentives in cruise terminal concession contracts. Res. Transp. Bus. Manag. 13, 36–42.

Wang, G., Notteboom, T.E., Pallis, A.A., 2014b. Incentives in cruise concession contracts. In: International Association of Maritime Economists Conference, US, pp. 1–22.

Wang, G., Pallis, A.A., Notteboom, T.E., 2015. Cooperation and vertical integration in cruise ports. In: International Association of Maritime Economists Conference, Malaysia, pp. 1–18.

Wang, G., Chang, W.-H., Yue, C., 2018. Responsible cruising and its economic impact. In: International Association of Maritime Economists Conference, Kenya.

Wang, G., Xiao, Y., Li, K., 2019a. Measuring environmental efficiency of cruise company considering corporate social responsibility. Mar. Policy, 99, 140–147.

Wang, G., Chang, W.-H., Yue, C., Qi, G., Li, K., 2019b. Sustainable Cruise Shipping With Environmental and Economic Considerations. Submitted.

Further reading

Wang, G., Chang, W.-H., Yue, C., 2017. Economic impacts and significance of the cruise ports in the United States. Int. J. Shipp. Transp. Logist. Accepted.

Chapter 6

How does the UK transport system respond to the risks posed by climate change? An analysis from the perspective of adaptation planning

Tianni Wang[a,b,c], Zhuohua Qu[a], Zaili Yang[b], Adolf K.Y. Ng[d]
[a]*Liverpool Business School, Liverpool John Moores University, Liverpool, United Kingdom,*
[b]*Liverpool Logistics, Offshore and Marine Research Institute, Liverpool John Moores University, Liverpool, United Kingdom,* [c]*Transport College, Shanghai Maritime University, Shanghai, China,*
[d]*Department of Supply Chain Management, Asper School of Business, University of Manitoba, Winnipeg, MB, Canada*

1 Introduction

In the United Kingdom, the transport sector is recognized as one of six key departments which is the most vulnerable to the risks posed by climate change (McKenzie Hedger et al., 2000). The country has a network covering 422,100 km of paved roads with different quality and capacity (Department for Transport, 2017; Department for Infrastructure, 2017). A unified road numbering system is used to classify and identify all the roads in the United Kingdom. Cooperated with the Department for Transport, Highways England (HE), for example, operates, maintains, and improves motorways and major "A" roads in England (Highways England, 2018a).

The United Kingdom opened locomotive-hauled public passenger railways in 1825. As the oldest railway system in the world, it has a network of 15,760 km of standard-gauge lines, including 5272 km electrified lines today (Wikinow, n.d.). The majority of railway track is managed and maintained by Network Rail (NR). Also, there are some services on public rail-based mass transit systems run by local authorities and an undersea rail link to France called the Channel Tunnel operated by Getlink. Some short tourist rail lines are managed by private railways.

Maritime Transport and Regional Sustainability. https://doi.org/10.1016/B978-0-12-819134-7.00006-X

The rapid rate of climate change challenges the infrastructure, operation, and policy-making in the context of transport systems. The transport-related activities are vulnerable to heterogeneous weather extremes, which include variations in temperature, precipitation, winds, sea-level/other-water levels, thunderstorms, fog period or visibility, frost, and thaw (e.g., Wang et al., 2019; Schweikert et al., 2014; Love et al., 2010). In the United Kingdom, the inland transport systems are threatened by four primary climate change hazards, namely high temperature, heavy precipitation and flooding, high wind and storms, and sea-level rise (SLR) (Wang et al., 2018a,b, 2019). The frequently occurring flooding events in Cumbria and heavy storms in Devon, for example, have caused catastrophic infrastructural and financial losses and casualties due to a variety of impacts including roads and rail line closure, bridges' deterioration, traffic disruption, service cancellation, and delays (e.g., BBC News, 2015a,b, 2017; Devon County Council, 2014a; Devon Maritime Forum, 2014).

Some reviews have been conducted to investigate the impacts of climate change on the British roads and railways in recent years (e.g., Wang et al., 2018b, 2019; Koetse and Rietveld, 2009). However, research on climate impacts on road and rail freight in the United Kingdom has remained relatively unexplored (Jaroszweski, 2015). It is only recently that more attention has been given to the impacts of climate change on roads and railways (e.g., Hooper and Chapman, 2012; Wang et al., 2018b, 2019). Current action plans in British roads have not been developed from a published, detailed, and official adaptation plan but mainly focus on internal technical documents within the relevant business areas (Committee on Climate Change, 2014). Likewise, the existing adaptation plan of NR mainly concerns with the identification of several climate thresholds and selection of the best risk scenario, owing to the uncertainties of long-term climate change risks and insufficiency of data on change rate and extreme events (Network Rail, 2015). Indeed, a comprehensive adaptation plan covering every aspect has not been published in either British road or rail networks.

The failure of implementing adaptation plans in the transport systems may potentially be attributed to the deficiency of precise data on climate change impacts and cost-benefit analysis of adaptation planning (e.g., Koetse and Rietveld, 2012). Accordingly, there have been considerable studies assessing climate risks and cost-effectiveness of adaptation measures in diverse transport modes (e.g., ports, roads, railways) (Ng et al., 2018; Yang et al., 2018; Wang et al., 2018b, 2019). Nevertheless, current published reports rarely cover the "hidden" problems in climate adaptation planning, such as planning methods, procedure, time horizon, and public participants. Understanding such, in this chapter, we conduct a comparative study on the UK road and railway networks to reveal the state-of-the-art understanding on how the two transport systems adapt to the risks of climate change, including the primary climate risks, adaptation options, and the implementation and development of adaptation plans.

This chapter performs an analysis of multicase studies via in-depth interviews with affiliated senior experts in the UK transport systems. The outcomes provide researchers, transport planners, and decision makers with an innovative thinking pattern from the identification of climate hazards to the implementation of climate adaptation planning, in order to bridge the research gaps and facilitate climatic adaptation in the inland transport industries.

2 Methodology

A qualitative approach is used to get access to a considerable amount of unpublished qualitative information, to analyze relationships and social process, which could be hard to achieve by only using quantitative methods (such as modeling) especially when data is scarce (Miles and Huberman, 1984). In addition to documental review, we conducted five semistructured, in-depth interviews with the associated domain experts from HE, NR, Transport for London (TfL), Environment Agency (EA), and Devon County Council (DCC) in early 2018. Their positions included policymakers, transport planners, environmental specialists, and climate change advisors.[a]

The interview, as a commonly used qualitative method, was used to collect data in four case studies. These data were expected to reveal some "hidden" problems (e.g., the key factors, processes, and references in a climate change adaptation plan) in adapting to the risks posed by climate change in the UK transport systems. Besides the primary threats posed by the climate change, the interviews attempted to figure out the current risk assessment and planning processes, as well as the crucial elements and dilemmas in current and future adaptation planning for climate change. The primary interview questions were

(1) What are the significant risks and uncertainties posed by climate change? Do road/rail stakeholders have an adaptation plan and measures to cope with such climate risks?

(2) How are these climate risks assessed in their entities? Is there a risk analysis system for climate change? What are the priorities and fundamental principles for adaptation planning in short- and long-terms?

(3) How do road/rail decision makers conceive the unique conditions of their entities? What kind of recourses, information, and references have been used or would be used for adaptation planning?

(4) What are the perceptions of road/rail decision makers on climate adaption planning? Who are the involved participants or will be involved in the climate adaptation planning process?

(5) What is the planning horizon of climate adaptation planning? What are the critical factors influencing the success of an effective climate adaptation plan?

These interview questions were semistructured and open-ended in terms of their expected answers. The five corresponding issues were: Part A, identifying

---

a. The basic information of interviewees can be found in the Appendix.

the vulnerabilities of rail and road posed by climate change; Part B, assessing risk and planning priorities; Part C, recognizing the characteristics and differences between rail and road's conditions; Part D, analyzing the preparation, environment, and stakeholders involving an adaptation plan; Part E, implementing an adaptation plan and developing adaptation strategies.[b]

All the interviewees were asked similar major questions based on a preset framework, which gave interviewees enough time to prepare their answers and allowed us to cover all questions adequately. However, we (as interviewers) did not strictly limit them only to answer the above questions but encouraged them to express their views freely to reflect their "real thinking" on climate risks and adaptation planning. After that, we integrated multisource evidence for triangulation, including the interviews, official reports, and local news and archival data, to enhance the validity of our understanding. The interview data were coded by a thematic coding analysis approach, which provided a practical and flexible approach to categorize and summarize the key characteristics of various qualitative data (Braun and Clarke, 2006).

Finally, both within-case (i.e., EA and TfL in London) and cross-case (e.g., NR and HE) analyses were undertaken to compare and contrast the similarities and differences of the organizations' adaptation plan, which also reinforced the external and internal validity (Yin, 2003). First, in a single case, within-case analysis allows us to recognize the existing and potential climate risks and adaptation strategies by hearing diverse voices from different entities in a specific region (i.e., London). Cross-case analysis facilitates the development of a comprehensive view on how the rail and road adapt to the risks posed by climate change respectively, so as to reveal the common issues and potential collaborative opportunities in an integrated inland transport system.

3 Adapting planning for climate change in the UK road and rail systems

3.1 Highways England

Formerly known as the Highways Agency, Highways England (HE) became a new government company in 2015 responsible for the operation, maintenance, and improvement of England's strategic road network covering more than 4300 of miles motorways and major A-class roads (Highways England, 2018a,b).

The Highways Agency initiated its first *Climate Change Adaptation Strategy and Framework* in 2009, aiming to recognize and assess the impacts posed by climate change on the road network to generate preferred adaptation options (Highways Agency, 2009). Through a comprehensive review, the *Highways Agency's Adaptation Framework Model (HAAFM)* was developed by setting up a detailed seven-stage adaptation process from "*define objectives*

---

b. The interview question framework can be provided as a request.

and decision making criteria" to "*adaptation programme review*". By 2014, a variety of climate risk assessment and adaptation action plans had been produced, with specific reports on flooding and winter adaptation, but few gaps in climate adaptation on roads remained unexplored (Highways England, 2016). The Committee on Climate Change (2014) commented that neither information on resilience spending plans nor reported progress of implementing resilience measures were publicly available. However, the *Transport Resilience Review* (Department for Transport, 2014) contained recommendations for improving climate resilience such as the management of high-sided vehicles due to high winds, drainage management for flooding, and roadside infrastructure for winter driving.

The latest *Climate Adaptation Risk Assessment Progress Update* (Highways England, 2016) set up an overview of climate adaptation in the 2015–2020 (Road Investment Strategy 1) period. HE identified the existing primary climate change hazards in its 2016 report. The trend of climate change included increased average and maximum temperature, more frequent and intense rainfalls in summer, high winds, and SLR. For instance, the variations in precipitation, such as flooding, storm surges, and groundwater level changes, could pollute and deteriorate transport asset as well as influence the design, operation, and maintenance of drainage, foundations, and skid. High temperature may modify bearings' layout and expansion joints. In addition, high winds may slightly affect structure and gantries but could result in severe disruption of construction work. Cascade failure risk is being talked about in infrastructure circles and will potentially be an issue in the future (Interviewee 1, January 26, 2018).

Still using the *HAAFM* methodology, HE's 2016 report highlighted current adaptation action plans. The existing climate risk evaluation takes account of four main factors, including the rate of climate change, the severity of disruption, uncertainties, and the extent of disruption (Conference of European Programme of Roads, 2010). The management of drainage, road pavements, and structures are still the primary focus, with the highest risk scores, which is the same as the results in the risk assessment 2011 (Highways Agency, 2011), whereas other potential climate vulnerabilities will be continuously monitored by HE.

Under the Adaptation Reporting Power in the Climate Change Act, HE is required to report to the British government on a 5-year basis. Current time horizons of road asset life/activity are assessed against two broad categories: short-term (less than 30 years) and long-term (more than 30 years). However, HE (Highways England, 2016) considers that the time horizon for climate change effects to become material can be divided into short-term (present-2020), mid-longer term (2020–2080), and longer term (beyond 2080).

Due to the uncertain nature of climate change, a longer time horizon might be required in future adaptation planning. This planning horizon can be referred to asset lifecycles up to 120 years (Interviewee 1, January 26, 2018). Road planning is a complicated procedure that involves geography, asset condition, and financial budgets. Also, different routes might have diverse time horizon because of its project-based feature. Hence, an appropriate time horizon for climate adaptation planning needs to be set up on a multifaceted basis (e.g., asset lifecycle, likelihood, frequency, severity of climate change and infrastructure resilience, route conditions, and adaptation costs, etc.).

In future adaptation planning, one of the critical challenges, as an interviewee mentioned ((Interviewee 1, January 26, 2018), is how to find an approach to embed climate change in standards. This might start with reviewing relevant road technical specifications (e.g., Design Manual for Roads and Bridges). A long-term plan with specific adaptation actions will be carried out, but how to deliver it to all the staff is yet to be addressed (Highways England, 2016). With the publication of UKCP18, a new-round review of derived products within the British road sector is required (Interviewee 1, January 26, 2018). Moreover, other mitigation measures should be supplementary with adaptation measures to reduce CO_2 emission owing to high temperature, which is a primary concern as stressed in *Highways England Delivery Plan (2015–2020)* (Highways England, 2015) and from our interviews. Still, risk analysis should be a significant component of road planning for climate adaptation. It will benefit from a standardized mechanism constructed by diverse road stakeholders, such as from the UKCP18 Government User Group.

3.2 Network Rail

Network Rail (NR) owns and operates the national railway infrastructure covering 20,000 miles of track, 30,000 bridges and viaducts, as well as thousands of tunnels, signals, level crossings, and points across England, Wales, and Scotland (Network Rail, 2018a). Its strategic national network has been divided into nine routes, including Anglia, Freight and National Passenger Operations, London North Eastern and East Midlands, London North Western, Scotland, South East, Wales, Wessex, and Western line since 2015. Although local train and freight operators run each route, they are supported by NR's national services and functions to maintain its safety and efficiency (Network Rail, 2018b).

Climate change adaptation and weather resilience are two mainstreams in environmental development in NR. Though climate adaptation and weather resilience initiatives have been prepared since 2012, an official *Climate Change Adaptation Report* was not published until 2015. Afterward, *Route Weather Resilience Plans* specialized for each route were produced in 2016 (Network Rail, 2018c,d). In the Western Route Plan, for instance, flooding, wind, and landslips were considered to be the highest priority risk and likely to cost a lot to repair.

The 2015 Adaptation Report (Network Rail, 2015) summarized the understanding of NR as to the existing and potential impacts posed by climate change on its rail performance and safety and the implementation of adaptation actions to deal with them. This included the identification of climate risks, thresholds and uncertainties, knowledge sharing, existing adaptation barriers and opportunities, and planned actions. A few significant climate hazards on rail infrastructure were recognized through an internal risk analysis supported by METeorological data EXplorer (METEX) and geographic information system (GIS) tools. These included changes in temperature and precipitation, increased flooding, high winds, SLR, extreme weather, lightning, and seasonal changes. For example, cold weather, such as snow and ice, would threaten overhead line equipment and block rail lines; heat may increase rail bucking and derailment risk; heavy rainfall and flooding could cause scour of embankment material and damage electricity equipment (Interviewee 2, April 6, 2018). Furthermore, *Tomorrow's Railway and Climate Change Adaptation Report*, as a part of the T1009 programme funded by the Rail Safety and Standards Board (RSSB), established an adaptation framework containing four action steps for the management of summer conditions, winter conditions, and flooding risk by drawing on the experiences of other countries in weather resilience and climate change adaptation (RSSB, 2016).

In the meantime, NR has been responding to the challenges of extreme weather in its daily operation (Network Rail, 2018c). The latest published *Weather Resilience & Climate Change (WRCC) Adaptation Policy* and *Weather Resilience and Climate Change Adaptation Strategy 2017–2019* (Network Rail, 2017a,b) laid solid foundation for the delivery of resilience plans of each route through setting up the context and funding values of specific adaptation actions. The *WRCC* reports (Network Rail, 2017a,b) set out NR's approach to creating a safer and more resilient network for future weather impacts. A four-pillared method included the following components: "analysis risk and costs," "integrate into business as usual," "streamline operational weather management," and "proactive investment" in its *2020 Review and Revise Strategy*.

Overall, NR has constructed a relatively comprehensive framework for adapting to climate change and extreme weather, with supplementary specific route plans and a professional resilience steering group (e.g., RSSB, 2016). The current adaptation report runs on a 5-year basis. However, the time horizon for rail adaptation planning may look at the next 30 years and beyond 2100 in a longer term depending upon the lifespan of specific assets and geographical conditions. There are four primary steps in real adaptation implementation, including risk assessment in place, data analysis, asset investment, and influence and discussion with stakeholders (Interviewee 2, April 6, 2018). Nevertheless, owing to the uncertainties of long-term climate change impacts and insufficiency of precise data on climate change rate and extreme events (Network Rail, 2015), the existing plan still focuses on the identification of several climate thresholds and selection of the best risk scenario (Interviewee 2, April 6, 2018). The quantification of climate risks and costs is still at an embryonic phase (Network Rail, 2017a).

Several gaps have been investigated in the last 5 years, including weather and climate-related thresholds, management of wet weather, and standards' design of uppertemperature thresholds (RSSB, 2016). Meanwhile, the need for asset investment funding to take account of the whole-life cycle of the rail network, as well as the cost-benefit analysis, was mentioned in the interview (Interviewee 2, April 6, 2018). As part of the process of preparing its climate adaptation strategy, NR has researched exceptional experience and suggestions from local transport authorities, such as TfL, as well as data analysis from EA, consultants, and scholars. To be successful, further plans should also incorporate stakeholder views, including public engagement. Besides, NR should continuously receive legislative and regulatory support from the Office of Road and Rail (ORR) and Department for Transport (DfT), and other relevant government bodies. Successful climate adaptation planning might mean that climate adaptation can be finally written into every business plan and become "business as usual" (Interviewee 2, April 6, 2018).

3.3   London (Transport for London & Environment Agency)

3.3.1   Transport for London

Transport for London (TfL), as a local transport authority, is responsible for the daily operations of the capital's public transport and road system (Transport for London, 2018a). Through delivering the transport strategy and policies from the Mayor of London, it commits to develop and maintain integrated, secure, efficient, and economical transportation infrastructure and various services mainly covering London Underground, Surface Transport, London Rail, and Emirates Air Line Cable Car (Transport for London, 2018a,b).

TfL started its climate risk assessment on London's transport networks based on the UK Climate Impact Programme's projections of the potential climate risks and opportunities due to flooding, drought, and overheating. It possesses mechanisms in managing extreme weather events and identifying the requirement to replace critical assets to make them more resilient to climate change (Greater London Authority, 2011). *Managing Extreme Weather at Transport for London* (Woolston, 2014) reviewed a series of local documents in climate adaptation in London, such as the London Underground's comprehensive flood risk review and EA's Thames Estuary 2100 Project, and attempted to establish a long-term flood risk management plan by a flexible "threshold" planning approach.

The recently published report *Providing Transport Service Resilience to Extreme Weather and Climate Change* (Transport for London, 2015) updated the findings in the 2011 report and provided an overview of existing risk assessment by TfL regarding operation and services. According to this report, the primary trend of climate change impacts to Greater London included the increased temperature in summer, flooding, and more frequent and intensive winter storms by 2080s. Flooding, high winds, and heating were deemed to be the main risks that affect the delays on the road network, the safety of surrounding buildings and infrastructure, as well as the comfort and health of passengers on trains, respectively.

Some extreme weather events were also reviewed by TfL, exemplified by a lightning strike at Docklands Light Railway Crossharbour in 2012, a hail storm at the Fore Street tunnel in 2013, and cloudburst flooding and localized rainfall in several spots in summer 2014 (Transport for London, 2015). Combined with the discussion results from an interviewee, the summary is that roads are less vulnerable to climate risk as they have alternative routes and modes to adapt to climate change; but rail and underground are usually more vulnerable due to lack of flexibility in asset construction. The wind is considered as one of the critical threats at present, while the published UKCP18 has changed the status quo by integrating additional factors, such as SLR (Interviewee 3, January 17, 2018; UK Climate Projections, 2018).

A critical scoring risk assessment method has been developed by TfL. The TfL Board initially develops a "top-down" risk appetite factor before each business area (London Rail, London Underground and Surface Transport) produces its own scoring scheme to reflect the local differences (the bottom up factor) (Transport for London, 2015). The strategic risk map primarily considers the likelihood and impacts of climate change. For instance, in London Underground, overheating was expected to pose "very high" impacts to key track, signals, and communications assets, as well as the comfort of staff and passengers. Current risk management of TfL is based on its day-to-day operations, asset management plans, and also infrastructure design and scheme planning in the long term. The forecast bulletins and daily real-time monitoring help identify temperature and precipitation changes to enable the corresponding adaptation options. For example, TfL would apply salt and grit to the road surface, bus station approaches, and platforms if cold weather and icy conditions are being forecasted. Meanwhile, its asset management framework sets out the high-level principles and specific strategies for every asset group with regard to required asset performance, conditions, and maintenance.

Established in 2003, the London Climate Change Partnership (LCCP) Transport Group committed to raising the awareness of the risks of climate change in the transport sector via the development of guidance and adaptation measures. Several projects, such as London Underground's cooling the tube and "Drain London," have offered pioneering trials in climate adaptation (Woolston, n.d.). Nevertheless, existing adaptation methods to climate change are still at an embryonic stage, and no comprehensive adaptation plans have been proposed to TfL. This could partially be due to uncertainties in forecasting and insufficient understanding of climate vulnerabilities and thresholds. Current risk assessment tends to rely on qualitative evidence rather than on a systematic quantitative method. Hence, priorities should be given to how to best utilize scientific data, as well as how to translate climate forecasts into meaningful scenarios.

As an integrated transport provider with financial support from the government, TfL has advantages in developing a holistic adaptation plan to make a resilient network in the future. Even so, gaining political interest is still a potential barrier. This would require TfL to continually provide substantial evidence on the level of risks alongside other factors to enable decision-making. In future adaptation plans, appropriate time horizons will vary for each project and different transport mode and its asset strategy. The future plan of climate adaptation can be led by TfL. The planners can draw on the adapting experience of NR and LCCP, while attracting the engagement of transport providers and utility providers (e.g., Thames Water, Environment Agency and Met Office) to further increase the likelihood of success of adaptation planning (Interviewee 3, January 17, 2018).

3.3.2 Environment Agency

In water transport, the Thames Barrier in London is one of the few moveable flood barriers in the world, which is run and maintained by the Environment Agency (EA). EA examines the barrier monthly and tests it at a high spring tide each year. Supported by its internal computer models and data from Met Office and the UK National Tide Gauge Network, it forecasts the risks up to 36 hours in advance to inform a decision on when the barrier should be closed. The closure of the Thames Barrier happens under a storm surge condition in order to protect London from the sea, depending upon the height of the tide and the tidal surge as well as the river flow entering the tidal Thames. Since 1982, the barrier has been closed over 170 times to protect against tidal and fluvial flooding (Environmental Agency, 2014).

The Thames Estuary 2100 (TE2100) Project, established by EA in 2002, is the first primary flood risk project in the United Kingdom to put climate change adaptation at its core. The plan mainly looked at tidal flooding, though other sources of flooding including high river flow as a result of heavy rainfall, and surface water flooding are simultaneously considered. Based on the prediction of SLR from 90 cm up to 2.7 m by 2100, the plan was designed to provide strategic guidance for adapting to flooding in the Thames Estuary over the next 100 years. A key driver is to consider how the tidal flood is likely to change in response to future change in climate, and how this would impact people and property in the floodplain. Additionally, there is a consensus that many existing flood walls, embankments, and barriers are getting older and would need to be raised or replaced to manage SLR (Interviewee 4, February 1, 2018).

The *Safeguarding London Transport* (Environmental Agency, 2008) comprehensively evaluated the risks that potential flooding pose to the London transport system and the Thames Estuary. Using Geographic Information System (GIS) and Key Performance Indicators (KPIs), it assessed the vulnerability of different assets in several transport networks (e.g., age of station, elevation, flood warning, and distance from the defenses). Generally, London Underground was most vulnerable to the risks of flooding as it was widely located in tunnels underneath the ground, though roads, generally at ground level, can also be extensively affected. However, the rail network had the lowest level of vulnerability because stations and rail tracks were usually located above the ground. Adaptation costs and network resilience were considered in responding to flooding risks.

A typical cost of installing a set of points and their related signals could be expected to be between British pounds (£) 175,000 and £250,000 (Environmental Agency, 2008). The resilience was measured by the recovery capacity of the transport network, including the scope to use other alternative routes or modes to bypass the partial closure of this system. London Underground, owing to its natural underground location and interconnected tunnels with a high possibility for water ingress, might be the least resilient to climate risks. In the worst flood risk scenario, a majority of the sections could be closed for an extended period, and the repair cost could be massive. With an updating requirement of a 5-year short-term review and a 10-year full review, the latest *TE2100 5-year review*, used historical data and report analysis to examine the results of ten indicators (e.g., sea level, peak surge level, asset condition, barrier operation, habitat, and public attitudes to flood risk, etc.) (Environmental Agency, 2016).

One of the significant challenges is the mismatch of aging flooding barrier infrastructure and a higher SLR rate where many flood defenses were built 30 years ago when SLR was 8 mm per year but now becomes 11 mm per year (Environmental Agency, 2017). Existing data are incapable of measuring wave conditions at a peak surge level and the amount of intertidal habitat in the Estuary. Meanwhile, the asset condition has declined in recent years in some areas, especially the outer Estuary. More funding is needed for asset improvement and maintenance and increasing the proportion of assets rating as fair and reasonable (Environmental Agency, 2016).

Nevertheless, the TE2100 Plan is believed to be on the right track with a broad range of stakeholders and public engagement, as an interviewee stated. Having had two earlier consultations in 2005 and 2008 on the critical findings of the project supported by a programme of public meetings and a web-based consultation, EA undertook its public consultation on the draft TE2100 Plan in 2009. These included 15 local workshops and public meetings across the Estuary, over 50 meetings with key organizations, to provide stakeholders (e.g., Greater Local Authority) with an opportunity to feedback and ask questions on any aspect of the Plan or its recommendations, as well as receiving 120 written responses (Interviewee 4, February 1, 2018). In future planning, it is expected to have more new and cost-effective barriers further downstream and tidal flood defenses for tackling more severe SLR and storm surges (Environmental Agency, 2017).

3.4 Devon County Council

Flooding is one of the critical issues for UK transport systems. A significant number of heavy storms in recent years have broken historical records since 2000 in the United Kingdom, and there are more frequent events projected in the future (Devon County Council, 2014a).

Dawlish Warren is a coastal spit on the south Devon coast of England. The cumulative effect of the rapid succession of over significant storms in winter

2012/2014 had the most severe impacts in the United Kingdom since the 1950s (Devon Maritime Forum, 2014). The South West main rail network was mainly affected with the collapse of the multisectional seawall at Dawlish, as well as a significant impact on transport resilience and the local economy of the South West Peninsula due to extreme weather (Devon County Council, 2014a; Devon Maritime Forum, 2014). Up to 46 m of railway track was swept away with part of the seawall in early February 2014, restricting the service linking Cornwall and much of Devon with the rest of the United Kingdom. Dawlish station was damaged, and the main rail line from Exeter to Newton Abbot was closed. In total, the storms had resulted in the 2-month closure of the mainline and over 7000 services canceled (Devon Maritime Forum, 2014).

NR estimated that the damage would take "at least" six weeks to recover, and an extra £100m was provided for flood repairs across the country (BBC News, 2014a, 2015c). A storm occurred again in early 2017, crashed into trains and over flood barriers as 50ft waves smashed the coasts. Boats, lighthouses, and seafront rail track were impaired by surges, and the gales caused temporary cancellation of some trains at Plymouth and between Newton Abbot and Exeter St David's (The Sun, 2017).

Devon County Council (DCC) is responsible for the maintenance and re-pair of 12,800 km of the public road network (not including Strategic Road Network) in Devon. On the basis of the UK Climate Impact Programme's projections, the potential impacts on Devon's roads include increased temperature, SLR and the changes in rainfall patterns, and humidity variations. DCC initiated its Weather Impacts Assessment in 2010 and introduced an Impact Assessment Tool (IAT) in 2011. The risks posed by climate change were evaluated through the "Devon Way for Risk Management" matrix, where the impact and likelihood of risks were identified as three scenarios ("low," "medium," and "high") in different timescales (the 2020s, 2050s, and 2080s).

In the *Extreme Weather Resilience Report* (Devon County Council, 2014a), a few risks on highways maintenance and connectivity posed by extreme weather events were documented after the 2013/2014 storm. These mainly contributed to the collapse of the sea wall at Dawlish, severe road deterioration, and road closures in multiple sections of "A" road, backlog in the carriageway, increases in potholes, fallen trees, and branches (Devon County Council, 2014a). £3 m initial clear-up was followed by more than £700 m for climate risk maintenance.

In April 2014, the main railway line through Dawlish in Devon was re-opened, rebuilt by a 300-strong team from NR at a cost of £35 m (BBC News, 2014b). By Dec. 2016, the government had commissioned NR to make a further £10 m plan to protect coastal lines from storms, which included moving the line and strengthening the cliffs above the line connecting Devon and Cornwall with the rest of the United Kingdom (BBC News, 2016). NR has outlined the ongoing maintenance for the regional rail network in Control Period 6 of its 5-year plan (2019–2024) (Devon Live, 2018).

Led by DCC, the Flood Recovery Coordination Group was established to provide operational and financial support for the affected communities threatened by flooding (Devon Maritime Forum, 2014). Since 2012, DCC has spent over £12 m on the storm-related emergency plan for highways, together with extensive drainage works implemented due to the 2013 storms. Nevertheless, a continual modification for existing design and operation and maintenance are required to adapt against further climate change (Devon County Council, 2014b).

More recently, an assessment of the risks posed by climate change to DCC's Highways Management Service was completed in April 2014 in coordination with Highways Agency and Department for Environment, Food & Rural Affairs (DEFRA). DCC, as a part of the South West partnership, including Somerset County Council and Wiltshire Council, has campaigned for government investment to enhance the strategic resilience of the A303/A30/A358 corridor (Devon County Council, 2014b). Meanwhile, the South West Peninsula Rail Task Force, made up of local authorities, enterprise, and academia, provided cross-sector support for guaranteeing a £7 m investment to develop a more resilient rail network (Devon Maritime Forum, 2014). Therefore, further collaboration between roads and railways is expected to deal with the potential risks posed by storms and other extreme weather events.

The modal shift solution from the road to rail may not fit the case of DCC, as pointed out by an interviewee (Interviewee 5, February 8, 2018). First, rail is more vulnerable to the variation of weather, as it is easier to identify alternative routes for an affected road. Second, the capacity of trains cannot meet the demand for emergency evacuation of cars on the road. For instance, the capacity of a train that can carry 500 people is only equivalent to 250 cars. Alternatively, as an emergency plan, the National Express provided five new "rail replacement" coach services, and Flybe had put on three extra flights from Newquay to Gatwick each day during that period (Transport Committee, 2014). Most importantly, since 2014, a solution proposed by NR to tackle storms is the reinstatement of the old Tavistock line, along the Great Western Railway Teign Valley route, and a new railway with five alternative routes to avoid the coastal section through Dawlish (BBC News, 2014c,d). More recently, the Peninsula Rail Task Force implemented the Dawlish Additional Line as a long-term priority in the 20-year plan, by reconnecting Okehampton to Plymouth route to make the network more resilient to extreme weather (Devon Live, 2018).

Although NR's efforts in storm adaptation are remarkable concerning rapid repairing capacity and replacement services, the condition of the coastlines and their connectivity to the diverse region are still inexplicit in the long term with the occurrence of more frequent and intense extreme weather events (Devon Maritime Forum, 2014). Overall, there is no comprehensive adaptation strategy for climate change at DCC, which may be associated with the kaleidoscopic nature of climate change itself. Currently, climate adaptation has been integrated into the risk management, by which DCC is primarily identifying the risks posed by climate change and working closely with NR to make specific adaptation measures for each risk. With more than £10m being put into drainage management, a near-sight plan (the 2020s) is to alleviate flooding and keep the water level as low as possible. One of the advantages in developing a holistic adaptation strategy in the future is that DCC is well aware of the risks of climate change via a bottom-up mechanism to collect local information and a top-down mechanism to deliver the governmental policy. Even so, a long-term adaptation plan still needs enough financial support and cross-party engagement to ensure its effective implementation (Interviewee 5, February 8, 2018).

4   Discussion and conclusion

This chapter explores the existing adaptation planning in the road and railway systems by exemplifying four typical case studies in the United Kingdom. A qualitative research method is utilized, including document review and five in-depth interviews with domain transport experts from HE, NR, EA, TfL, and DCC. By doing so, the evolvement of climate risk assessment and adaptation actions, current advantages, and potential challenges are dissected for each organization. To compare the similarities and differences of adaptation plans among different organizations, within-case (e.g., London) and cross-cases analyses (e.g., NR and HE) are further explained to strengthen the external and internal validity.

Table 1 summarizes the primary progress of the UK road and rail sectors in adapting to climate change based on the Committee on Climate Change's latest report (2017b) and the new findings from the studied cases.

In the cases of HE and NR, a series of climate risk and adaptation reports have been published on the basis of the UK Climate Projections (UKCP09) and relevant legal guidance (e.g., Committee on Climate Change, 2017a,b; Department for Transport, 2014). Although fewer weather-related delays in England occurred in recent years, regional extremes are still witnessed and will potentially trigger significant costs due to the uncertainties of climate change. In the road and rail sectors, the facilitation of flood resilience is a shared priority with over £100 million funding being allocated by the government (HM Treasury, 2016). Meanwhile, the Committee on Climate Change's ARP2 and DfT's Transport Resilience Review have provided HE, NR, and other local authorities with specific guidance for improving climate resilience on their transport networks.

Simultaneously, the road and rail systems face many challenges in adapting to climate change risks. Overall, rail and underground are more vulnerable to the impacts of climate change due to the limited flexibility and complexity in rail infrastructure construction. Compared with roads, a strategic rail adaptation plan covering all the nine national routes (exclude the Freight and National Passenger Operators route but add the West of Exeter route) has been prepared (Network Rail, 2017a,b; Rail Safety and Standards Board, 2016). However,

TABLE 1 Primary progress of the UK road and rail sectors in climate adaptation.

Similarities		
	Risks to infrastructure	Increased frequency and severity of flooding (will double the number of assets exposed to climate change by 2080s); temperature and precipitation changes; increased maximum wind speeds; other uncertainties such as fog, storms, and lightning (Dawson et al., 2016)
	Vulnerability	Fewer weather-related delays in England in recent years
		The road is less vulnerable to climate risk as having alternative routes and modes to adapt to climate change; rail and underground are more vulnerable due to the limited flexibility and complexity in rail infrastructure construction (e.g., in cases of London and Devon)
	Risk assessment	*TfL—Providing Transport Service Resilience to Extreme weather and Climate Change* (2015): A scoring risk assessment method considering likelihood and impacts of climate change at each business area in London Rail, London Underground and Surface Transport has been developed
		EA—TE2100 5-Year Review (2016): Evaluated the flooding risks based on identified 10 KPIs
	Funding	*The Autumn Statement 2016 transport projects*: Announced £150m governmental funding for flood resilience improvement with £10m on roads and £50m on rails (HM Treasury, 2016)
		London: several £million has invested for Docklands Light Railway, and an estimated cost of at least £1m is required to carry out a sustainability assessment for pathways in TfL (Transport for London, 2015); delivering £308m of investment on tidal flood defense improvements across the Tidal Thames for the Thames Estuary Asset Management programme (Institution of Civil Engineering, 2017)
		Devon: Over £12m storm emergency plan in highways from DCC and a further £10m plan for protecting the coastal line from storms for NR (BBC News, 2016); £7m investment for establishing a resilient rail network from The South West Peninsula Rail Task Force (2016)
	Guidance	*The UK Climate Projections* (*UKCP09*) provided comprehensive evidence for risk assessment; the new *UKCP18* projections may change the level of climate risks
		The second round of the Adaptation Reporting Power (ARP2) for 2015-2020 (2017a): HE and NR's reports
		DfT—Transport Resilience Review: A review of the resilience of the transport network to extreme weather events (2014): provided HE and NR with the specific recommendation for improving climate resilience

Continued

TABLE 1 Primary progress of the UK road and rail sectors in climate adaptation—cont'd

Railways	*RRSB—Tomorrow's Railway and Climate Change Adaptation Report (2016)*: Established an adaptation framework for climate change
	NR—An internal audit of weather resilience and climate change (2016): Recognized the demand for setting up strategic targets, and standardizing risk management and decision making.
	NR—Weather Resilience and Climate Change Strategy (2017): Covered all national routes (including West of Exeter); risk analysis and site-specific actions (focusing on embankments, bridge stability, and coastal defenses)
	Implementation: NR has cooperated with the Energy Network Association to investigate the electricity substations; aging railway infrastructure is a challenge
Roads	*HE—Climate Change Adaptation Strategy and Framework (2009)*: Initiated the HHAFM by setting up a seven-stage adaptation process
	HE—Climate Adaptation Risk Assessment Progress Update (2016): Embarked a flood risk analysis through utilizing EA's flood risk maps and other data; recognized high risk and very high-risk hotspots and culverts and reduced floods at 124 flooding hotspots and culverts
	DCC—Service Resilience in a Changing Climate Highways Management (Devon County Council, 2014b): Developed a "Devon Way for Risk Management" matrix for evaluating the impact and likelihood
	Implementation: HE has improved drainage and flood resilience to climate change on some regional routes; some local authorities have increased its strategic planning and investment in resilience

Based on "Progress in preparing for climate change 2017 Report to Parliament" (Committee on Climate Change, June 2017a).

the existing adaptation plan of NR mainly focuses on the identification of various climate thresholds (Network Rail, 2015). Sometimes climate adaptation is regarded as part of risk management, where the attention focuses on risk assessment for the road system with specific extreme weather adapting plans (Interviewee 5, February 8, 2018). The absence of a holistic adaptation strategy might reflect the deficiency of scientific data (e.g., SLR for TE2100), cost-benefit analysis, and understanding of climate vulnerabilities and thresholds (Interviewee 3, January 17, 2018), and these gaps reflect the findings in the literature (e.g., Koetse and Rietveld, 2012). Hence, a new set of climate projections (UKCP18) published in November 2018 by the UK government is believed to offer clear guidance for dealing with the challenges of estimation and selection of risk scenarios under various climate conditions.

One of the significant challenges revealed by the case of the TE2100 is that the aging flooding infrastructure might not be able to catch up with the higher SLR (Environmental Agency, 2017), while aging infrastructure could be a standard issue for the whole railway industry owing to the increasing rate of climate change. Furthermore, there is an ambiguous time horizon for road and rail adaptation planning. For the majority of cases, adaptation reports run on a 5-year basis (Interviewee 1, January 26, 2018; Interviewee 2, April 6, 2018; Interviewee 4, February 1, 2018); thereafter, the time horizon for a long-term adaptation plan is undetermined: it may look at next 30 years and beyond 2100 (Interviewee 2, April 6, 2018) or 100 years (Interviewee 3, January 17, 2018) for railways and up to 120 years for roads. It can be linked to the diverse lifespan of specific assets, geographic conditions, climate change prediction, financial budgets, etc. In the future, climate adaptation planning needs to be regularized and written into every business plan (Interviewee 3, January 17, 2018), embedded in technical standards and delivered to all staff seamlessly (Interviewee 1, January 26, 2018). A successful adaptation plan must be aware of budgetary constraints and strike a balance between corporate priorities and technical requirements (Wang et al., 2019).

The establishment of several partnerships, on behalf of the London Climate Change Partnership (LCCP) Transport Group and the South West Peninsula Rail Task Force, has offered a chance to deal with regional climate change issues. However, in practice, owing to the project-based characteristics in most adaptation cases, as per several interviewees, road and rail stakeholders usually only consult each other but undertake projects separately. A modal shift strategy has been successfully applied into practice, for instance, where road traffic was converted to rail by establishing a rail platform and offering a new rail service in Workington, Cumbria in a quick response (Ace Geography, n.d.), and a rail replacement service by increasing buses and flights from Newquay to London due to seawall damage in Dawlish, Devon (Transport Committee, 2014). More extensive and efficient cooperation between roads and railways is expected, but the development of an integrated inland transport system requires thorough consideration of multiple factors, such as mode capacity, the severity of consequences, and geographic conditions. The trans-mode and cross-sectoral collaborations in the future should enable planners to create a new blueprint for climate adaptation, effectively facilitated by governmental regulation (e.g., ORR, DfT), broader stakeholder management, and public engagement from decision-making to adaptation implementation.

Finally, we believe that this chapter contributes to further studies in climate change, risk assessment, transport planning, policy making, and other interdisciplinary areas. The all-around data analysis from interviews will facilitate interviewees in the case studies to better understand the impacts posed by climate change in decision-making and to recognize their strengths and barriers in future adaptation planning. Also, an integrated thinking pattern concerning roads and railways in an integrated inland transport will enlighten transport planners to consider the consistency and resiliency of diverse modes in a systematic transport network in future planning.

Appendix. Basic information of interviewees

Interviewee	Position	Organization	Interview date
Interviewee 1	Middle	Highways England	January 26, 2018
Interviewee 2	Middle	Network Rail	April 6, 2018
Interviewee 3	Senior	Transport for London	January 17, 2018
Interviewee 4	Middle	Environmental Agency	February 1, 2018
Interviewee 5	Senior	Devon County Council	February 8, 2018

Remarks: "Senior" means policy maker, transport planner, etc.
"Middle" means environmental specialist, climate adaptation advisor/manager, etc.

References

Ace Geography (n.d.). Flooding Case Studies: Cockermouth, UK - Rich Country (MEDC). Retrieved from: http://www.acegeography.com/flooding-case-studies-gcse.html. 20 December 2017.

BBC News, 2014a. Devon and Cornwall Storm Causes 'Devastation'. Retrieved from: http://www.bbc.com/news/uk-england-26044323. (Accessed June 10, 2018).

BBC News, 2014b. Storm-Hit Dawlish Rail Line Compensation Payout Revealed. Retrieved from: https://www.bbc.co.uk/news/uk-england-devon-27055780. (Accessed June 10, 2018).

BBC News, 2014c. UK Storms Destroy Railway Line and Leave Thousands Without Power. Retrieved from: http://www.bbc.com/news/uk-26042990. (Accessed June 10, 2018).

BBC News, 2014d. How Do You Fix the Dawlish Problem?. Retrieved from: https://www.bbc.co.uk/news/uk-26068375. (Accessed June 10, 2018).

BBC News, 2015a. Cumbria Hit by Flooding After Heavy Rain. Retrieved from: http://www.bbc.com/news/uk-england-cumbria-35003649. (Accessed August 9, 2017).

BBC News, 2015b. Storm Desmond: Defences Against Indefensible Floods. Retrieved from: http://www.bbc.com/news/science-environment-35028180. (Accessed December 20, 2017).

BBC News, 2015c. Dawlish Rail Line: Closure 'Costs Economy up to £1.2bn'. Retrieved from: http://www.bbc.com/news/uk-england-devon-31140192. (Accessed June 11, 2018).

BBC News, 2016. Dawlish Railway Line: £10m Pumped Into Saving Coastal Route. Retrieved from: https://www.bbc.co.uk/news/uk-england-devon-38012833. (Accessed June 11, 2018).

BBC News, 2017. Flooding Closes Cumbria Roads and Schools. Retrieved from: http://www.bbc.com/news/uk-england-cumbria-41582974. (Accessed December 20, 2017).

Braun, V., Clarke, V., 2006. Using thematic analysis in psychology. Qual. Res. Psychol. 3 (2). 77–101, 474, 486.

Committee on Climate Change, 2014. Managing climate risks to well-being and the economy. Adaptation Sub-Committee Progress Report 2014.

Committee on Climate Change, 2017a. Progress in preparing for climate change. . 2017 Report to Parliament.

Committee on Climate Change, 2017b. The Second Round of the Adaptation Reporting Power (ARP2) for 2015-2020.

Conference of European Directors of Roads, 2010. Retrieved from http://www.fehrl.org/index.php?m=32&mode=download&id_file=10736, Accessed 20 May 2018.

Dawson, R.J., Thompson, D., Johns, D., Gosling, S., Chapman, L., Darch, G., Watson, G., Powrie, W., Bell, S., Paulson, K., Hughes, P., Wood, R., 2016. UK Climate Change Risk Assessment Evidence Report: Chapter 4, Infrastructure. Report prepared for the Adaptation Sub-Committee of the Committee on Climate Change, London.

Department for Infrastructure, 2017. Northern Ireland Transport Statistics 2016-2017. . UK Government (Crown copyright).

Department for Transport, 2014. Transport Resilience Review: A Review of the Resilience of the Transport Network to Extreme Weather Events. Retrieved from: https://assets.publishing.service.gov.uk/government/uploads/system/uploads/attachment_data/file/335115/transport-resilience-review-web.pdf. (Accessed July 4, 2017).

Department for Transport, 2017. Road Lengths in Great Britain 2016. DfT Statistical Release. UK Government (Crown copyright).

Devon County Council, 2014a. Extreme Weather Resilience Report. Devon County Council, Devon. May 2014.

Devon County Council, 2014b. Service Resilience in a Changing Climate Highways Management. Environment Group, Devon County Council.

Devon Live, 2018. It's Been Four Years Since the Dawlish Rail Disaster—What Has Happened Since? Retrieved from: https://www.devonlive.com/news/its-been-four-years-dawlish-1152802. (Accessed June 11, 2018).

Devon Maritime Forum, 2014. Holding the line? Reviewing the Impacts, Responses and Resilience of People and Places in Devon to the Winter Storms of 2013/2014. Retrieved from: http://www.devonmaritimeforum.org.uk/images/stories/DMFdocuments/DMFmeetingArchives/2014Autumn/DMF%20Storms%2013-14%20Summary%20Report.pdf. (Accessed June 11, 2018).

Environmental Agency, 2008. Thames Estuary 2100 Safeguarding Transport Phase 3 Studies—Work Element 6.2.

Environmental Agency, 2014. The Thames Barrier: How the Thames Barrier Works, and When It Is Scheduled to Close. Retrieved from: https://www.gov.uk/guidance/the-thames-barrier. (Accessed June 4, 2018).

Environmental Agency, 2016. TE2100 5 Year Review, Non-Technical Summary. July 2016.

Environmental Agency, 2017. Policy Paper: Thames Estuary 2100 (TE2100). Updated 2 March 2017. Retrieved from: https://www.gov.uk/government/publications/thames-estuary-2100-te2100/thames-estuary-2100-te2100. (Accessed June 5, 2018).

Greater London Authority, 2011. Managing Risks and Increasing Resilience: The Mayor's Climate Change Adaptation Strategy.

Highways Agency, 2009. Climate Change Adaptation Strategy and Framework. Revision B—issued in November , p.2009.

Highways Agency, 2011. Climate Change Risk Assessment August 2011. Retrieved from: http://assets.highways.gov.uk/about-us/climate-change/HA_Climate_Change_Risk_Assessment_August_2011_v2.pdf. (Accessed May 22, 2018).

Highways England, 2015. Highways England Delivery Plan 2015–2020.

Highways England, 2016. Climate Adaptation Risk Assessment Progress Update 2016.

Highways England, 2018a. Retrieved from: https://www.gov.uk/government/organisations/highways-england/about. (Accessed May 20, 2018).

Highways England, 2018b. Retrieved from: https://highwaysengland.co.uk/highways-england-about-us/. (Accessed May 22, 2018).

HM Treasury, 2016. Policy Paper: Autumn Statement 2016 Transport Projects. Retrieved from: https://www.gov.uk/government/publications/autumn-statement-2016-transport-announce-ments/autumn-statement-2016-transport-projects. (Accessed May 30, 2018).

Hooper, E., Chapman, L., 2012. Chapter 5. The impacts of climate change on national road and rail networks. In: Transport and Climate Change. Emerald Group Publishing Limited, pp. 105–136.

Institution of Civil Engineering, 2017. Thames Estuary Asset Management 2100 Programme (TEAM2100). Retrieved from: https://www.ice.org.uk/getattachment/events/thames.../TEAM2100-flyer.pdf.aspx. (Accessed 14 June 2018.)

Jaroszweski, D., 2015. Chapter 6. The impacts of climate change on the National Freight Sector. In: Transport and Climate Change. Emerald Group Publishing Limited, pp. 137–173.

Koetse, M.J., Rietveld, P., 2009. The impact of climate change and weather on transport: an overview of empirical findings. Transp. Res. Part D: Transp. Environ. 14 (3), 205–221.

Koetse, M.J., Rietveld, P., 2012. Adaptation to climate change in the transport sector. Transp. Rev. 32 (3), 267–286.

Love, G., Soares, A., Püempel, H., 2010. Climate change, climate variability and transportation. Procedia Environ. Sci. 1, 130–145.

McKenzie Hedger, M., Gawith, M., Brown, I., Connell, R., Downing, T.E., 2000. Climate change: assessing the impacts—identifying responses. In: The First Three Years of the UK Climate Impacts Programme. UKCIP and DETR, Oxford.

Miles, M.B., Huberman, A.M., 1984. Qualitative data analysis: a sourcebook of new methods. In: Qualitative Data Analysis: A Sourcebook of New Methods. Sage Publications.

Network Rail, 2015. Climate Change Adaptation Report 2015. Retrieved from: http://16cbgt3sbwr8204sf92da3xxc5m-wpengine.netdna-ssl.com/wp-content/uploads/2016/11/Climate-Change-Adaptation-Report-2015_FINAL.pdf. (Accessed April 7, 2017).

Network Rail, 2017a. Weather Resilience and Climate Change Adaptation Strategy 2017-2019.

Network Rail, 2017b. Weather Resilience & Climate Change Adaptation Policy.

Network Rail, 2018a. Retrieved from: https://www.networkrail.co.uk/who-we-are/. (Accessed May 25, 2018).

Network Rail, 2018b. Retrieved from: https://www.networkrail.co.uk/running-the-railway/our-routes/. (Accessed May 25, 2018).

Network Rail, 2018c. Retrieved from: https://www.networkrail.co.uk/communities/environment/climate-change-weather-resilience/weather-resilience/. (Accessed May 25, 2018).

Network Rail, 2018d. Retrieved from: https://www.networkrail.co.uk/communities/environment/climate-change-weather-resilience/climate-change/. (Accessed May 25, 2018).

Ng, A.K., Wang, T., Yang, Z., Li, K.X., Jiang, C., 2018. How is business adapting to climate change impacts appropriately? Insight from the Commercial Port Sector. J. Business Ethics 150 (4), 1029–1047.

Rail Safety and Standards Board, 2016. Tomorrow's Railway and Climate Change Adaptation: Final Report. Rail Safety and Standards Board Ltd.

Schweikert, A., Chinowsky, P., Espinet, X., Tarbert, M., 2014. Climate change and infrastructure impacts: comparing the impact on roads in ten countries through 2100. Proc. Eng. 78, 306–316.

The South West Peninsula Rail Task Force, 2016. 'Closing the Gap'—The South West Peninsula Strategic Rail Blueprint. Retrieved from: https://peninsularailtaskforce.co.uk/closing-the-gap-the-south-west-peninsula-strategic-rail-blueprint/. (Accessed June 11, 2018).

The Sun, 2017. Doris Pray as 50ft Waves Batter Britain's Coasts, Met Office Warns Brits to Brace Themselves as Worst of Storm Doris Will Hit TODAY. Retrieved from: https://www.thesun.co.uk/news/2765380/storm-doris-uk-weather-warning/. (Accessed May 31, 2018).

Transport Committee, 2014. Oral evidence: transport's winter resilience: rail flooding, HC 1087, Meeting on Tuesday 25 February 2014. Retrieved from: http://data.parliament.uk/writtenevidence/WrittenEvidence.svc/EvidenceHtml/6845. (Accessed June 12, 2018).

Transport for London, 2015. Providing Transport Service Resilience to Extreme Weather and Climate Change. Update Report following the last report to Government in 2011.

Transport for London, 2018a. Retrieved from: https://tfl.gov.uk/corporate/about-tfl/what-we-do?intcmp=2582. (Accessed May 31, 2018).

Transport for London, 2018b. Retrieved from: https://tfl.gov.uk/corporate/about-tfl/how-we-work/how-we-are-governed?intcmp=2724. (Accessed May 31, 2018).

UK Climate Projections, 2018. UKCP18 Science Overview report, November 2018. Retrieved from https://www.metoffice.gov.uk/pub/data/weather/uk/ukcp18/science-reports/UKCP18-Overview-report.pdf. (Accessed February 20, 2019).

Wang, T., Qu, Z., Yang, Z., Nichol, T., Dimitriu, D., Clarke, G., Bowden, D., 2018a. Impacts of climate change on rail systems: a new climate risk analysis model. In: Haugen, S., Barros, A., Gulijk, C.V., Kongsvik, T., Vinnem, J. (Eds.), Safety and Reliability—Safe Societies in a Changing World. CRC Press, London.

Wang, T., Qu, Z., Yang, Z., Nichol, T., Dimitriu, D., Clarke, G., Bowden, D., 2018b. How can the UK road system be adapted to the impacts posed by climate change? By creating a climate adaptation framework. In: Proceedings of the International Conference on Project Logistics (PROLOG), University of Hull, UK, 28-29 June 2018.

Wang, T., Qu, Z., Yang, Z., Nichol, T., Dimitriu, D., Clarke, G., Bowden, D., 2019. How can the UK road system be adapted to the impacts posed by climate change? By creating a climate adaptation framework. Transp. Res. D: Transp. Environ. Available from: https://doi.org/10.1016/j.trd.2019.02.007.

Wikinow (n.d.). British Railway System. 7 April 2017, Retrieved from: http://www.wikinow.co/topic/british-railway-system

Woolston, H. (2014). Managing Extreme Weather at Transport for London Climate Change. Transport for London. 30 May 2018, Retrieved from: https://www.arcc-network.org.uk/wp-content/pdfs/ARCC-2014-2b-TfL-Woolston.pdf.

Woolston, H. (n.d.). Climate Change—Impact on Transport Network User. LCCP, Transport for London. 30 May 2018, Retrieved from: http://sd-research.org.uk/sites/default/files/publications/Helen%20Woolston%20CC%20Adaptation.pdf

Yang, Z., Ng, A.K., Lee, P.T.W., Wang, T., Qu, Z., Rodrigues, V.S., … Lau, Y.Y., 2018. Risk and cost evaluation of port adaptation measures to climate change impacts. Trans. Res. D: Trans. Environ. 61, 444–458.

Yin, R.K., 2003. Case Study Research: Design and Methods. Sage, Thousand Oaks, CA.

Further reading

Getlink (n.d.). 20 May 2018, Retrieved from: https://www.getlinkgroup.com/uk/home/.

Network Rail, 2014. West of Exeter Route Resilience Study. Retrieved from: https://cdn.net-workrail.co.uk/wp-content/uploads/2016/11/West-of-Exeter-Route-Resilience-Study.pdf. (Accessed June 11, 2018).

Network Rail, 2016. Great Britain National Rail Network Diagram. Retrieved from: http://www.nationalrail.co.uk/static/documents/content/routemaps/nationalrailnetworkmap.pdf. (Accessed April 7, 2017).

Peterson, T.C., McGuirk, M., Houston, T.G., Horvitz, A.H., Wehner, M.F., 2008. Climate Variability and Change With Implications for Transportation. Transportation Research Board.

UK Climate Projections, 2009. Retrieved from: http://ukclimateprojections.metoffice.gov.uk/21687. (Accessed June 1, 2018).

Part III

Improving environmental practice

Chapter 7

Green port initiatives for a more sustainable port-city interaction: The case study of Barcelona

Marta Gonzalez-Aregall, Rickard Bergqvist
Department of Business Administration, School of Business, Economics and Law, University of Gothenburg, Gothenburg, Sweden

1 Introduction

Shipping is the most cost-effective way to transport goods across long distances (Butt, 2007; Papaefthimioua et al., 2016). This is supported by the fact that over 80% of global trade is handled by seaports worldwide (UNCTAD, 2017). Therefore, port infrastructure has the capacity to connect global logistics with national and regional industries through its integration with cities and the hinterland (Akhavan, 2017; Monios et al., 2018; Suykens, 1989; Xiao and Lam, 2017; Zhao et al., 2017).

Traditionally, cities with harbors have coped with regular adaptation to political, social, and maritime changes, such as cargo-handling technology, trading patterns (Grossmann, 2008; Hoyle, 1989, 1992, 2000; McCarthy, 2003; Proudfoot, 1996), and inner-city regeneration (Hoyle, 2000). Consequently, the expansion due to increasing vessel size and hinterland connections (Asteris and Collis, 2010) has required appropriate urban planning between the seaport and the city institutions (Akhavan, 2017; Hoyle, 2000; Daamen and Vries, 2013; Debrie and Raimbault, 2016). This relationship between institutions aims to create economic opportunities for the areas surrounding seaports (Hoyle, 1992; Lee et al., 2014; Schipper et al., 2017; Suykens, 1989) and to manage negative impacts, such as coastal water level (Hanson et al., 2011), congestion, noise, and pollution (Monios et al., 2018; Del Saz-Salazar et al., 2013; Xiao and Lam, 2017).

Although shipping is an essential source for maritime cities' economies (UNCTAD, 2017), it can also generate atmospheric pollutants[a] and have adverse effects on human health and the marine environment in coastal residential

---

a. The main air emissions are carbon dioxide (CO_2), nitrogen oxides (NOx), carbon monoxide (CO), volatile organic compounds (VOC), sulfur oxide (SO_2), and particulate matter (PM).

Maritime Transport and Regional Sustainability. https://doi.org/10.1016/B978-0-12-819134-7.00007-1
109

areas (Cullinane and Cullinane, 2013; Ros Chaos et al., 2018; Mat et al., 2016; Na et al., 2017; OECD, 2011; Papaefthimioua et al., 2016).

Nowadays, more than 50% of the world's population lives in urban regions located along rivers and coastlines (Mat et al., 2016). Although approximately 70% of shipping emissions occur at sea (Corbett et al., 2007), ships' fuel combustion can spread to the land (Tichavska and Tovar, 2015) and can consequently cause severe illnesses and diseases (Corbett et al., 2007; Merico et al., 2017; Papaefthimioua et al., 2016; Tian et al., 2013). According to Lack and Corbett (2012), shipping produces around 3.3% of global CO_2 emissions and around 2% of black carbon emissions. Moreover, other studies state that vessels' emissions contribute to increased atmospheric pollution in port cities (Saxe and Larsen, 2004).

Consequently, several actors have proposed initiatives for improving air quality in coastal regions affected by shipping.[b] Although CO_2 emissions from shipping are not covered by the Kyoto Protocol (Merico et al., 2017) or the Copenhagen Accord (UNCTAD, 2010), the International Maritime Organization (IMO) has implemented several initiatives through the International Convention for the Prevention of Pollution from Ships, or MARPOL. For instance, Emission Control Areas (ECAs) limit the concentration of sulfur content from ships in specific regions of the sea to 0.1%. In non-ECA regions, the limit is 3.5%, and that limit will be reduced to 0.5% in January 2020 (Gonzalez Aregall et al., 2018; Merico et al., 2017; UNCTAD, 2018). The current regulation on nitrogen oxides (NOx) is limited to new vessels with specific engine characteristics (Merico et al., 2016).

In addition to international bodies, regional institutions like the European Sea Port Organization (ESPO) aim to enhance environmental policies in European ports, through the EcoPorts Foundation (Gonzalez Aregall et al., 2018). Local port authorities have also had to develop environmental port management and green policies, identifying the negative externalities of shipping activities and transportation (Bergqvist and Egels-Zandén, 2012; Davarzani et al., 2016; Gonzalez Aregall et al., 2018; Puig et al., 2015).

As a result, one of the most common policies for ports has been the implementation of monitoring programs to control the sources of air emissions (Gonzalez Aregall et al., 2018; Puig et al., 2015). Relatedly, some studies have evaluated the emissions ensuing from the movement of containers at terminals (Geerlings and Van Duin, 2011).

Undoubtedly, regulatory frameworks can play a significant role in reducing air pollution. However, in order to combat environmental impacts on port cities, port stakeholders and city managers must plan sustainable port-city interfaces in order to guarantee sustainable development and to optimize the economic, environmental, and social benefits of ports despite challenges caused by urban congestion and pollution (Monios et al., 2018).

---

b. For a detailed description of different initiatives, see Christodoulou et al. (2019).

Given these trends, we examined port-developed strategies that have reduced the environmental impacts by ports and that have promoted sustainable actions to combat climate change in large urban areas. Guided by an extensive literature review on green initiatives at ports, we selected the Port of Barcelona in Spain as a case study to analyze green initiatives that could facilitate the resilience and growth of Barcelona's port-city complex.

Located on the Mediterranean Sea, Barcelona's port experiences high volumes of both freight and cruise traffic, which make it an interesting source for examining sustainability challenges associated with freight transport routes around urban areas as well as possible solutions. According to Monteiro et al. (2018), in terms of shipping emissions, northern Europe and the Mediterranean area are the most damaged regions. Thus, this study considered the perspectives of all actors involved in managing Barcelona's port-city interface, particularly regarding hinterland initiatives. In terms of methodology, this study considered monitoring stations' data because, according to Dalsøren et al. (2009), emission inventories are an important factor in evaluating the effect of emissions on the environment and human health and are thus a helpful tool for policy makers.

This chapter is organized as follows: Section 2 outlines the description of the specific case study of Barcelona. Section 3 provides an in-depth analysis of the specific hinterland initiative enacted here. Section 4 provides the main results of this specific case study focused on reducing air pollution. Finally, Section 5 is devoted to the main conclusions of this chapter.

2 Relevance of the Port of Barcelona

The Port of Barcelona is not only a gateway for the city to the sea but also a driver of economic growth. The recovery of the port's seafront for citizens' use, and its consolidation as a seaport with diversified traffic, has encouraged different government agencies to coordinate in promoting ecofriendly practices at the port.

The primary goal of the Port of Barcelona has been to boost its competitiveness, service quality, and productivity by integrating maritime and land transport via logistics (The Port of Barcelona, 2017a). To that purpose, the commercial strategy of the Port of Barcelona has revolved around becoming a benchmark logistics hub in the Mediterranean. Consequently, the capacity expansion of the seaport and its logistics has made the Port of Barcelona one of the most advanced ports in southern Europe[c] (The Port of Barcelona, 2017a).

In terms of container traffic, measured in TEU (20-foot equivalent units), the Port of Barcelona is ranked as the 12th busiest port in Europe (IAPH Database, 2018) with the third highest level of traffic in Spain (Puertos del Estado, 2019).

---

c. In terms of container terminals, the two most important expansions were: the new BEST container terminal, operated by TERCAT-Hutchinson since 2014; and the expansion of the terminal operated by TCB (Barcelona Container Terminal) in 2013 (Gonzalez-Aregall, 2014).

Furthermore, because of tourism, the port is fast becoming the leading turnaround cruise port in Europe and is the world's fourth most-important base port in cruise ship transit, with a total of 2.7 million cruise passengers in 2017 (The Port of Barcelona, 2017b). In terms of environmental impact, The Port of Barcelona (2017c) has estimated that cruise vessel emissions contribute to about 1.2% of the city's air pollution, 0.23% of NOx levels, and 0.23% of PM_{10} levels.

In the context of economic growth in the region,[d] the Port of Barcelona handled more than 60 million tons in 2017 (The Port of Barcelona, 2017b). Fig. 1 shows the evolution of goods through the port from 2000 to 2018, considering the types of cargo (left axis) and container traffic by TEUs (right axis). Following a period of markedly less traffic from 2010 to 2013 (due to the global economic crisis), the port has since experienced growth in traffic as the economy has recovered, particularly due to the effects of the port's infrastructure investment program[e] (The Port of Barcelona, 2017a).

Another characteristic of the Port of Barcelona is its port authority's management system. Adhering to a landlord-based port structure, the port authority does not directly manage the port's different terminals but rather leases them to private companies by public tender. As such, the port is part of an autonomous economic and financial structure, with centralized management regulated by law.[f] Aiming to promote competition among ports and increase private ownership and investment, the port authority may apply a policy of port charge discounts for companies that operate the terminals and for shipping companies, according to several criteria, including the type of traffic and the environmental practices[g] (Fageda and Gonzalez-Aregall, 2018; Gonzalez-Aregall, 2014).

Correspondingly, the city of Barcelona has been subject to different processes of urbanization, particularly on its waterfront.[h] Although the successful integration between the marine and urban areas has promoted culture and

---

d. According to Villalb and Demisse Gemenchu (2011), in 2008 the Port of Barcelona was responsible for approximately 77% of foreign maritime trade in Catalonia and for approximately 23% of Spain's foreign maritime trade. In this regard, according to the The Port of Barcelona (2019a), the port is responsible for 6% of Catalonia's Gross Added Value.

e. The benchmark model is based on several measures focused on increasing the port's capacity to receive larger vessels, enhancing hinterland strategies, and improving relations with clients and port-related agencies.

f. The Spanish port system, owned by the state, consists of 28 port authorities that oversee 44 ports of general interest scattered around the country. For further detail on the Spanish port system and the process of Spanish port reform, see Castillo-Manzano et al. (2016) and Fageda and Gonzalez-Aregall (2018).

g. The last Spanish port reform has provided greater autonomy for all port authorities to set tariffs. However, the regulation limited the modification of seaports' port charges to their traffic forecasts and cost structures. In practice, it seems that this "price liberalization" benefits the biggest Spanish ports (Algeciras and Valencia), which have a high proportion of transhipment traffic, and has a limited impact on ports like Barcelona, which is very dependent on the demand generated by the local hinterland (Fageda and Gonzalez-Aregall, 2018).

h. For further detail and historical urbanization planning, see Jauhiainen (1995), Marti et al. (2018), and Ressano Garcia (2008).

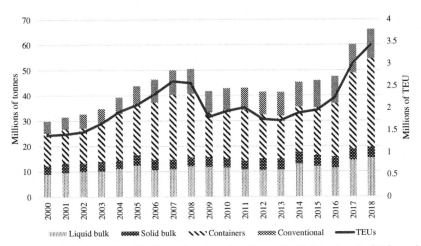

FIG. 1 Evolution of traffic in the Port of Barcelona from 2000 to 2018. *(Based on the database of Spanish Ministry of Transport.)*

tourism (Griffin and Hayllar, 2006; McCarthy, 2003; Kostopolou, 2013), the geographical layout of Barcelona in relation to the port and its logistic area has led port-generated pollution to affect the city's residents (Pérez et al., 2016).

3 Environmental impact of the Port of Barcelona on the city

Port cities have to coordinate a variety of governance processes, often by negotiating complex conflicts of interest among stakeholders in the public and private sectors, in which regulations for urban space and different management models complicate the establishment of new policies (Debrie and Raimbault, 2016; Frantzeskaki et al., 2014; Proudfoot, 1996).

The impact of institutions on the complexity of spatial governance and on the active planning of policies (Daamen and Vries, 2013) is especially significant in Barcelona's case. While the port authority (managed by the Spanish Ministry of Transport) focuses on green initiatives for the port, the city's municipal government and the regional government that administers the Barcelona Metropolitan Area[i] implement different policies to reduce the air pollution in the city and its vicinity.

In the case of regulation, different government bodies have established air emissions' reduction measures through several air quality plans, and they focus on the impact of the port on the city of Barcelona.

---

i. The metropolitan area of Barcelona has more than 3.2 million inhabitants and occupies an area of 636 km^2 in the Catalonia region. This area includes Barcelona and 37 municipalities (Àrea Metropolitana de Barcelona, 2019).

First, from a national perspective, in 2013, the Ministry of Agriculture, Food and Environment established the National Plan for Air Quality and Protection of the Atmosphere for executing several action plans. Some of these actions focused on reducing pollution of different port activities by controlling ships' emissions in port areas as well as promoting facilities for alternative fuels in shipping and road transport (The Port Authority of Barcelona, 2016).

Second, from a regional perspective, the Catalan Regional Governance has promoted the Air Quality Monitoring and Control Service through measuring stations to evaluate air quality and collect data with the Atmospheric Pollution Surveillance and Forecast Network (XVPCA) and according to current legislation.[j] Furthermore, the Catalan Governance published the Action Plan for the Improvement of the Air Quality in the Horizon 2020 (PAMQA). Specifically, some actions described in this air quality plan have a direct impact on port activity such as enhancing the rail network for goods inside and outside the port area and monitoring air pollution emission (The Port Authority of Barcelona, 2016).

Finally, Barcelona has 11 monitoring stations located throughout the city that collect specific data about major air pollutants on daily, monthly, and annual bases (Barcelona City Council, 2016). Along with the stations, in 2016 the Barcelona City Council and the Barcelona Metropolitan Area unveiled "the Government Means Program of Anti-Air Pollution Measures" in order to establish specific actions to reduce atmospheric pollutants, as well as a specific protocol for managing occasional episodes of environmental problems.[k] Furthermore, the Barcelona City Council has launched its Urban Mobility Plan to promote safe, sustainable, equitable, and efficient types of transport, as well as "the Air Quality Improvement Plan: 2015–2018," a report that shows all of the initiatives, plans, and measures aimed at reducing air pollution levels in the city from 2015 to 2018. This plan has specific measures to reduce pollution for the Port of Barcelona (The Port Authority of Barcelona, 2016; Barcelona City Council, 2018).

Several studies have evaluated the sources of atmospheric pollutants in Barcelona and the port area as well as the effects of air emissions from the Port to the city's residential areas.

From a city perspective, the principal source of air pollution is road transportation emission. However, nearly 7.6% NOx levels and 1.5% PM_{10} levels of the city stem from emissions from the Port of Barcelona[l] (The Port Authority of Barcelona, 2016). Specifically, although the limits for SO_2 emissions are quite low, such emissions in the city seem to mostly relate to the port's activities (APICE, 2013; Barcelona Regional, 2015; IDAEA-CSIC, 2015).

---

j. European regulation: Directive 2008/50/EC and Spanish regulation: Royal Decree 102/2011.
k. In particular, these measures aim to enhance the use of nonpolluting vehicles, enacting policies to promote the use of less polluting vehicles, regulating taxis, funding studies to evaluate the negative effects of pollution on health, and enhance cooperation with other agencies (Barcelona City Council, 2018).
l. In this regard, since 2007, ambient PM concentrations in Barcelona have decreased significantly due to the success of different initiatives addressing air quality (Pérez et al., 2016).

In the port area, cargo vessels, container vessels, and cruise vessels are the greatest sources of pollution. Thus, container vessels are responsible for 1.7% of nitrogen dioxide (NO_2) emissions and 0.38% of PM_{10} levels. Cargo vessels contribute 2% of NO_2 levels and 0.48% of PM_{10} levels (Barcelona Regional, 2015). In all cases, marine vessels operating in the port terminal show higher levels of emissions (The Port Authority of Barcelona, 2016). Consequently, according to the literature, several port authorities have been adapting their shore-side electrification at berth for reducing vessels' atmospheric pollution (Papaefthimioua et al., 2016). In the port area, road transportation is most responsible for air pollution, with 89% of NOx levels and almost 77% of PM_{10} emissions, followed by diesel passenger cars, with 5.8% of emissions.

Although the port has a limited autonomy to set prices (Fageda and Gonzalez-Aregall, 2018) and it is not located in an environmentally restricted area, it has still implemented several environmental initiatives, mainly thanks to EU Funds as well as to specific Green Plans and particular fields of actions.

3.1 Environmental initiatives at the Port of Barcelona

In recent years, the Port of Barcelona has implemented an environmental management system that, since 2018, has focused on managing the port area, maritime activities, and maintenance works (The Port of Barcelona, 2019b).

From a general perspective, in terms of climate strategy, the port has adopted different strategies to reduce greenhouse gas emissions and focus on logistics aspects and Port Authority structure (The Port of Barcelona, 2019c). First, from a logistics perspective, since 2014, the port has collaborated with terminals and service operators to develop the "BCN Zero Carbon." This project aims to establish an inventory of greenhouse gas emissions from freight activities at the port area (The Port of Barcelona, 2019d). Second, the Port of Barcelona has developed software to calculate the CO_2 emissions of customers' transport routes and identify which is the less atmospheric pollutant alternative. With this tool, port customers can decide on a transport route considering environmental aspects (The Port of Barcelona, 2019e). Finally, specific agreements with private companies involving exclusive rail services,[m] as well as with government agencies aiming to reduce electric consumption,[n] have helped to reduce road transportation.

With regard to air emissions, the port authority has implemented several green policies to reduce air pollution from port activities and to influence the atmospheric quality in the port area. Although meteorological conditions are an uncontrollable factor in the dispersion of pollution into the atmosphere, the port

---

m. For instance, since 2008, the SEAT automotive industry has had a long-term agreement with the port in order to enhance the direct railway line (Autometro/Cargometro) from the SEAT Factory in Martorell to the port (distance: 40 km) (Bestfact, 2019).

n. In 2012, the Port Authority of Barcelona and the Catalan Climate Change Office (CCCO) signed voluntary agreements to reduce direct and indirect emissions caused by fuel and electrical consumption (The Port of Barcelona, 2019f).

has developed specific actions to reduce and control the concentration of contaminants in ambient air. These actions include monitoring stations,[o] specific action plans (like the "Improvement Plan for Air Quality," published in 2016), and European Union (EU) Projects (The Port of Barcelona, 2019g).

Based on information reported in the "Improvement Plan for Air Quality," published by the Port Authority of Barcelona in 2016, the port has designed 53 programs, which are grouped into seven fields of action (The Port Authority of Barcelona, 2016). The fields of action are[p]:

1. The port aims to invest in new infrastructure to encourage the use of alternative fuels for cargo transport. Based on the EU's Alternative Fuel Strategy approved in 2012, the Port of Barcelona considers Liquefied Natural Gas (LNG) to be the most appropriate fuel alternative for marine vessels, because it can remove SO_2 and PM and reduce NOx emissions up to 85%, and it is the cheapest energy source with stable prices.

2. Since almost 71% of NOx emissions are generated when the vessel is moored on the quay, the port aims to evaluate the feasibility of providing an electric connection for marine vessels.

3. Based on the port reform of 2010, that allows price discounts to be applied for green port practices, the port wants to incentivize shipping companies to improve environmental performance by reducing port charges.

4. In terms of land transport, the port is replacing diesel-engine vehicles with electric vehicles. Furthermore, the port is running a project with container transport trucks called "RePort" to convert diesel-engine trucks to a dual system with Natural Gas (NG).

5. The port authority is also concerned with port terminal equipment. In this regard, the port aims to increase the electrification and gasification of port terminal machinery.

6. Considering the port's commercial strategy to become an efficient logistic hub in the Mediterranean, the port authority aims to invest in better infrastructure in order to improve the use of rail and short sea shipping (SSS) as an alternative to road transportation.

7. The port aims to enhance collaboration with port customers and external actors to promote sustainable mobility among port companies.

Tables 1 and 2 summarize the main atmospheric pollution initiatives developed over the last few years. Fig. 2 summarizes the different initiatives in chronological order and considers whether they are based on infrastructure investment or focused on energy development.

---

o. The port monitors concentrations of pollutants in the ambient air using a network of manual networks to evaluate samples for particulate matter and a network of automatic analyzers for gaseous air pollutants. Note that one of the manual stations (Port Vell) is part of the Catalonia Air Pollution Monitoring and Forecast Network (the XVPCA), managed by the Catalan government.

p. All the information is based on the The Port Authority of Barcelona (2016) and the The Port of Barcelona (2019g).

TABLE 1 Current green initiatives on the Port of Barcelona.

Initiative	Area of action	Period	Actors affected	Co-funded	Aim	Field of action
Autometro/cargometro	Hinterland (Port ZAL-Martorell)	Since 2008	Port authority and SEAT Factory	Port authority and private company	Transporting new cars and automotive parts by rail service	6
						7
BCN Zero Carbon	Port terminals	Since 2014	Terminals and service operators	Port authority and port customers	Inventory of GHG emissions from port freight activities	7
CarEsmatic	Port terminals	2016–18	Terminals and vessel operator	EU project	Increase the transport of electric cars by SSS and improve port access and railway connections	4
						6
CCCO agreements	Port area	Since 2012	Port authority and service operators	Port authority and Catalan Regional Government	Reduce the electrical and the fuel consumption of the Port fleet	7
CleanPort	Port terminal	2014–18	Terminals and vessel operator	EU project	Installation an auxiliary power system based on NG and LNG on a ferry vessel. Construction of an on-shore bunkering infrastructure at the berth	1
						2
Core LNGas Hive	Port terminal	2014–20	Terminal facilities and vessels	EU project	Supporting the implementation of LNG infrastructure for maritime transport and port terminal operations	1
						2
						7

Continued

TABLE 1 Current green initiatives on the Port of Barcelona—cont'd

Initiative	Area of action	Period	Actors affected	Co-funded	Aim	Field of action
EcoCalculator	Hinterland (Europe)	Since 2012	Logistics operators	EU project	Software that calculates the CO_2 emissions of different transport routes	7
New Rail access TEN-T	Port area	2015–18	Terminals and railway infrastructure	The Spanish Government, the port authority and EU project	Implement the TEN-T network by the construction of a new rail access to the port	6
RePort	Port Area	2016–20	Truck transportation	EU project	Convert truck diesel engines into dual-fuel system	4
Sea traffic management	Hinterland (Europe)	2015–18	Maritime stakeholders	EU project	Create a European consortium to communicate and share information between stakeholders in the maritime transport industry	7

Based on the information available on the Port of Barcelona's webpage.

TABLE 2 Past green initiatives on the Port of Barcelona.

Initiative	Area of action	Period	Actors affected	Co-funded	Aim	Field of action
B2MOS	Hinterland (Europe)	2013–15	Logistics stakeholders	EU project	Implement the TEN-T network by promoting the SSS and cooperation with different countries	6 7
CLYMA	Hinterland (Europe)	2013–15	Logistics stakeholders	EU project	Implement the TEN-T Network by connecting the Madrid-Lyon axis	6
MOS4MOS	Hinterland (Europe)	2011–12	Logistics stakeholders	EU project	Implement the TEN-T Network by information technology to improve multimodal SSS Services	6
New terminal for SSS	Port area	2013–14	Port Authority	EU project	Construction of a new roll-on/roll-off and passenger terminal	6
Railway UIC gauge	Port area	2012–13	Terminals and railway infrastructure	EU project	Adaptation of the port railway network to standard (UIC) gauge	6

Based on the information available on the Port of Barcelona's webpage.

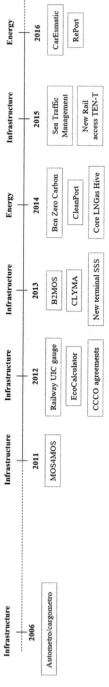

FIG. 2 Summary of chronological green initiatives in the Port of Barcelona. (*Based on the information available on the Port of Barcelona's webpage.*)

4   Results from the evaluation of port initiatives

To evaluate the impact of these port initiatives on the Barcelona urban area, monitoring stations have been considered, which are distributed over Barcelona and the two cities bordering the port area, Hospitalet de Llobregat[q] and El Prat de Llobregat.[r] It is important to remember that the Port of Barcelona is close to the airport (El Prat) and to a logistics distribution area (Port ZAL[s]); consequently, results related to atmospheric pollutants in the different urban areas are sensitive to the airport's infrastructure and to logistics-related transport. To control these effects, the analysis has considered monitoring stations close to the airport (El Prat) and port monitoring stations (Port ZAL and Port Dàrsena Sud) close to the logistic distribution area. Fig. 3 shows the municipalities' locations and the monitoring stations in urban and port areas. Note that not all the monitoring stations calculate the same atmospheric pollutants.

In the case of air pollutant analysis, this study considered the most relevant atmospheric pollutants related to maritime transport and road transportation, in order to control the impact of port activity.

In terms of atmospheric emissions, this analysis considered sulfur oxide (SOx), nitrogen oxide (NOx),[t] and particulate matter (PM_{10} and $PM_{2.5}$).[u] According to Viana et al. (2014), in European coastal regions, the average contribution of NO_2 levels ranges from 7% to 24%, while in terms of particulate matter, shipping emissions account for 1%–7% of PM_{10} levels and 1%–14% of $PM_{2.5}$ levels.

This section is divided into four subsections, which describe the different atmospheric pollutants in detail and consider the atmospheric pollutants in cities around the port as well as in specific neighborhoods. Note that in urban areas, this study only reports on data from 2011 or 2012 to 2017, due to a lack of data before 2011.

4.1   Analysis of NO_2

This section analyzes nitrogen dioxide (NO_2) levels from two perspectives. First, it analyzes the annual mean of this atmospheric pollutant in the cities

---

q. L'Hospitalet de Llobregat is the second largest city in Catalonia, with almost 260,000 inhabitants. This municipality has an area of 12 km^2 and is located 8 km from Barcelona (Generalitat de Catalunya, 2019).

r. El Prat de Llobregat (El Prat) has a total of 64,000 inhabitants and is located in the delta of the Llobregat River, next to Barcelona. The area of El Prat is about 31 km^2 (Generalitat de Catalunya, 2019).

s. ZAL means "Zona Activitat Logística," or Logistics Activity Zone.

t. Nitrogen dioxide is mainly produced by road traffic of trucks and light vehicles.

u. According to WHO (2006), in terms of SO_2, the maximum level of exposure is 20 $\mu g/m^3$ daily mean level. In terms of NO_2 emission, the limit is 40 $\mu g/m^3$ annual mean level. Finally, in terms of particulate matter, the limits are 20 $\mu g/m^3$ annual mean level for PM_{10} and 10 $\mu g/m^3$ annual mean level for $PM_{2.5}$.

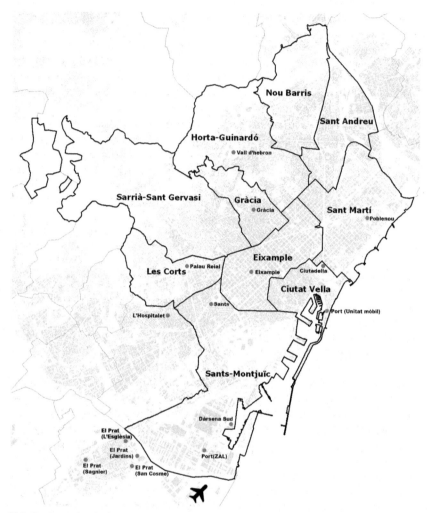

FIG. 3 Area of analysis and location of monitoring stations.

around the port from 2007 to 2017. Second, based on disaggregated data, this section describes the evolution of different urban neighborhoods parallel to the evolution of port emissions from 2012 to 2017 (Fig. 4).

In terms of the implementation of different environmental policies, it seems that the initiatives established in 2012, 2014, and 2016 had a major impact on the reduction of NO_2. In this regard, these initiatives mainly focused on new infrastructure that would encourage the use of alternative fuels (LNG and NG) for marine vessels, trucks, and light vehicles. Furthermore, the success of these initiatives is related to a clear increase in collaborations with port customers and service operators.

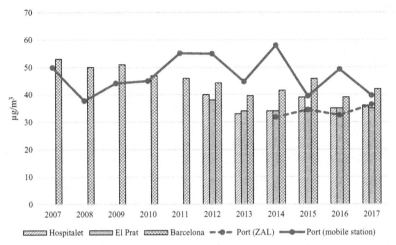

FIG. 4   Annual mean of NO₂ emissions.

Furthermore, it seems that the approach to reduce road congestion through an increase in short sea shipping and rail facilities had a positive effect on the reduction of NO_2. In this regard, the reduction of NO_2 levels in 2008 is potentially related to the effects of the economic crisis and to a subsequent reduction in freight traffic and port vehicles. At the same time, the gradual recovery of the economy increased levels of NO_2 emissions (The Port of Barcelona, 2019h; Fig. 5).

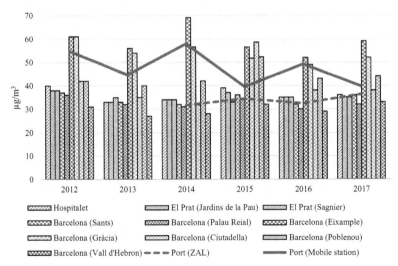

FIG. 5   Annual mean of NO₂ emissions by specific areas.

In terms of the impact of this atmospheric pollutant on all areas, NO_2 levels are similar to levels in the port area. In this regard, the increase in NO_2 emission levels in the port (ZAL) was in response to economic growth and to its close location to the airport (El Prat).[v]

In terms of location, it seems that the more distant monitoring stations of the port area (Vall d'Hebron, Gracia, Ciutadella, and Poblenou) present higher values. Thus, it is possible that the NO_2 emitted at the port has not had a direct effect on the NO_2 levels of the city.

4.2 Analysis of SO_2

This section analyzes sulfur dioxide (SO_2) levels from two perspectives. First, it analyzes the annual mean of this atmospheric pollutant in the cities around the port from 2003 to 2017. Second, based on disaggregated data, this section describes the evolution of different urban neighbors parallel to the evolution of port emissions from 2011 to 2017 (Fig. 6).

According to the Port Authority of Barcelona, the reduction in SO_2 levels from those in 2008–09 is mainly due to the closing of an industrial sulfur processing plant outside the port, located near the monitoring station. Furthermore, in 2010, an obligation was established for ships to use low sulfur content fuel during mooring activities at the port berth.

In relation to the port initiatives previously described, after energy efficient policies for marine vessels were implemented in 2014, the levels of SO_2 have remained quite constant (Fig. 7).

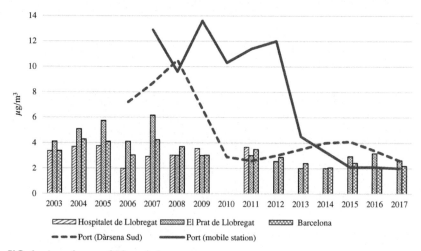

FIG. 6 Annual mean of SO_2 emissions.

v. According to the EEA (2018), aviation has a higher effect on NOx emissions than other atmospheric pollutions, such as SO_2, PM_{10}, or $PM_{2.5}$.

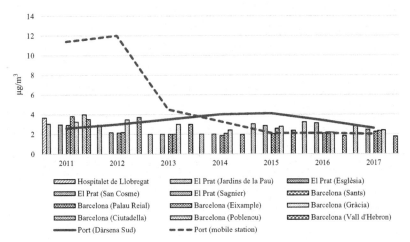

FIG. 7 Annual mean of SO_2 emissions by specific areas.

In terms of urban areas, the effect of the emissions from the port is higher than that of the emissions in residential areas. This result could be related to the source of emissions: marine vessels. Thus, it is important to keep in mind the reduction of SO_2 emissions after energy policies were implemented on vessels in 2013–14. Overall, it seems that the trend and effects are similar to the previous figures.

4.3 Analysis of PM_{10}

This section analyzes particulate matter (PM_{10}) levels from two perspectives. First, it analyzes the annual mean of this atmospheric pollutant in the cities around the port from 2004 to 2017. Second, based on disaggregated data, this section describes the evolution of different urban neighborhoods parallel to the evolution of port emissions from 2012 to 2017 (Fig. 8).

PM_{10} levels have continuously decreased over the last ten years; consequently, this atmospheric pollutant is below the 40 $\mu g/m^3$ limit values for urban areas (The Port of Barcelona, 2019h; WHO, 2006).

The gradual decrease in PM_{10} levels is due to several causes. First, the reduction in 2008 coincides with the growing completion of construction works for expanding the terminal. Second, with the different green policies described above taken into account, it seems that promoting the implementation of TEN-T network and focusing on enhancing intermodal transportation[w] (which has increased 3% from 2016 to 2017) have had a positive effect on the reduction of particulate matter in the port area (Fig. 9).

---

w. Intermodal transport units (UTIs) are defined as "*thos e means of unitization, self-propelled or otherwise, that are directly or indirectly used as a means of land transport (e.g. trailers, platforms, trucks, freezer trucks, etc.)*" (The Port of Barcelona, 2017b, p. 14).

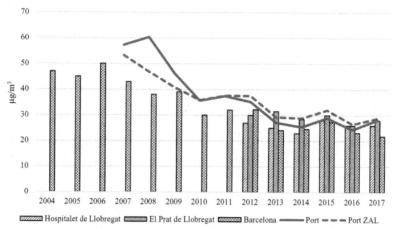

FIG. 8 Annual mean of PM_{10} emissions.

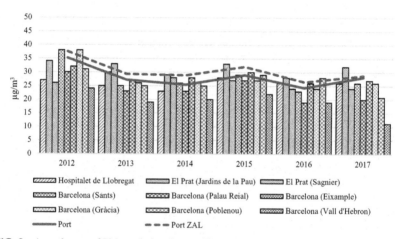

FIG. 9 Annual mean of PM_{10} emissions by specific areas.

Considering the evolution of particulate matter in different urban areas, it seems that the values of PM_{10} levels are higher than in the port area. This is in line with the literature that evaluates the sources of air pollutants in urban areas; in Barcelona, it is related to private vehicles. The evolution is quite analogous between the port and different neighborhoods.

4.4 Analysis of $PM_{2.5}$

This section analyzes the particulate matter ($PM_{2.5}$) levels from two perspectives. First, it analyzes the annual mean of this atmospheric pollutant in the cities around the port from 2007 to 2017. Second, based on disaggregated data, this section describes the evolution of different urban neighborhoods parallel to the evolution of port emissions from 2007 to 2017 (Fig. 10).

The evolution of $PM_{2.5}$ is quite similar to the PM_{10} progress in its reduction of levels, although it has increased since 2016.

In terms of the initiatives taken by the port, policies established in 2011, 2013, and 2015 that are focused on increasing intermodal transport have had a positive effect on the reduction of this atmospheric pollutant (Fig. 11).

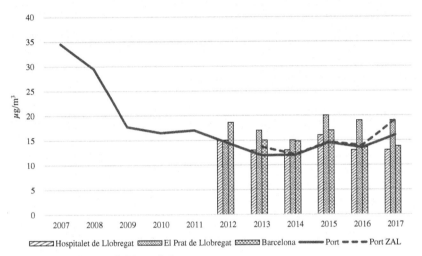

FIG. 10   Annual mean of $PM_{2.5}$ emissions.

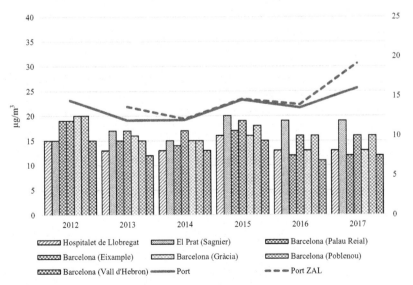

FIG. 11   Annual mean of $PM_{2.5}$ emissions by specific areas.

In terms of the distribution over urban areas, the levels of $PM_{2.5}$ show a similar trend to the levels of $PM_{2.5}$ in the Port of Barcelona. Thus, it seems that the coordination among different institutions has helped to control the increase of this type of emission. Still, since 2014, the emissions in the port area have increased. This result could be related to the opening of the new BEST terminal and TCB Terminal in 2014 and 2016, respectively. Note that these terminals are located close to El Prat (Sagnier), which shows higher values during those years.

5 Discussion and conclusion

To evaluate the impact of port initiatives on the Barcelona urban area, we have considered data reported by monitoring stations distributed over the city of Barcelona and the two cities bordering the port area. Thus, this study considered the most relevant atmospheric pollutants (SO_2, NO_2, PM_{10}, and $PM_{2.5}$) related to maritime transport and road transportation in order to determine the impact of port activity.

Based on a case study of Barcelona, this research has examined port-developed strategies that have reduced the environmental impact caused by the port and that have promoted sustainable actions to combat climate change in a large urban area. With regard to regulatory framework and port-city stakeholders, the port has implemented several atmospheric pollution initiatives in order to combat negative environmental impacts on the urban area.

Therefore, considering that the Port of Barcelona has experienced growth in traffic, it should be necessary to encourage institutions and policymakers in Barcelona to continue working toward robust environmental performance in the port-city area.

Specifically, it seems that the policies concerning alternative fuels, i.e., Clean Port and Core LNGas Hive initiatives, had beneficial effects on shipping emissions like SO_2. In addition to these policies, the main results suggest that the approach to reduce road congestion through an increase of rail facilities, like new rail access to the port, as well as improving Short Sea Shipping services, for instance MOS4MOS or CarEsmatic initiatives, had positive effect on the reduction of NO_2. Similarly, this study suggests that enhancing intermodal transportation, i.e., implementing the TEN-T network, has had a positive effect on the reduction of the values of particular matter (PM_{10} and $PM_{2.5}$) in the port and urban areas.

Although the Port of Barcelona has implemented many successful measures for reducing detrimental environmental effects on the larger urban area, other examples from other ports might be useful to consider for the future, such as the "Clean Truck Replacement Incentive Program" of Port New Orleans[x] or the "Mode Shift Incentive Scheme" of Port of Melbourne.[y]

---

x. Since 2016, the Port of New Orleans provides competitive grants to truck and fleet owners to voluntary invest in replacement trucks with cleaner models (year 2012 or newer). This initiative is similar as the Port of Barcelona's initiative: RePort.

y. This program aims to support port industry to shift more containerized freight from road to rail. This project is similar as the port of Barcelona's TEN-T initiatives.

In conclusion, this study indicates that actions focus on alternative fuels for marine vessels and enhance intermodal transport through Short Sea Shipping and rail services have positive effects on reducing atmospheric pollutants in cities around the port. Thus, this research recommends to implement similar policies that could benefit the environment by reducing air pollution at the port and improving air quality in the city.

References

Akhavan, M., 2017. Development dynamics of port-cities interface in the Arab Middle Eastern world—the case of Dubai global hub port-city. Cities 60, 343–352.

APICE, 2013. Pla APICE Barcelona: Mitigació d'emissions marítimes i portuàries per a la millora de la qualitat de l'aire (In English: APICE Plan: Mitigation of Maritime and Port Emissions for the Improvement of Air Quality). APICE (Actions for the Mitigation of Port, Industries and Cities Emissions) Project.

Àrea Metropolitana de Barcelona, 2019. Available at: http://www.amb.cat/s/web/territori/territori.html.

Asteris, M., Collis, A., 2010. UK container port investment and competition: impediments to the market. Transp. Rev. 30 (2), 163–178.

Barcelona City Council, 2016. Mesura de Govern: Programa de mesures contra la contaminació atmosfèrica (In English: The Government Means Program of Anti-Air Pollution Measures). Barcelona City Council.

Barcelona City Council, 2018. Pla de Millora de la Qualitat de l'Aire 2015-2018 (In English: The Air Quality Improvement Plan: 2015-2018). The Barcelona City Council.

Barcelona Regional, 2015. Anàlisi de la contribució en emissions i immissions del port de Barcelona (In English: Analysis of the Contribution in Emissions and Immissions of the Port of Barcelona). Barcelona Regional.

Bergqvist, R., Egels-Zandén, N., 2012. Green port dues—the case of hinterland transport. Res. Transp. Bus. Manag. 5, 85–91.

Bestfact, 2019. Transport by short distance train in automotive industry—SEAT. Bestfact Practices in Green Logistics and Co-Modality Number 2/141. EU Seventh Framework Program. Available at: http://www.bestfact.net/.

Butt, N., 2007. The impact of cruise ship generated waste on home ports and ports of call: a study of Southampton. Mar. Policy 31 (5), 591–598.

Castillo-Manzano, J.I., Castro-Nuño, M., Fageda, X., Gonzalez-Aregall, M., 2016. Evaluating the effects of the latest change in Spanish port legislation: another "turn of crew" in port reform? Case Stud. Transp. Policy 4 (2), 170–177.

Christodoulou, A., Gonzalez-Aregall, M., Linde, T., Vierth, I., Cullinane, K., 2019. Targeting the reduction of shipping emissions to air: a global review and taxonomy of policies, incentives and measures. Marit. Bus. Rev. 4 (1), 16–30.

Corbett, J.J., Winebrake, J.J., Green, E.H., Kasibhatla, P., Eyring, V., Lauer, A., 2007. Mortality from ship emissions: a global assessment. Environ. Sci. Technol. 41 (2), 8512–8518.

Cullinane, K.P.B., Cullinane, S.L., 2013. Atmospheric emissions from shipping: the need for regulation and approaches to compliance. Transp. Rev. 33 (4), 377–401.

Daamen, T.A., Vries, I., 2013. Governing the European port-city interface: institutional impacts on spatial projects between city and port. J. Transp. Geogr. 27, 4–13.

Dalsøren, S.B., Eise, M.S., Endesen, Ø., Mjelde, A., Gravir, G., Isaksen, I.S.A., 2009. Update on emissions and environmental impacts from the international fleet of ships: the contribution from major ship types and ports. Atmos. Chem. Phys. 9, 2171–2194.

Davarzani, H., Fahimnia, B., Bell, M., Sarkis, J., 2016. Greening ports and maritime logistics: a review. Transp. Res. D 48, 473–487.

Debrie, J., Raimbault, N., 2016. The port-city relationships in two European inland ports: a geographical perspective on urban governance. Cities 50, 180–187.

Del Saz-Salazar, S., García-Menéndez, L., Merk, O., 2013. The port and its environment: methodological approach for economic appraisal. OECD Regional Development Working Papers, 2013/24, OECD Publishing. https://doi.org/10.1787/5k3v1dvb1dd2-en.

EEA, 2018. Emissions of Air Pollutants From Transport. Available at: https://www.eea.europa.eu/data-and-maps/indicators/transport-emissions-of-air-pollutants-8/transport-emissions-of-air-pollutants-6.

Fageda, X., Gonzalez-Aregall, M., 2018. Port governance reform in Spain. In: Pettit, S., Bredsford, A. (Eds.), Port Management: Cases in Port Geography, Operations and Policy. Kogan Page Edition, Great Britain and The United States.

Frantzeskaki, N., Wittmayer, J., Loorbach, D., 2014. The role of partnerships in 'realising' urban sustainability in Rotterdam's City Ports Area, The Netherlands. J. Clean. Prod. 65, 406–417.

Geerlings, H., Van Duin, R., 2011. A new method for assessing CO2 emissions from container terminals: a promising approach applied in Rotterdam. J. Clean. Prod. 19 (6), 657–666.

Generalitat de Catalunya, 2019. Municipality Information. Available at: http://municat.gencat.cat/ca/inici/.

Gonzalez Aregall, M., Bergqvist, R., Monios, J., 2018. A global review of the hinterland dimension of green port strategies. Transp. Res. D Transp. Environ. 59, 23–34. https://doi.org/10.1016/j.trd.2017.12.013.

Gonzalez-Aregall, M., 2014. The Port of Barcelona: prospects and challenges in an independent Catalonia. In: White Paper: Barcelona, the Capital of New State 334-341. Barcelona City Council. November 2014.

Griffin, T., Hayllar, B., 2006. Historic waterfronts as tourism precincts: an experiential perspective. Tour. Hosp. Res. 7 (1), 3–16.

Grossmann, I., 2008. Perspectives for Hamburg as a port city in the context of a changing global environment. Geoforum 39 (6), 2062–2072.

Hanson, S., Nicholls, R., Ranger, N., Hallegatte, S., Corfee-Morlot, J., Herweijer, C., Chateau, J., 2011. A global ranking of port cities with high exposure to climate extremes. Clim. Chang. 104, 89–111.

Hoyle, B.S., 1989. The port-city interface: trends, problems and examples. Geoforum 20 (4), 429–435.

Hoyle, B.S., 1992. Waterfront redevelopment in Canadian port cities: some viewpoints on issues involved. Marit. Policy Manag. 19 (4), 279–295.

Hoyle, B., 2000. Global and local change on the port-city Waterfront. Geogr. Rev. 90 (3), 395–417.

IAPH Database, 2018. World Container Traffic Data 2018. International Association of Ports and Harbors.

IDAEA-CSIC (2015) "Avaluació de l'impacte d'emissions de l'àrea portuària a la qualitat de l'aire de la zona urbana de Barcelona "Evaluation of the Impact of Emissions From the Port Area on the Air Quality of the Urban Area of Barcelona" Published by The Institute of Environmental Assessment and Water Research – Consejo Superior de Investigaciones Cientificas (IDAEA-CSIC) (in English).

Jauhiainen, J.S., 1995. Waterfront redevelopment and urban policy: the case of Barcelona, Cardiff and Genoa. Eur. Plan. Stud. 3 (1), 3–23.

Kostopolou, S., 2013. On the revitalized Waterfront: creative milieu for Creative Tourism. Sustainability 5, 4578–4593.

Lack, D.A., Corbett, J.J., 2012. Black carbon from ships: a review of the effects of ship speed, fuel quality and exhaust gas scrubbing. Atmos. Chem. Phys. 12, 3985–4000.

Lee, T., Yeo, G.-T., Thai, V., 2014. Environmental efficiency analysis of port cities: Slacks-based measure data envelopment analysis approach. Transp. Policy 33, 82–88.

Marti, P., Garcia Mayor, C., Melgarrejo, A., 2018. Waterfront landscapes in Spanish cities: regeneration and urban transformations. WIT Trans. Built Environ. 179, 45–56.

Mat, N., Cerceau, J., Sh, L., Park, H.-S., Junqua, G., Lopez-Ferber, M., 2016. Socio-ecological transitions toward low-carbon port cities: trends, changes and adaption processes in Asia and Europe. J. Clean. Prod. 114, 362–375.

McCarthy, J., 2003. The cruise industry and port regeneration: the case of Valleta. Eur. Plan. Stud. 11 (3), 341–350.

Merico, E., Donateo, A., Gambaro, A., Cesari, D., Gregoris, E., Barbaro, E., Dinoi, A., Giovanelli, G., Masieri, S., Contini, D., 2016. Influence of in-port ships emissions to gaseous atmospheric pollutants and to particulate matter of different sizes in a Mediterranean harbor in Italy. Atmos. Environ. 139, 1–10.

Merico, E., Gambaro, A., Argiriou, A., Alebic-Juretic, A., Barbaro, E., Cesari, D., Chasapidis, L., Dimopoulous, S., Dinoi, A., Donateo, A., Giannaros, C., Gregoris, E., Karagiannidis, A., Konstandopoulos, A.G., Ivosevic, T., Liora, N., Melas, D., Mifka, B., Orlic, I., Poupkou, A., Sarovic, K., Tsakis, A., Giua, R., Pastore, T., Nocioni, A., Contini, D., 2017. Atmospheric impact of ship traffic in four Adriatic-Ionian port-cities: comparison and harmonization of different approaches. Transp. Res. D 50, 431–445.

Monios, J., Bergqvist, R., Woxenius, J., 2018. Port-centric cities: the role of freight distribution in defining the port-city relationship. J. Transp. Geogr. 66, 53–64.

Monteiro, A., Russo, M.M., Gama, C., Borrego, C., 2018. Impact of SO2 shipping emissions on air quality: the airship project. WIT Trans. Ecol. Environ. 230, 429–437.

Na, J.-H., Choi, A.-Y., Ji, J., Zhang, D., 2017. Environmental efficiency analysis of Chinese container ports with CO2 emissions: an inseparable input-output SBM model. J. Transp. Geogr. 65, 13–24.

OECD, 2011. Environmental Impacts of International Shipping: The Role of Ports. OECD. https://doi.org/10.1787/9789264097339-en.

Papaefthimioua, S., Maragkogianni, A., Andriosopoulos, K., 2016. Evaluation of cruise ships emissions in the Mediterranean basin: the case of Greek ports. Int. J. Sustain. Transp. 10 (10), 985–994.

Pérez, N., Pey, J., Reche, C., Cortés, J., Alastuey, A., Querol, X., 2016. Impact of harbour emissions on ambient PM_{10} and $PM_{2.5}$ in Barcelona (Spain): evidences of secondary aerosol formation within the urban area. Sci. Total Environ. 571, 237–250.

Proudfoot, P.R., 1996. Government control in urban waterfront renewal: a comparative review. J. Urban Des. 1 (1), 105–114.

Puertos del Estado, 2019. Monthly Traffic Statistics. Available at: http://www.puertos.es/en-us/estadisticas/Pages/estadistica_mensual.aspx.

Puig, M., Wooldridge, C., Michail, A., Mari Darba, R., 2015. Current status and trends of the environmental performance in European ports. Environ. Sci. Pol. 48, 57–66.

Ressano Garcia, P., 2008. The role of the port authority and the municipality in port transformation: Barcelona, San Francisco and Lisbon. Plan. Perspect. 23 (1), 49–79.

Ros Chaos, S., Pino Roca, D., Saurí Marchán, S., Sánchez-Arcilla Conejo, A., 2018. Cruise passenger impacts on mobility within a port area: case of the Port of Barcelona. Int. J. Tour. Res. 20, 147–157.

Saxe, H., Larsen, T., 2004. Air pollution from ships in three Danish ports. Atmos. Environ. 38, 4057–4067.

Schipper, C.A., Vreugdenhil, H., de Jong, M.P.C., 2017. A sustainability assessment of ports and port-city plans: comparing ambitions with achievements. Transp. Res. D 57, 84–111.

Suykens, F., 1989. The city and its port—and economic appraisal. Geoforum 20 (4), 437–4445.

The Port Authority of Barcelona, 2016. Pla de Millora de la Qualitat de l'Aire del Port de Barcelona (In English: Improvement Plan for Air Quality). Port Authority of Barcelona. July.

The Port of Barcelona, 2017a. Annual Report 2017. The Port of Barcelona. Available at: http://www.portdebarcelona.cat/en/web/autoritat-portuaria/memoria_vigent.

The Port of Barcelona, 2017b. Port of Barcelona Traffic Statistics. Accumulated data December 2017, Port of Barcelona's Statistics Service.

The Port of Barcelona, 2017c. Heading for Sustainability: The Environmental Impact of Cruise Ships on Barcelona. The Port of Barcelona.

The Port of Barcelona, 2019a. Economic Motor. Available at: http://www.portdebarcelona.cat/en/web/economic/servicios.

The Port of Barcelona, 2019b. Environmental Management System. Available at: http://www.portdebarcelona.cat/en/web/el-port/sistema.

The Port of Barcelona, 2019c. Climate Strategy. Available at: http://www.portdebarcelona.cat/en/web/el-port/estrategia-climatica.

The Port of Barcelona, 2019d. BCN Zero Carbon. Available at: http://www.portdebarcelona.cat/en/web/el-port/114.

The Port of Barcelona, 2019e. EcoCalculator. Available at: http://www.portdebarcelona.cat/en/web/el-port/115.

The Port of Barcelona, 2019f. Atmospheric Environment. Available at: http://www.portdebarcelona.cat/en/web/el-port/medi-atmosferic.

The Port of Barcelona, 2019g. Voluntary CCCO Agreements. Available at: http://www.portdebarcelona.cat/en/web/el-port/113.

The Port of Barcelona, 2019h. EU Projects. Available at: http://www.portdebarcelona.cat/en/web/el-port/accessos-ferroviaris.

The Port of Barcelona, 2019i. Atmospheric Environment—AIR Quality. Available at: http://www.portdebarcelona.cat/en/web/el-port/108.

Tian, L., Ho, K.-F., Louie, P.K.K., Qiu, H., Pun, V.C., Kan, H., Yu, I.T.S., Wong, T.W., 2013. Shipping emissions associated with increased cardiovascular hospitalizations. Atmos. Environ. 74, 320–325.

Tichavska, M., Tovar, B., 2015. Port-city exhaust emission model: an application to cruise and ferry operations in Las Palmas Port. Transp. Res. A 78, 347–360.

UNCTAD, 2010. Review of Maritime Transport 2010. UNCTAD, Geneva.

UNCTAD, 2017. Review of Maritime Transport 2017. UNCTAD, Geneva.

UNCTAD, 2018. Review of Maritime Transport 2018. UNCTAD, Geneva.

Viana, M., Hammingh, P., Colette, A., Querol, X., Degraeuwe, B., de Vlieger, I., van Aardenne, J., 2014. Impact of maritime transport emissions on coastal air quality in Europe. Atmos. Environ. 90, 96–105.

Villalb, G., Demisse Gemenchu, E., 2011. Estimating GHG emissions of marine ports—the case of Barcelona. Energy Policy 39, 1363–1368.

WHO, 2006. WHO Air Quality Guidelines for Particulate Matter, Ozone, Nitrogen Dioxide and Sulfur Dioxide: Global Update 2005—Summary of Risk Assessment. World Health Organization.

Xiao, Z., Lam, J., 2017. A systems framework for the sustainable development of a Port City: a case study of Singapore's policies. Res. Transp. Bus. Manag. 22, 255–262.

Zhao, Q., Xu, H., Wall, R.S., Stavropoulos, S., 2017. Building a bridge between port and city: improving the urban competitiveness of port cities. J. Transp. Geogr. 59, 120–133.

Chapter 8

Climate change adaptation and mitigation in ports

Advances in Colombia

Gordon Wilmsmeier

School of Management, Universidad de los Andes, Bogota, Colombia, University of Applied Sciences Bremen, Bremen, Germany

1   Context

Colombia is the second most biodiverse country in the world and borders on the Pacific Ocean and Caribbean Sea. The country's marine-coastal zones display a great variety of strategic ecosystems coral reefs, mangroves, sea grass areas, and beaches, among others. These landscapes and ecosystem present a key role in Colombia's exposure to climate change, as they provide protection against sea-level change, climate regulation and climate change hydrology, and erosion control (UNDP, 2014). However, these ecosystems have been subject of degradation due to unplanned development of economic activities. This has resulted that erosive processes are affecting a quarter of the Colombian coastline. Coastal areas on the Caribbean Sea (23%) and the Pacific (25%) have been categorized as critical, where erosion is affecting coastal ecosystems and infrastructure.

This chapter describes the relevance of climate change adaption and mitigation for the Colombian port system and discusses the identified threats and general adaptation and mitigation needs in the national port climate change action plan. Colombia's climate action plan, or Intended Nationally Determined Contribution (INDC), includes the goal to reduce its greenhouse gas emissions by 20% by 2030, as compared to a projected business-as-usual scenario. Colombia's INDC document stresses that climate action is fundamentally a development issue (Colombian Government, 2015). Thus, innovative and strong development in the various sectors of the economy will support efforts to reach this goal. Therefore, the second part of this chapter focuses on mitigation efforts by presenting results for current baseline measures for implementing and monitoring mitigation matters in the port sector.

Maritime Transport and Regional Sustainability. https://doi.org/10.1016/B978-0-12-819134-7.00008-3

2   Climate change impacts and ports

An analysis of INVEMAR-MADs (2016) revealed that Colombia's seaports are exposed to a variety of threats associated with the climate variability and change. The analysis identified 13 threats for the 12 coastal regions. All coastal regions are affected by gales, floods, erosion, and sea surge. Four regions are at least exposed to 10 of the threats. The least affected regions Cordoba and Sucre are still affected by 6 and 5 threats, respectively (see Table 1).

Studies have also estimated the aggregate or macroeconomic and sectoral impacts of climate change in the General Equilibrium Model Climate Change Computable (MEG4C) developed by the National Planning Department (DNP). The model, applying IDEAM's future climate scenarios, finds a negative aggregate impact of climate change on the economy. Taking only the impacts on a number of sectors between 2011 and 2100, the average impact would be annual GDP loss of 0.49%. This means that each year GDP would be 0.49% lower than in a macroeconomic scenario without climate change. It is important to bear in mind that the analysis was only carried out on subsectors of the economy which together account for 4.3% of total GDP (BID-CEPAL-DNP, 2014).

TABLE 1 Exposure to threats from climate variability and change by coastal department in Colombia.

Threat	Bolívar	Antioquia	Chocó	Magdalena	S. Andrés	Atlántico	Guajira	Nariño	Cauca	V. Del Cauca	Córdoba	Sucre	Total number of coastal departments affected by threat
Gale													12
Flood													12
Erosion													12
Cam sea													12
Tropical storm													10
Earthquake													10
Slip													9
Hurricane													8
Drought													6
Sea level rise													5
Salinization													4
Tsunami													3
Tornado													3
Number of threats per coastal department	12	11	10	9	10	9	8	9	7	7	6	5	

Source: author based on Invemar-MADS 2016.

The sum of the losses, not considering inflation, would be equivalent between 3.6 and 3.7 times the GDP value of 2010. However, despite this exemplified impact for the country as a whole, the identified impacts by sectors and regions are very heterogeneous. In general, forestry could be one of the sectors that could actually benefit from climate change as an increase in temperature in combination with higher precipitation in some regions could have a positive effect on the growth of certain species, while livestock, agriculture, aquacultures, and fisheries can be expected to present losses in their production (BID-CEPAL-DNP, 2014).

Turning toward the transport sector, this sector contributes 4.07% to the national GDP, a figure that has remained relatively constant over the last 15 years (Mintransport, 2018). Across all modes the greatest contribution remains with the road transport sector (69%, considering constant prices) of the whole transport sector. While the contribution of the maritime sector to GDP does not seem of high relevance in terms of direct contribution to the country's GDP, over 93% of Colombia's trade in terms of volume are moved via the maritime mode. In 2017 the Colombian port system transferred 205 million tons, a volume that doubled in comparison to 2006. 178 million tons of the total cargo handled in ports correspond to international trade.

The transport sector has two strategic priorities: the improvement and development of logistics and infrastructure; and the consolidation of transport systems in cities. However, the level of progress in meeting these objectives may be affected by climate effects. Climate variability and change in precipitation patterns can have negative effects on the assets and operation of the transport sector. Floods and droughts can cause damage to roads and ports, and increase closures and disruptions in land, rail, sea, and air traffic leading to increases in freight rates, reduced reliability, shortages, and damming of cargo with impacts on the rest of the economy. On the other hand, rising sea levels along with storm surges can cause flooding of transport infrastructure, causing costly damage and closures for the sector (BID-CEPAL-DNP, 2014).

3 Climate change, national strategies and ports

An analysis of the literature also reveals that a "first wave" of climate change risk studies in ports was realized around 2010 (e.g., Scott et al, 2013; UNFCC, 2007). Relevant impact on the awareness of port authorities and increased visibility of this topic probably relates back to the survey of Becker et al. (2011) and continued efforts of international institutions such as UNCTAD (UNCTAD, 2018) and associations, e.g., IAPH and the World Ports Sustainability Programme, set up in 2017 following the World Ports Climate Initiative from 2008.

In 2017 Colombia published the "Plan de Gestión del Cambio Climatico para los Puertos Marítimos de Colombia" [Climate change plan for maritime ports in Colombia]. As numerous similar reports, the work presents a first stocktaking and guide how to assess climate risks and vulnerability assessments. At the same time risk assessment in Colombian ports has a certain tradition. One of the first risk assessments of climate changes was realized by the IFC for the Terminal Maritimo Muelles el Bosque in Cartagena. This analysis included legal, financial, environmental, local community, operational, health and safety, reputational and external stakeholder criteria (Stenek et al., 2011). Furthermore, in 2015 the City of Cartagena was one of the pioneers to develop a climate change action plan (inveMAr-MAds-Alcaldía Mayor de Cartagena de indias-CdKn, 2012).

What can be found in the Colombian national report as in similar local, regional, or other works are guidelines and risk assessment frameworks based on other national or individual experiences. Therefore, this case like others leads to the invariable first question: what are the most appropriate actions a port should be taking now to deal with climate change. Second, what of the plans has been put into reality? In the case of the Cartagena plan advances toward implementation have been very limited until today (Ortiz, 2017).

4 The Colombian port system

There are nine main port management zones defined by the Integrated Port Management Plan—PIOP and port expansion plans (INVEMAR, 2016; Conpes 3611 of 2009 and 3744), of which seven are located on the Caribbean Coast (La Guajira, Santa Marta, Ciénaga, Barranquilla, Cartagena, Golfo de Morrosquillo, Urabá, and San Andrés) and two in the Pacific (Buenaventura and Tumaco). Overall more than 85% of all cargo in terms of volume is moved on the Caribbean coast. In 2018 43% of all cargo in terms of volume was coal, liquid bulk (25%), dry bulk (not coal) 8%, general cargo 3%, and containerized cargo 20%. Santa Marta-Ciénaga, Morrosquillo, and Guajira are the ones in which more volume is handled. These mobilize 65% of the total cargo entering or leaving the country in terms of volume (Fig. 1).

However, given the Colombian cargo structure this picture is somewhat skewed due to the dominance of bulk cargo on the Caribbean coast. Therefore, it is necessary to take a separate look at geographical distribution of the container port activity, which reached around 4 million TEU in 2018. The two principal container port activity zones are Cartagena (58.5%) and Buenaventura (34.3%). These are followed in distance by Barranquilla (4.0%), Santa Marta (2.6%), San Andres (0.4%), and Guajira (0.1%) (Superintendencia de Puertos y Transporte, 2019). About 1 million containers were transshipped in Colombia, 267 thousand in Buenaventura, and 802 thousand in Cartagena.

Figures from the National Infrastructure Agency (ANI) show that between 2010 and 2018 the investment made in the country's public port terminals reached 2558 billion dollars. According to the ANI, these investments have increased the installed port capacity by 55%. The capacity in 2010 was approximately 286 million tons per year; by 2018 it reached 444 million tons per

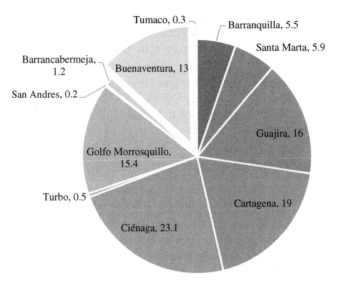

Source: Superintendencia de Transporte, 2019

FIG. 1   Share of port activity in tons by Colombian port region.

year of installed capacity. According to projections of the entity, the installed capacity of the Colombia's port infrastructure could reach 514 million tons per year by 2021. No figures exist on the volume of investment considering or preparing the port system for climate change impacts.

In this sense, given the volume of coal exports the respective capacity in specialized ports (in addition to Puerto Bolívar de Cerrejón the direct loading ports of Drummond and Prodeco were opened) increased from 69 million tons per year in 2010 to 157 million tons in 2018, and by 2021 it is expected to reach 201 million tons.

Further, estimations indicate that the capacity of 124 million tons per year for petrochemical products will remain stable, container capacity is planned to increase from 164 million tons per year in 2018 to 189 million tons in 2021, which implies an increase of 89% over the 2010 capacity level, which was 100 million tons per year.

In terms of the regional port system Cartagena and Buenaventura aim to develop their role as transshipment and logistics hubs. Beyond the pure transshipment activity, the expansion of the two ports as regional logistics centers is part of this strategy, thus capturing value-added services and using the attractiveness of an emerging national market of 49 million inhabitants, the second largest in South America. Example, of first successes, is the establishment of Cartagena as the Latin American distribution center for Decathlon (La Semana, 2018). On the Pacific coast, the development of the Centro Logístico del Pacífico (CELPA) has made the port of Buenaventura attractive for international companies as a distribution center for the West Coast of South America; at the same time the development of national distribution centers in adjacency to the port becomes a more attractive option (CELPA, 2019; Calivia, 2016). The strategies are accompanied by increased expectations in the functioning and reliability of the Colombian port and logistics infrastructure as these will lead to a greater integration of the country in global and regional value chains. Given the identified climate change-related risks, infrastructure adaptation and mitigation move up in the level of importance.

5   Climate change adaptation

Climate change poses new challenges to infrastructure. On the one hand, customers demand reduction of greenhouse gases due to corporate environmental goals (carbon neutrality) or national targets in the countries where they operate. On the other hand, the effects of the climate change are impacting the port infrastructure and its functioning, requiring the development of higher investments for maintenance and generating incident losses. Given this situation ports need to implement measures to mitigate their emissions and to adapt to climate change.

Adaptation measures must include indicators to assess their impact on the environment and on the determinants of vulnerability. In Colombia IDEAM developed indicators and models for this purpose and across all sectors. IDEAM also is responsible for the upcoming National System of Adaptation Indicators, developed by the National Government, and led by the Ministry of Environment and Sustainable Development. Colombia's commitment to GHG emission reductions by 2030 is approximately equivalent to 670 million tons of CO_2 equivalent over the period 2015–30.

Regarding GHG mitigation, the measures considered must be formulated with methodologies validated by the Intergovernmental Panel on Climate Change (IPCC), such as the GHG Protocol or ISO 14064.

The figure below shows GHG adaptation and mitigation measures suggested in Colombia's Port climate change management plan, based on national and international case studies. These are proposed as guidance for decision-makers (Ministry of the Environment and Sustainable Development, 2017) (Fig. 2).

The Colombian government has identified key adaptation, mitigation, and transversal actions (see Tables 2 and 3) to lead the country's port sector in the fight against climate change impacts and to depict possibilities to contribute to the set targets of greenhouse gas emissions.

In the key actions for adaptation and mitigation the relevance of the mangroves as coastal protection becomes evident. Thus, the protection of the current mangroves and reforestation of damaged areas are a central part of the strategy. The listing also shows that the identified actions apply for the Caribbean as for the Pacific coast of the country.

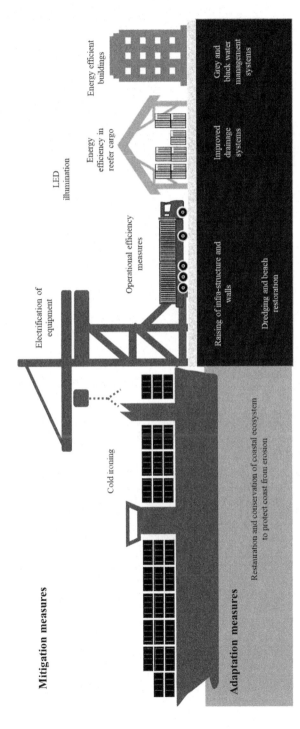

FIG. 2   Proposed mitigation measures in Colombia's port climate change strategy.

TABLE 2 Key actions for climate change adaptation in Colombian seaports.

Threat	Action	Port zone
Sea level rise, flooding, erosion	Protection through management of coastal ecosystems, e.g., mangrove plantation	Gulf of Morrosquillo, Cartagena, Turbo, Buenaventura and Tumaco
	Adaptation of port facilities to minimize impacts from floods, droughts, and coastal erosion, e.g., protective walls, elevation of infrastructure	All
	Articulation of actions of the Road-CC Plan with climate change strategies for seaports in order to reduce vulnerability of ports in terms of connectivity, e.g., consider results of pilot studies of the Road-CC Plan on port access roads	
	Use of the dredging material in accordance with the guidelines of the National Dredging Plan, e.g., beach landfills	
Rainfall increase, temperature increase	Maintenance planning and rehabilitation of port infrastructure taking into account the effects of climate variability and climate change, e.g., with IDEAM climate forecasts	All
	Planning and design of new works considering the risks derived from climate variability and climate change, e.g., guide to incorporate climate change in projects, works or activities of the Ministry of Environment and Sustainable Development	
	Integration of climate change variables into port operating procedures or plans, e.g., ships, terminals, and depots, may require more efficient refrigeration systems in drought seasons, relocation of port facilities to areas not susceptible to sea level rise	
	Reforestation of mangroves and riparian forest and revegetation of areas adjacent to the terminal and in terminal buildings to reduce temperature increase	

TABLE 2 Key actions for climate change adaptation in Colombian seaports.—cont'd

Threat	Action	Port zone
Sedimentation	Planning of dredging activities according to the guidelines of the National Plan	Barranquilla, Dragados Cartagena, Urabá, Buenaventura, Tumaco
Floods and droughts	Implement rainwater capture, management, and storage systems, e.g., rainwater gardens, bio-infiltration or plant trenches, and permeable pavement to infiltrate as much runoff water as possible	All
	Technologies to reduce water consumption, e.g., improved irrigation systems for bulk solids, water recycling systems for carbon wetting	
All	Review and update of standards, codes, and regulations applicable to infrastructure to incorporate the effects of climate variability and change	All
	Identification of future climate change maintenance needs (e.g., tools, plans)	

Source: Ministry of the Environment and Sustainable Development., 2017. Plan de Gestión del Cambio Climático para los Puertos Marítimos de Colombia. Documento de Trabajo. http://www. minambiente.gov.co/images/cambioclimatico/pdf/Plan_nacional_de_adaptacion/Plan_CC_Puertos_ version_trabajo.pdf (accessed May 2019).

TABLE 3 Key CO_2 mitigation actions in Colombian seaports.

		Port zones
CO_2 capture	Protection or restoration of ecosystems with CO_2 capture functions, e.g., Mangroves (highest capacity), seagrasses, coral reefs	Golfo de Morrosquillo, Cartagena, Urabá, La Guajira, Buenaventura and Tumaco
Renewable energies	Use of the wind season to generate and supply alternative energy	La Guajira, Santa Marta—Ciénaga, Cartagena, Barranquilla, San Andres
	Installation of photovoltaic panels to energize small infrastructure units (e.g., bathrooms, to gradually increase its capacity. participation	All

Continued

TABLE 3 Key CO_2 mitigation actions in Colombian seaports.—cont'd

		Port zones
Energy efficiency	Adaptation of port facilities to optimize energy and water consumption, e.g., reuse of gray water, and maximize the use of energy and natural lighting	All
	Cold ironing	
	Replacement of technologies by more efficient ones, e.g., reconversion of cranes, forklifts, vehicles, and other equipment to electrical systems	
	Replacement of light bulbs and luminaires in the entire port/terminal	
	Switching off light sensors in buildings	
Transport	Encourage employees to use active travel modes (bicycle, walking), public transport, or car pooling	All
	Establish a policy for minimizing idling tines in port facilities: vehicles turn off engines (eco driving)	
	Promotion of renovation or reconversion of the vehicle fleet to electric, hybrid, or natural gas vehicles	

Source: Ministry of the Environment and Sustainable Development., 2017. Plan de Gestión del Cambio Climático para los Puertos Marítimos de Colombia. Documento de Trabajo. http://www.minambiente.gov.co/images/cambioclimatico/pdf/Plan_nacional_de_adaptacion/Plan_CC_Puertos_version_trabajo.pdf (accessed May 2019).

In terms of mitigation actions, the report focuses on actions that can directly be employed by the sector in the country, renewable energy and energy efficiency building a central axis.

The transversal actions (Table 4) go beyond the "traditional" adaptation and mitigation actions. Here the focus lies on seizing competitive advantage by implementing environmental standards and optimization of the port and logistics system. A crucial element is the development of financing strategies and leveraging public and private investment resources. Further capacity building and constructing of information systems are seen as essential to visualize progress and raise the awareness of the climate change actions in the port sector. The success

TABLE 4 Key transversal actions in the Colombian port sector.

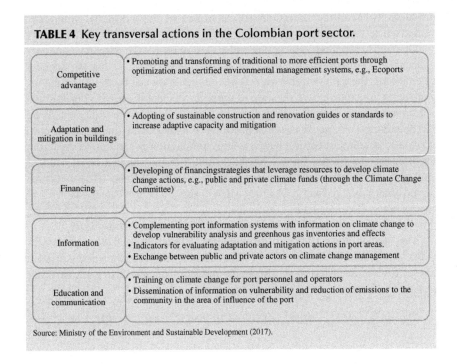

Competitive advantage	• Promoting and transforming of traditional to more efficient ports through optimization and certified environmental management systems, e.g., Ecoports
Adaptation and mitigation in buildings	• Adopting of sustainable construction and renovation guides or standards to increase adaptive capacity and mitigation
Financing	• Developing of financingstrategies that leverage resources to develop climate change actions, e.g., public and private climate funds (through the Climate Change Committee)
Information	• Complementing port information systems with information on climate change to develop vulnerability analysis and greenhous gas inventories and effects • Indicators for evaluating adaptation and mitigation actions in port areas. • Exchange between public and private actors on climate change management
Education and communication	• Training on climate change for port personnel and operators • Dissemination of information on vulnerability and reduction of emissions to the community in the area of influence of the port

Source: Ministry of the Environment and Sustainable Development (2017).

of the implementation of the latter is directly related with the possible success in the search for climate finance, as donors will request for clear success and progress indicators.

5.1   Mitigation efforts and constructing energy consumption and emissions baseline data

Given the focus of mitigation key actions on changing energy sources, energy consumption, and efficiency this section presents and discusses efforts from the Colombian government to create a baseline for mitigation efforts in ports. In a first step the Colombian government has adopted and further developed the energy consumption measuring methodology described in Wilmsmeier and Spengler (2016), Wilmsmeier et al. (2014), and Spengler and Wilmsmeier (2019).

The methodology establishes the system boundaries at terminal level in order to capture differences between terminal types (container, general cargo, dry and liquid bulk). The information has been collected regularly in a joint effort of the Transport Ministry and the School of Management of the Universidad de los Andes. The goal of the data collection is to establish and develop baseline data for climate change mitigation actions as described in Table 3. Beyond energy consumption the collected data include traditional productivity measures, emissions (if calculated by the terminal), water consumption, as well as energy and water costs used.

Given the complexity to capture full emission profiles, energy data are only collected for emissions calculation of scope 1 and scope 2. Thus, for scope 1 calculations, the data on fuels burned in equipment owned by the reporting organization are collected. In the case of scope 2, all electric energy consumption at equipment level as well as the overall purchased electricity is collected. Since the boundaries are set to the organizational boundaries of the terminal, conversion losses that occur in electric energy production in the conversion from fossil energy to electric energy are neglected. For the calculation of scope 2 emissions the form of electric energy production is of relevance. The databases of the International Energy Agency (IEA)[a] and the Latin American Energy Organization [Organización Latinoamericana de Energía] (OLADE) database provide the necessary information and aggregated data at country level.

The methodology allows to disaggregate defined process clusters to individual equipment level. The following equation exemplifies the energy process cluster for container terminals (Spengler and Wilmsmeier, 2019), which is adjusted for other terminal types, respectively:

$$TC_{ij} = \sum_{z=1}^{n}\left(QCC_{ij} + HOC_{ij} + CRC_{ij} + BC_{ij} * LC_{ij} + OC_{ij} + GEN_{ij}\right) + UC_{ij} \quad (1)$$

where z is the type of energy; TC_{ij} is the total energy consumption in terminal i in period j; QCC_{ij} is the energy consumption within the process cluster of quay cranes; HOC_{ij} is the energy consumption within the process cluster of horizontal operations; CRC_{ij} is the energy consumption within the process cluster of reefer cooling; BC_{ij} is the energy consumption within the process cluster of buildings; LC_{ij} is the energy consumption within the process cluster of lighting; OC_{ij} is the energy consumption within the process cluster of others; GEN_{ij} is the energy consumption within the process cluster of generators; and UC_{ij} is the undefined consumption.

Since individual energy consumption of different equipment types and activity clusters might not always be the same as the indicated total consumption, the category the "Undefined consumption" is introduced. This category captures the cases where the terminal's total consumption exceeds the sum of the individual activity clusters (Eq. 2). It has however rather to be understood as a mathematical necessity than a process cluster. The formula for undefined consumption can be seen in the equation below:

$$UC = TC - \left(QCC + HOC + CRC + BC + LC + OC + GEN\right) \quad (2)$$

where UC is the undefined consumption; TC is the total energy consumption from all sources; QCC is the energy consumption from all sources within

---

a. International Energy Agency: Colombia: Balances for 2012. (URL: http://www.iea.org/statistics/statisticssearch/report).

the process cluster of quay cranes; *HOC* is the energy consumption from all sources within the process cluster of horizontal operations; *CRC* is the energy consumption from all sources within the process cluster of reefer cooling; *BC* is the energy consumption from all sources within the process cluster of buildings; *LC* is the energy consumption from all sources within the process cluster of lighting; *OC* is the energy consumption from all sources within the process cluster of others; and *GEN* is the energy consumption from all sources within the process cluster of generators.

The baseline data are collected using a semistructured questionnaire that is sent to Colombian terminals. The process is iterative with the aim to improve data over time, since terminal operators still have relatively little knowledge about the subject of energy consumption as well as no experience in recording historic energy consumption data in their terminals. In several cases, specific energy consumption source monitoring is not installed. The latter is particularly present in smaller ports and terminals, as they are not acquainted with energy consumption measures. Therefore, some questionnaires are not filled out completely in a first round, but in personal discussion during follow-ups it was possible to obtain the required in most cases. This iterative process also aims to measure the awareness and knowledge of the topic in the sector. Since data on resource consumption (energy and water), productivity, as well as expenses in general are considered confidential, no terminal names or specific locations are published. Terminals receive individual feedback on their data in terms of completeness and quality, including a benchmark exercise to indicate them where they are positioned in comparison to their peers. The latter, however, is dependent on the availability of sufficient data.

In 2018 the database included 23 bulk, liquid bulk, general cargo, and container terminals. The data for a couple of terminals is far back as 2012. The latest data are available for 2017.

The following calculations showcase the advances in Colombia and are based on data provided from six container terminals representing a throughput of 2.66 million containers in 2017. These terminals spent over 936 million current USD on energy sources, consuming 12.48 megaliters of diesel and 83.12 GWh of electricity, emitting an estimated 43.5 thousand tons of CO_2 (Table 5).

Diesel is the main energy source in container terminals across the globe (74% of overall energy consumption) (ECLAC and MTT, 2016; Gonçalves de Souza Viera et al., 2010). Colombia is no exception. In Colombia the share of electricity has been oscillating around 40% between 2014 and 2017. The current dependency on diesel marks a significant potential toward electrification.

The variation of consumption patterns across terminals depends significantly on the equipment configurations in each terminal. In the case of Colombia advances in electrification are not only dependent on the terminals but also requires significant improvement of electricity provision. The current network

TABLE 5 Container terminal climate change baseline data for Colombian container terminals, 2017.

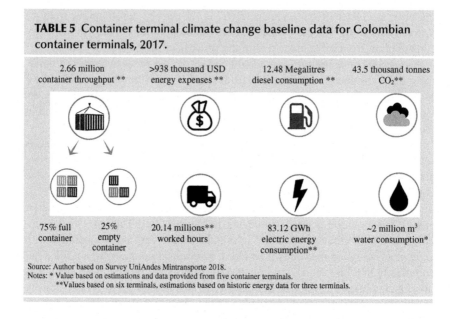

| 2.66 million container throughput ** | >938 thousand USD energy expenses ** | 12.48 Megalitres diesel consumption ** | 43.5 thousand tonnes CO$_2$** |

| 75% full container | 25% empty container | 20.14 millions** worked hours | 83.12 GWh electric energy consumption** | ~2 million m^3 water consumption* |

Source: Author based on Survey UniAndes Mintransporte 2018.
Notes: * Value based on estimations and data provided from five container terminals.
    **Values based on six terminals, estimations based on historic energy data for three terminals.

in certain regions does not provide the necessary energy security to make an investment in electric equipment.

The use of different energy sources also requires the transformation of the different energy types into one comparable unit to derive comparable indicators. In the following the different energy sources are converted into diesel liters equivalent to enable such comparison (Spengler and Wilmsmeier, 2019).

The results from the latest survey show that the greatest energy consumption continues to be related to horizontal movements in the terminals (Fig. 3). Further, despite concerted efforts to improve the knowledge and increase the relevance of the topic the share of "undefined consumption" remains relevant. This indicates that more concerted efforts will be necessary to explain the process cluster approach and to convince operators on the relevance in enabling measurement by activity clusters.

Spengler and Wilmsmeier (2019) show the variation in diesel liters equivalent used to handle a single dry container in different countries. The median consumption per dry container across their global benchmark was 10.4 L diesel equivalent. In the case of Colombia, the average energy used to move one container has improved significantly between 2014 and 2017. In 2014 the average consumption was 10.7 L of diesel equivalent. This value reduced to 7.9 L diesel equivalent by 2017. In Latin America and the Caribbean Spengler and Wilmsmeier (2019) calculated an average consumption of 8 L (Fig. 4).[b]

b. Based on 41 terminals in 17 countries with a total throughput of over 37 million TEU.

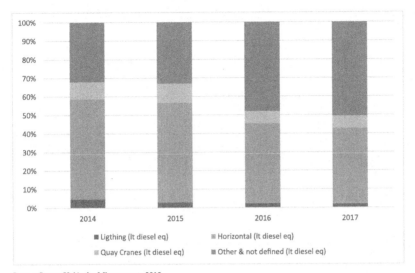

Source: Survey UniAndes Mintransporte 2018
Notes: For the years 2014-15 values based on data provided from six container terminals. For the years 2016-17, values based on data provided from three container terminals and estimations based on partial energy data for three terminals

FIG. 3 Share of energy consumption by activity cluster in container terminals, Colombia 2014–17.

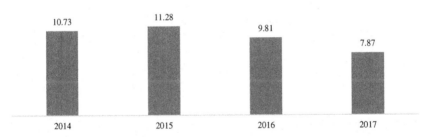

Source: Survey UniAndes Mintransporte 2018
Notes: For the years 2014-15 values based on data provided from six container terminals. For the years 2016-17, values based on data provided from three container terminals and estimations based on partial energy data for three terminals

FIG. 4 Average liters of diesel equivalent consumed for handling one dry box (excluding reefer consumption) in Colombia, 2014–17.

Considering the different energy sources, CO_2 emissions (scope 1 and scope 2) per container varied between 12.4 and 42.7 kg in 2017. This difference is significant and reveals the gap between terminals in the same country. It also sets the benchmark for the less efficient terminals. A key element to incentivize terminals, whether public or private, to invest in strategies to reduce emissions is the cost of energy. Thus, the data on energy expenses reveal the economic dimension of energy consumption in general, substituting energy sources, and potential challenges in the current energy tariff structure when implementing

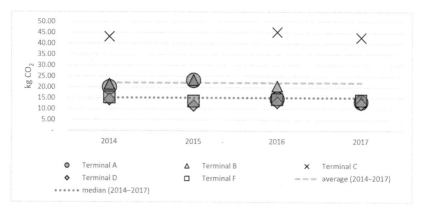

Source: Survey UniAndes Mintransporte 2018.
Notes: For the years 2014–2017 values based on data and estimated data from four container terminals.

FIG. 5   Estimated CO_2 emissions (scope 1 and scope 2) per container and terminal in Colombia, 2014–17. *From Survey UniAndes Mintransporte, 2018. Notes: For the years 2014–2017 values based on data and estimated data from 5 container terminals.*

electrification strategies. In 2017 the average total energy costs for the terminal for moving a container through a Colombian container terminal are estimated to sum up to 37 current USD, 22 current USD for diesel, and 15 for electricity.[c] Consequently, any increase in energy efficiency will not only have a significant impact on the terminal's emissions but also improve its economic performance. The economic dimension in the climate change discussion should not be underestimated since it can bring key arguments for terminals to invest in new technology, change processes, or operating pattern (Fig. 5).

6   Conclusions

This chapter presents an overview of Colombia's strategy for adaptation and exemplifies the efforts to create and develop a baseline for CO_2 emissions in the country's port sector. While the country is clear on the identified actions for mitigation and adaptation (INVEMAR, 2016; INVEMAR-MADs, 2016), the financing of these actions and concrete implementation remain a major challenge. Further, agency to implement proposed actions or guidelines is not always clear. The role of government of private actor responsibilities remains fuzzy and no incentive of investment programs has been developed to incentivize change.

Nevertheless, Colombia shows clear advances in building baseline data for climate relevant indicators, which will facilitate monitoring progress of climate

---

c. Energy costs will vary by container type, e.g., reefer container as well as type of movement, e.g., transshipment, import, export as these require different handling operations in the terminal.

action in the future. Yet, existing results should not hide that data coverage focuses only on the largest terminals in the country. The current data base on sustainability and productivity of terminals includes 23 terminals. However, according to the Superintendencia de Puertos y Transporte (2016) at least 207 terminals exist in the country, of which 121 are maritime and 86 are inland waterway terminals. Continued efforts to increase the participating number and types of terminals and, thus, to also include a significant number of bulk and river terminals is required.

Acknowledgment

The author would like to thank Magda Buitraga, Transport Ministry of Colombia and all participating terminals in Colombia for their collaboration and support in this research.

References

Becker, A., Inoue, S., Fischer, M., Schwegle, B., 2011. Climate change impacts on international seaports: knowledge, perceptions, and planning efforts among port administrators. Clim. Change. https://doi.org/10.1007/s10584-011-0043-7.

BID-CEPAL-DNP, 2014. Impactos Económicos del Cambio Climático en Colombia—Síntesis. In: Calderón, S., Romero, G., Ordóñez, A., Álvarez, A., Ludeña, C., Sánchez, L., … Pereira, M. (Eds.), Banco Interamericano de Desarrollo. Monografía, No. 221 y Naciones Unidas, LC/L.3851, Washington, DC.

Calivia, 2016. Celpa S.A. en Buenaventura. La plataforma logística con beneficios de zona franca. https://www.caliviva.com/informes/item/1531-celpa-s-a-en-buenaventura-la-plataforma-logistica-con-beneficios-de-zona-franca. (accessed June 2019).

CELPA, 2019. https://www.celpazonafranca.co/. (accessed June 2019).

Colombian Government, 2015. Contribución Prevista y Determinada a Nivel NacionaliNDC. http://www.minambiente.gov.co/images/cambioclimatico/pdf/colombia_hacia_la_COP21/iNDC_espanol.pdf. (accessed June 2019).

ECLAC and MTT, 2016. Consumo y Eficiencia Energética en los Principales Terminales Portuarios de Chile. https://www.cepal.org/sites/default/files/events/files/boletin_ee-puertos-chile-cepal-mtt.pdf. (accessed June 2019).

Ortiz, F., 2017. Cartagena Struggles to Get Pioneering Climate Plan Into Action, Reuters. https://www.reuters.com/article/us-colombia-climatechange-cartagena/cartagena-struggles-to-get-pioneering-climate-plan-into-action-idUSKCN1BK00N. (accessed June 2019).

Gonçalves de Souza Viera, P., et al., 2010. Hito 2.2—Diagnóstico de la Situación Energética Actual en el Ámbito Portuario Estatal, EFICONT (EFIciencia Energética en Terminals Portuaria de CONTenedores), Project Report.

INVEMAR, 2016. Guía Ambiental de Terminales portuarios. http://www.invemar.org.co/documents/10182/43044/Version+Preliminar+Terminales+Portuarios+V1.pdf/53124700-911d-4265-82e1-85ee847e1f14. (accessed June 2019).

INVEMAR-MADs, 2016. Informe del estado de los ambientes y recursos marinos y costeros de Colombia 2015. 1692-5025.

inveMAr-MAds-Alcaldía Mayor de Cartagena de indias-CdKn, 2012. Lineamientos para la adaptación al cambio climático de Cartagena de indias. Proyecto integración de la Adaptación al Cambio Climático en la Planificación territorial y Gestión sectorial de Cartagena de indias. In:

Rojas, G.X., Blanco, J., Navarrete, F. (Eds.), Cartagena. Serie de Documentos Generales del inveMAr n° 55. 40 pp.

La Semana, 2018. La multinacional Decathlon entra a Colombia por Cartagena. https://www.se-mana.com/contenidos-editoriales/la-cuarta-oportunidad/articulo/la-multinacional-decathlon-entra-a-colombia-por-cartagena/592860. (accessed June 2019).

Ministry of the Environment and Sustainable Development., 2017. Plan de Gestión del Cambio Climático para los Puertos Marítimos de Colombia. Documento de Trabajo. http://www.minambiente.gov.co/images/cambioclimatico/pdf/Plan_nacional_de_adaptacion/Plan_CC_Puertos_version_trabajo.pdf. (accessed May 2019).

Mintransport, 2018. https://www.mintransporte.gov.co/documentos/15/estadisticas/. (accessed June 2019).

Scott, H., McEvoy, D., Chhetri, P., Basic, F., Mullett, J., 2013. Climate change adaptation guidelines for ports. Enhancing the resilience of seaports to a changing climate report series. National Climate Change Adaptation Research Facility, Gold Coast. 28 pp.

Spengler, T., Wilmsmeier, G., 2019. Sustainable performance and benchmarking in container termi-nals—the energy dimension. In: Bergqvist, R., Monios, J. (Eds.), Green Ports. Elsevier, ISBN: 9780128140543, pp. 125–154. https://doi.org/10.1016/B978-0-12-814054-3.00007-4.

Stenek, V., Amado, J.C., Connell, R., Palin, O., Wright, S., Pope, B., Hunter, J., McGregor, J., Morgan, W., Stanley, B., Washington, R., Liverman, D., Sherwin, H., Kapelus, P., Andrade, C., Pabon, J.D., 2011. Climate Risks and Business Ports; Terminal Maritimo Muelles el Bosque, Cartagena. International Finance Corporation, Colombia.

Superintendencia de Puertos y Transporte, 2016. Boletín Estadístico. Tráfico Portuario en Colom-bia. Ministerio de Transporte, Bogotá. Primer Trimestre 2016.

Superintendencia de Puertos y Transporte, 2019. http://www.supertransporte.gov.co/index.php/superintendencia-delegada-de-puertos/. (accessed June 2019).

UNCTAD, 2018. Risk to trade if ports not climate change proofed. https://unctad.org/en/pages/newsdetails.aspx?OriginalVersionID=1949. (accessed May 2019).

UNDP, 2014. PNUD 2014, V Informe Nacional de Biodiversidad de Colombia ante el Convenio de Diversidad Biológica, Programa de las. Naciones Unidas para el Desarrollo. http://www.undp.org/content/dam/colombia/docs/MedioAmbiente/undp-co-informebiodiversidad-2014.pdf. (accessed June 2019).

UNFCC, 2007. Vulnerability and Adaptation to Climate Change in Small Island Developing States—Background Paper for the Expert Meeting on Adaptation for Small Island Develop-ing State. https://unfccc.int/files/adaptation/adverse_effects_and_response_measures_art_48/application/pdf/200702_sids_adaptation_bg.pdf. (accessed June 2019).

Wilmsmeier, G., Froese, J., Zotz, A.-K., Meyer, A., 2014. Energy consumption and efficiency: emerging challenges from reefer trade in South American container terminals. FAL Bull. 329 (1). https://repositorio.cepal.org/bitstream/handle/11362/37283/Bolet%C3%ADn%20FAL%20329_en.pdf?sequence=1&isAllowed=y. (accessed June 2019).

Wilmsmeier, G., Spengler, T., 2016. Energy consumption and container terminal efficiency. FAL Bull. 350 (6). https://repositorio.cepal.org/bitstream/handle/11362/40928/S1601301_en.pdf. (accessed June 2019).

Part IV

Opening up the Arctic seas

Chapter 9

Navigational risk factor analysis of Arctic shipping in ice-covered waters

Mingyang Zhang[a,b,c], Di Zhang[a,b], Chi Zhang[a,b], Wei Cao[a,b]

[a]*Intelligent Transportation Systems Research Center, Wuhan University of Technology, Wuhan, China,* [b]*National Engineering Research Center for Water Transport Safety, Wuhan University of Technology, Wuhan, China,* [c]*Aalto University, School of Engineering, Department of Mechanical Engineering, Maritime Technology, Espoo, Finland*

1 Introduction

Recently, with global warming and a large amount of sea ice melting, the Arctic sea ice has started to decline in volume, extent, and thickness, and the extremely valuable Northern Sea Route (NSR) has led to an increased interest in Arctic shipping activities, as this is the shortest track between northern Europe and northeast Asia (Beveridge et al., 2016; Fu et al., 2017; Zhang et al., 2018; Schøyen and Bråthen, 2011; Zhang et al., 2017a,b,c). The latter is the reason why many shipping companies aim to use this sea route to decrease trip times and costs (Fu et al., 2016; Zhang et al., 2016, 2017a,b,c, 2018). Moreover, hydrocarbon resources in Arctic areas can be exploited and transported during such trips.

Merchant ships typically navigate Arctic waters for approximately four to five months, depending on ice conditions. It is very difficult to ensure the safety of navigation in Arctic ice-covered waters when vessels sail independently and face harsh conditions, such as the presence of sea ice, low temperatures, electromagnetic interference, and other complex environmental conditions (Stoddard et al., 2015; Goerlandt et al., 2016; Fu et al., 2017; Khan et al., 2018; Ostreng et al., 2013; Zhang et al., 2018). At the same time, ordinary vessels lack icebreaking capability, meaning that they cannot ensure ship safety when sailing independently in a harsh ice environment; this can, in turn, easily lead to ice accidents (Kum and Sahin, 2015; Zhang et al., 2017a,b,c; Fu et al., 2016).

In order to expand the navigational window, many ships are escorted or convoyed by an icebreaker. Thus, two shipping mode scenarios in Arctic ice-covered waters can be conceptualized: *independent navigation and navigation*

Maritime Transport and Regional Sustainability. https://doi.org/10.1016/B978-0-12-819134-7.00010-1

153

with icebreaker assistance. The latter can be seen as organized into four identified icebreaker operations: *escort, convoy, breaking a ship loose,* and *towing*
(Goerlandt et al., 2017; Valdez Banda et al., 2015). Despite these measures,
Arctic ice-covered waters continue to be regarded as harsh environments for
shipping. There is an overall high degree of uncertainty about this navigational
environment and ship safety management within it. As an example, accident
statistics in ice-covered waters in the Russian sea area (Goncharov et al., 2011;
Lobanov, 2013), and in the gulf of the Finnish sea area (Valdez Banda, 2017),
are presented in Fig. 1.

There are two types of ship accidents, from which three typical accident scenarios in Arctic ice-covered waters can be derived. As can be seen from Fig. 1,
in the Finnish sea area 22% of all accidents are collisions. That percentage increases to 48% for an ice-ship collision (most hull damage is caused by ice-ship

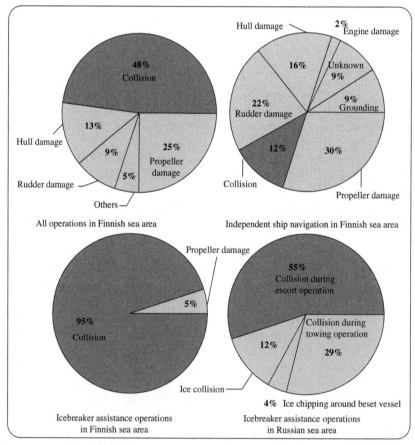

FIG. 1   Accident statistics for ice-covered waters in the Russian Sea and Finnish Sea areas (Valdez
Banda, 2017; Zhang et al., 2018).

collisions, and most propeller and rudder damage by ice-ship collisions during backing operations when ships get stuck in ice), and the percentage of grounding on ice is 9% out of all accidents when ships are independently navigated. In the icebreaker assistance scenario, i.e., in the Finnish sea area, it can be seen that 48% of all accidents are collisions, and 95% of all accidents occur under icebreaker assistance. In the Russian sea area, collisions during the escort operation can be seen to account for 55% under similar assistance conditions. Despite different ways of statistics in the Finnish and Russian sea areas, the statistics indicate that ship collisions (ice-ship collision and ship-ship collision) constitute the most typical accident type in ice-covered waters, in turn implying that collision accidents should be avoided in ice-covered waters.

In Arctic ice-covered waters, it is difficult to ensure navigational safety when ships sail independently and face harsh conditions (Stoddard et al., 2015; Goerlandt et al., 2016; Fu et al., 2017; Khan et al., 2018; Ostreng et al., 2013). At the same time, ordinary vessels lack ice-breaking capability, making them unable to sail independently in a harsh ice environment, which can easily lead to becoming stuck in ice (Kum and Sahin, 2015; Zhang et al., 2017a,b,c; Fu et al., 2016). Ship accidents frequently result in propeller or rudder damage. Meanwhile, ships navigating with an icebreaker reduce the risk of ice-ship collisions, becoming stuck in ice, and grounding on ice. Nevertheless, collision accidents also occur between icebreakers and assisted ships. Therefore, navigational risk factors should be analyzed in order to improve the safety level of ice navigation.

Above all, the navigational risks of ice-going ships cannot be ignored in Arctic ice-covered waters. The literature contains some studies that have been carried out with a focus on independent ship navigation in ice-covered waters. The risks of ship collision and grounding have been analyzed using a root cause analysis method in Arctic ice-covered waters (Kum and Sahin, 2015). Another study analyzed an Arctic shipping accident scenario in order to identify essential accident risk factors in a potential accident scenario (Afenyo et al., 2017). Risk analysis models of ships stuck in ice have also been proposed (Fu et al., 2015, 2016; Montewka et al., 2015). Another line of work has focused on the application of risk-based design principles to Arctic shipping (Bergström et al., 2016; Ehlers et al., 2017). Other studies have been conducted with a focus on icebreaker ships in ice-covered waters. The navigational risks of collision under icebreaker assistance have been found to be different from other ship collision accidents, estimated to be higher in ice-covered waters than in open waters (Zhang et al., 2014; Franck and Holm Roos, 2013; Sulistiyono et al., 2015).

Accordingly, navigational risk factors should be investigated under extreme conditions. Furthermore, the analysis of the risks of navigational operations in ice-covered waters suggests that escort and convoy operations under icebreaker assistance are quite dangerous when performed in ice-covered waters. Overall, whether navigating independently or under icebreaker assistance, ship accidents present the most significant risk in Arctic ice-covered waters (Valdez Banda et al., 2016; Goerlandt et al., 2017; Zhang et al., 2018).

However, the existing studies are limited in terms of their risk analysis of typical operational conditions or accidents, such as collisions between ships, between a ship and ice, or grounding accidents.

In particular, a systematic and multifactorial analysis of navigational factors is presented in the chapter, which aims at identifying and classifying collision risk factors. The research relies on the HFACS and text mining approaches, which are utilized to identify and classify the collision risk factors mentioned in accident reports and research papers on ice-covered waters. A Fault Tree model is proposed, employing a cause-consequence analysis of navigational risk factors according to accident reports and expert knowledge. Following this, a qualitative analysis is carried out to analyze collision risk factors using structural importance degree coefficients and minimum cut sets, thus providing a theoretical basis for the formulation of risk control strategies.

The rest of this chapter is organized as follows: Section 2 describes the research methodology and materials regarding navigational risk factors in Arctic ice-covered waters. The model development and results, which identify navigational risk factors, are presented in Section 3. The navigational risk factors analysis is carried out in Section 4, and Section 5 presents the discussion and conclusions.

2   Method and data

In this chapter, in order to identify and classify the factors contributing to accidents, and according to the characteristics of three typical accidents in Arctic ice-covered waters, the appropriate methods and models are proposed based on the available data. According to accident statistics and the scientific literature, accidents caused by human and organizational factors (HOFs) account for 90% of the total number of maritime accidents (Chauvin et al., 2013). Thus, with a focus on ship collision under icebreaker assistance, the HFACS and fault tree models are used here to analyze contributory collision factors with reference to historical accident reports and expert knowledge. So with a focus on accident scenarios occurring during independent ship navigation, a text mining approach is used to identify the contributory accident factors based on scientific research papers, while the fault tree model is applied to analyze the risk factors. A flowchart of the navigational risk factor analysis of typical accident scenarios in Arctic waters is shown in Fig. 2.

2.1   The HFACS framework

The Human Factors Analysis and Classification System (HFACS) framework, initially proposed by Wiegmann and Shappell (2003), consists of four layers: *organizational factors*, *unsafe supervision*, *preconditions for unsafe acts*, and

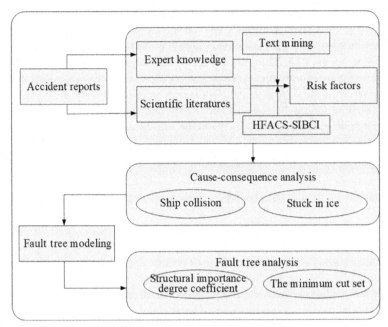

FIG. 2 Flowchart of navigational risk factor analysis of typical accident scenarios in Arctic ice-covered waters.

unsafe acts. Zhang et al. (2018) proposed a fifth layer by which to analyze the human and organizational factors involved in ship collisions under icebreaker assistance. This modified, five-layer HFACS framework was applied to the current study in order to classify and identify accident-contributing factors, with reference to accident and incident reports. In these applications of the HFACS framework to specific contexts, the contributing factors of each layer were interpreted in specific accident situations. Overall, the HFACS framework may be seen as a valid one for risk assessment and risk analysis, where the factors of each layer change continuously according to the research object. In addition, the HFACS-Ship-Icebreaker Collision in Ice-covered waters (HFACS-SIBCI) is utilized to identify and classify the collision risk factors during icebreaker assistance.

2.2 Fault tree analysis (FTA)

The FTA is a typical method and an accidental evolutionary logic analysis tool used to estimate the safety and reliability of a complex system. It has been used conclusively to establish the relationships among accident risk factors (Zhou et al., 2017; Kum and Sahin, 2015; Wang et al., 2013). Moreover, the FTA also has the capacity to reproduce the evolution of underlying factors and high-level accidents, thus aiding understanding of the development course of collision accidents.

In the FT model, following the identification and classification of collision risk factors, this chapter further explored the causes and consequences of accidents by analyzing the relationship between the latter and navigational risk factors, using the FTA. The collision accidents were also qualitatively analyzed based on the structural importance degree coefficient and the minimum cut sets. The essence of the FTA is to establish logical relationships among certain factors based on the mathematical logic theory. The logical relationships are denoted by OR and AND gates. The formulas are shown in Eqs. (1), (2)

$$\varphi(x) = \sum_{i=1}^{n} x_i = \{x_1 + x_2 + \cdots + x_n\} \tag{1}$$

$$\phi(x) = \prod_{i=1}^{n} x_i = \{x_1 \times x_2 \times \cdots \times x_n\} \tag{2}$$

where $\varphi(x)$ and $\phi(x)$ denote the top event used to describe the complex system state, and x_i, the basic factors of i. The output event $\varphi(x)$ presented by the OR gate occurs when at least one input factor occurs, and the output event $\phi(x)$ presented by the AND gate occurs when both input factors occur (Lee et al., 1985).

In the qualitative analysis, the probabilities of the occurrence of different navigational risk factors were assumed to be equal. At the same time, the institutional importance of all risk factors was calculated using the minimum cut sets of the proposed FT model. The structural importance degree of the FT was also calculated. The latter denotes the importance of each basic event based on the structure of the fault tree. For the purposes of the current chapter, the probabilities of all of the navigational risk factors of the three accident scenarios were assumed to be equal in qualitative analysis using FT. Then, the influence of the upper event on the top event was analyzed and the results sorted according to the structural importance degree. Finally, the risk control options (RCOs) were developed according to the structural importance degrees obtained through the qualitative analysis, using the proposed FT for ship collision accidents under icebreaker assistance in ice-covered waters.

Generally speaking, there are two ways to calculate the structural importance degree of the FT: (i) calculate the structural importance degree coefficient, and (ii) use the minimum cut sets for judgment, as now explained in further detail.

(a) Calculate the structural importance degree coefficient of each basic event.

Assume that the state is (0) when the system operates normally, and (1) when the system fails. Then, when the state of the basic event changes (usually from normal operation to failure), the system may experience the following four changes:

- The system changes from normal operation to failure $\{(0) \rightarrow (1)\}$.
- The system remains in a normal working condition $\{(0)\}$.
- The system remains in a state of failure $\{(1)\}$.
- The system changes from a state of failure to normal operation $\{(1) \rightarrow (0)\}$.

The structural importance degree coefficient can be obtained from Eq. (3):

$$I(i) = \begin{cases} \dfrac{1}{2^{n-1}} \sum \left[\varphi(1_i,x) - \varphi(0_i,x) \right] \text{OR gate,} \\ \dfrac{1}{2^{n-1}} \sum \left[\phi(1_i,x) - \phi(0_i,x) \right] \text{AND gate} \end{cases} \tag{3}$$

where $I(i)$ denotes the structural importance degree coefficient, $\varphi(x)$ and $\phi(x)$ represent the top event used to describe the complex system state, and x_i are the basic factors of i (Rausand, 2013).

(b) Determine the order of the degrees of structural importance by using the minimum cut sets.

2.3 Textual data mining approach

To construct ship accident scenarios in Arctic ice-covered waters using a text mining approach, the data first need to be categorized according to their type, and appropriate data mining tools need to be considered. In terms of the above analysis, the focus of this chapter lay in mining the scientific research papers found on the Web of Science, in particular. The most common text mining system is that of the Institute of Computing Technology, Chinese Lexical Analysis System, which can segment and count Chinese text data. However, this system has obvious merit in that it cannot conclusively identify the terminologies in one specific field. Compared with this system, the R language is an open software platform, which can be added according to the user's needs in order to undertake functions such as statistical analysis, analytical processing, and visualized results. As a result, the analysis and extraction of navigational risk factors in the current chapter were realized by utilizing the R language and related software package to extract and analyze the scientific literature regarding independent navigation in Arctic ice-covered waters.

2.4 Accident reports and scientific literature

In order to analyze navigational risk factors, official accident reports and scientific literature pertaining to accident analysis were used in this research. Official accident reports play an essential role in risk factor analysis because they present valuable and detailed information of said accidents that is compiled by an accident investigation board (Mazaheri et al., 2015). Examples of the latter include the Swedish Accident Investigation Board and the Marine Accident Investigation Branch (MAIB) from the United Kingdom, and the Russian FleetMon and Arctic database. Specifically, 17 accidents that had occurred during 1989–2017—and information about which was freely accessible to the public—were considered for the current study and analyzed so as to uncover the collision risk factors under icebreaker assistance. The collision risk factors mentioned in the accident reports are considered and further classified based on the proposed model. For their part, ship accident reports and the scientific literature were analyzed using a text mining approach to identify and classify the navigational risk factors in the context of ships making independent voyages in Arctic ice-covered waters, such as Web of Science, Scopus, and conference proceedings.

3 Model and results

In order to analyze the navigational risk factors, the analysis procedure was divided into three stages: (1) The navigational risk factors were identified using the HFACS-SIBI model (Zhang et al., 2018), based on official accident reports and experts' knowledge, with the focus on collision accidents occurring in the context of icebreaker assistance. (2) The navigational risk factors were identified using the text mining approach based on scientific papers and experts' knowledge, with the focus on ships being stuck in ice and collisions with ice during independent ship navigation. (3) The fault tree was used to establish the fault model for accident risk analysis in Arctic ice-covered waters; the model's development and results are described in the following section.

3.1 Ship collision factors analysis under icebreaker assistance

3.1.1 HFACS-SICI

The HFACS-based ship collision risk analysis model of the HFACS-SIBCI (HFACS-Ship-Icebreaker Collision in Ice-covered waters) was applied in this study to classify and identify ship collision factors, which consisted of five risk factor analysis levels and 28 classification categories, as shown in Fig. 3.

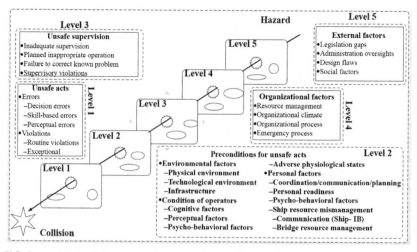

FIG. 3 Accident risk factors classification model based on HFACS-SIBCI.

The HFACS-SIBCI was thus established as a five-level framework (Zhang et al., 2018). In particular, the 28 classification categories contain fundamental collision risk factors affecting collision accidents in the context of icebreaker assistance.

The contributing factors mentioned in the accident reports were identified as risk factors. To begin with, ship collision risk factors under icebreaker assistance were identified using the five-layer HFACS-SIBCI model. The classification model was then utilized to classify ship collision factors based on the categories containing the five collision risk analysis levels and 28 classification categories, such as *decision errors*, *technical errors*, and *legislation gaps*. The HFACS-SIBCI accident risk factors classification model is shown in Fig. 3, which also describes the classification categories in detail (Zhang et al., 2019).

3.1.2 Ship to icebreaker collision risk factors identification

For the purposes of the current chapter, ship collision accidents were classified as occurring between icebreakers and assisted ships, and a hierarchical structure of ship collision risk factors was established. The collision risk classification procedure was as follows: First, ship collision-contributing factors that were mentioned in the accident reports were selected. Second, additional ship collision factors were identified by experts in the field, such as *wrong icebreaker course* and *engine failure*. The latter step was employed so as to ensure that key collision factors were not missed in the case of a lack of accessible accident reports. At the same time, the relevant literature was referenced in order to verify the results regarding ship collisions in open water (Chauvin et al., 2013) and Arctic ice-covered waters (Kum and Sahin, 2015), as shown in Zhang et al. (2018). The ship collision factors in the context of icebreaker assistance were classified as shown in Table 1.

3.2 Analysis of ice-ship collision factors in the context of independent navigation

3.2.1 Text mining to identify risk factors

The text mining approach was used to analyze the scientific literature and identify the navigational risk factors during independent ship navigation. As previously stated, the R language platform was used to extract and analyze the relevant literature. The text mining approach was used to analyze the scientific literature from 2000 to 2018 in international journals indexed on the Web of Science platform. The visualization of the relationships between the authors of the papers found is shown in Fig. 4, and the visualization of major keywords in this research field is shown in Fig. 5.

3.2.2 Identifying ice-ship collision risk factors

68 papers were collected from Web of Science to undertake the risk analysis for ships in ice-covered waters. 51 of these published papers pertained to

TABLE 1 Navigational risk factors during icebreaker assistance.

No.	Risk factors		No.	Risk factors	
1	Unsafe acts	Maneuver failures of the assisted ship [L1]	13	Preconditions for unsafe acts	Communication equipment failure [L13]
2		Maneuver failures of the icebreaker [L2]	14		Poor communication between ships [L14]
3		Lack of situational awareness [L3]	15		Improper route selection [L15]
4		Negligence [L4]	16	Unsafe supervision	Wrong icebreaker course [L16]
5		Judgment failures [L5]	17		Exceeding safe speed limit [L17]
6	Preconditions for unsafe acts	Ice conditions [L6]	18		Unmaintained safety distance [L18]
7		Ice ridge [L7]	19		Deviation from suggested route [L19]
8		Poor visibility [L8]	20	Org. factors	Lack of emergency operation [L20]
9		Snow or rain [L9]	21	External factors	Lack of icebreaking ability [L21]
10		Engine failure [L10]	22		Lack of engine power [L22]
11		Steering gear failure [L11]	23		Anticollision rule gap [L23]
12		Anticollision system failure [L12]			

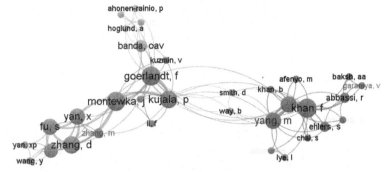

FIG. 4 Visualization of relationships between authors.

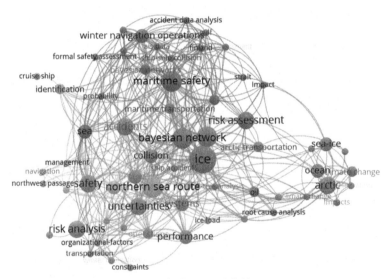

FIG. 5 Visualization of major keywords in the research field.

independent ship navigation, focusing on indication analyses of ships stuck in ice and ice-ship collision accident analyses. The scientific literature was analyzed using the proposed approach, as presented in Section 2.3. The collision risk identification procedure was as follows: First, the scientific literature was collected from Web of Science. This involved a preliminary selection of ship accident risk factors mentioned in the scientific literature as the word dictionary. Second, ship accident-contributing factors were selected using the text mining approach, based on the word dictionary. Finally, the relevant literature was referenced in order to verify the results, with the focus on ship accidents during independent navigation in ice. The navigational risk factors found in this context are shown in Fig. 6 and Table 2.

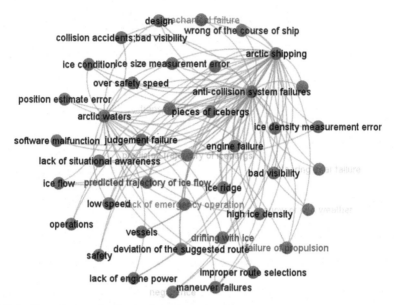

FIG. 6 Network view of navigational risk factors for ships navigating independently in ice.

TABLE 2 Navigational risk factors during independent ship navigation.

No.	Risk factors	No.	Risk factors
1	Drifting with ice [C1]	17	Snow or rain weather [C17]
2	Ice size measurement error [C2]	18	Steering gear failure [C18]
3	Pieces of icebergs [C3]	19	Wrong course [C19]
4	Predicted trajectory of icebergs [C4]	20	Engine failure [C20]
5	High ice density [C5]	21	Ice ridge [C21]
6	Ice density measurement error [C6]	22	Lack of engine power [C22]
7	Low speed [C7]	23	Lack of icebreaking ability [C23]
8	Predicted trajectory of ice flow [C8]	24	Improper route selections [C24]
9	Ice flow [C9]	25	Failure of propulsion [C25]
10	Bad visibility [C10]	26	Maneuver failures [C26]
11	Deviation from suggested route [C11]	27	Position estimate error [C27]
12	Ice conditions [C12]	28	Strong winds [C28]
13	Judgment failures [C13]	29	Software malfunction [C29]
14	Lack of emergency operation [C14]	30	Mechanical failure [C30]
15	Lack of situational awareness [C15]	31	Anticollision system failure [C31]
16	Negligence [C16]	32	Over safety speed [C32]

4 Navigational risk factors analysis using fault tree model

According to the statistics of accidents in Arctic ice-covered waters, shipping accidents constitute the most typical accident scenarios. Thus, this chapter focuses on the following three typical accident scenarios in Arctic ice-covered waters: a ship-icebreaker collision, a ship being stuck in ice, and an ice-ship collision. In order to establish the FT model for navigational risk analysis in ice-covered waters, the proposed statistical analysis procedure was carried out to analyze accident procedures (Zhang et al., 2018). In order to analyze the collision accident procedure, the Cause-Consequence Analysis approach was employed in order to draw the accident procedure map (Chen et al., 2013), an example of which is presented as follows: On 20th January 2011, at 0057 LT, a collision occurred between an icebreaker and an assisted ship during an icebreaker assistance operation at 65°05.1'5N, 026°41.0'1E. This ship collision accident was analyzed based on the HFACS-SIBCI model, in line with the Cause-Consequence Analysis approach shown in Fig. 7.

After identifying the navigational risk factors from the accident reports, the scientific literature, and experts' knowledge, the navigational factors could be operationalized using the proposed model, which is a basis for the FT of collision accidents.

Many navigational risk factors can lead to an accident in Arctic waters, as argued in Sections 3.1.2 and 3.2.2. Lower events connect navigational risk factors.

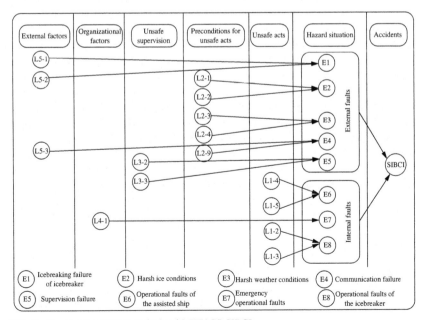

FIG. 7 Cause-consequence analysis with HFACS-SIBCI.

Thus, all branches can be terminated with the risk factors and intermediate factors using the FT model. Following this, this study analyzed the causes and consequences among the navigational risk factors. At the same time, hazard situations and events were defined as intermediate according to the causes and consequences analysis, as shown in Fig. 7. Accordingly, the preliminary FT was established using the causes and consequences analysis based on the accident reports. Finally, the FT model was established according to the hierarchical structure model based on the causes and consequences analysis, as detailed in the following section.

4.1 FT modeling for ship collision accidents under icebreaker assistance

For the purposes of this chapter, *ship collision risk under icebreaker assistance* was selected as the top event. Next, a ship collision risk analysis in the context of icebreaker assistance, due to both external and internal failures, was considered, according to the icebreaker assistance operation mode deemed suitable for the procedure of ship collision accidents. On one hand, icebreaker assistance is a complex navigation system involving the external environment, ships (the assisted ships and the icebreaker), crews, and coordination. This chapter analyzed *external operation errors* and *unsafe management* in line with the statistical HFACS-SIBCI analysis described in Section 3.1.1, and then established the FT of the external failure of icebreaker assistance operations, as shown in Fig. 8. Here, M5 denotes the failure of the icebreaker, M6 represents harsh navigational conditions, M6 contains two aspects (harsh ice conditions and harsh weather conditions), M7 denotes communication failure, and M15 supervision failure. In addition, the collision factors were connected by lower events; the relationships between the factors and events are shown in Fig. 7.

On the other hand, the specifics of icebreaker assistance imply its operation forms different from those of ships in open waters, in that tacit cooperation between the icebreaker and the assisted ship is required. Failures not only test the crews' skills but also require a high level of navigational performance under extreme conditions in ice-covered waters. At the same time, a detailed emergency operation plan needs to be developed.

Furthermore, icebreaker assistance involves many internal failures, in the sense that human errors can easily cause accidents. Accordingly, the FT of the internal failure of icebreaker assistance was established based on *internal operational faults* and *technical operational faults*, according to the Cause-Consequence analysis, as shown in Fig. 8. Here, M10 represents the operation faults of the assisted ship, M11 denotes emergency operation faults, and M12 stands for the icebreaker's operational faults. The FT model contains 23 risk factors influencing collision accidents under icebreaker assistance.

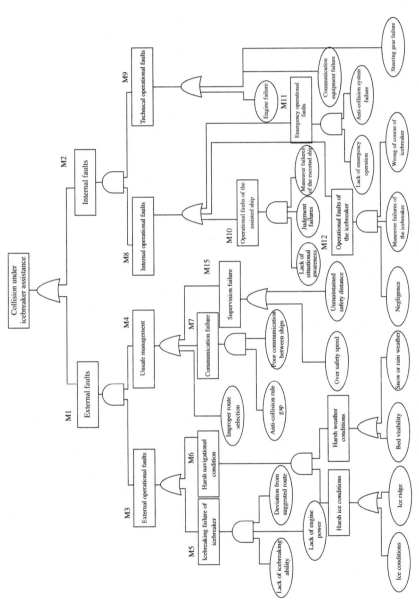

FIG. 8 Ship collision risk FT under icebreaker assistance.

4.2 FT modeling for ice-ship collision in the context of independent navigation

According to the results of the text mining of the scientific literature, the navigational risk factors during independent ship navigation in Arctic ice-covered waters were collected. The FTs of *stuck in ice* and *ice collision* were established based on accident procedure and experts' knowledge.

- FT of *ice-collision*—ice-ship accident

In this study, the FT modeling of ice-ship collisions in Arctic ice-covered waters selected *ice-ship collision* as the top event. Then, an ice-ship collision risk analysis during independent ship navigation was undertaken, taking into account navigational failures, unsafe management, and operational failures. The latter were considered according to the accident scenario of independent ice-going ships, based on the procedure of ice-ship collision accidents. The latter FT model is shown in Fig. 9.

- FT of *stuck in ice* of ice-ship accidents

To operationalize the FT modeling of ships stuck in ice in Arctic ice-covered waters, *ship stuck in ice* was chosen as the top event. Then, the risk analysis of ships being stuck in ice during independent ship navigation was conducted, considering navigational failures, unsafe management, and operational failures, based on the scenario of such independent ice-going ships undergoing the procedure consequent to being stuck in ice. The latter FT is shown in Fig. 10, highlighting 26 risk factors influencing the accident of ships being stuck in ice during independent ship navigation.

4.3 Qualitative analysis of collision risk

Based on the FT model proposed in Section 4.2, the qualitative analysis of navigational risk factors was carried out in light of the qualitative analysis of the FT. First, the minimum cut sets of the FT were determined. Next, the structural importance degree coefficients based on the minimum cut sets were calculated and sorted according to their sizes. Finally, the relationship analysis of different accident scenarios is conducted.

(1) Minimum cut sets of the proposed FT model

The cut set in a fault tree is a set of basic events whose (simultaneous) occurrence ensures that the TOP event occurs, where a cut set is said to be minimal if the set cannot be reduced without losing its status as a cut set. According to the logical relationships among the various factors of the FT for ship collisions under icebreaker assistance, in this study the minimum cut sets were obtained using the Boolean algebra method (Gupta and Agarwal, 1983) (V2.19) (Easy Draw, 2013). The results containing minimum cut sets for different accident scenarios in Arctic ice-covered waters are shown in Appendix.

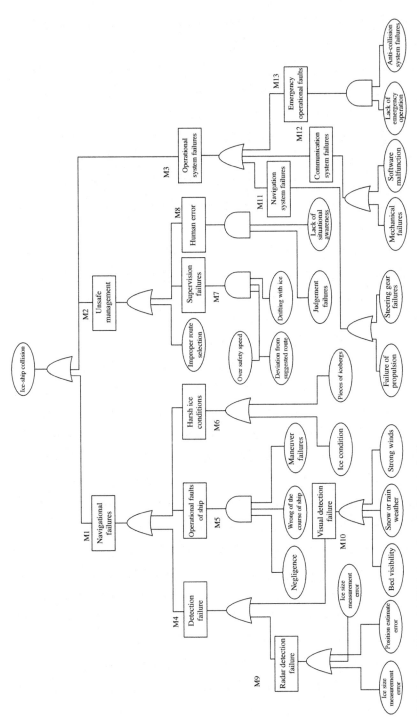

FIG. 9   Ice-collision risk FT during independent ship navigation.

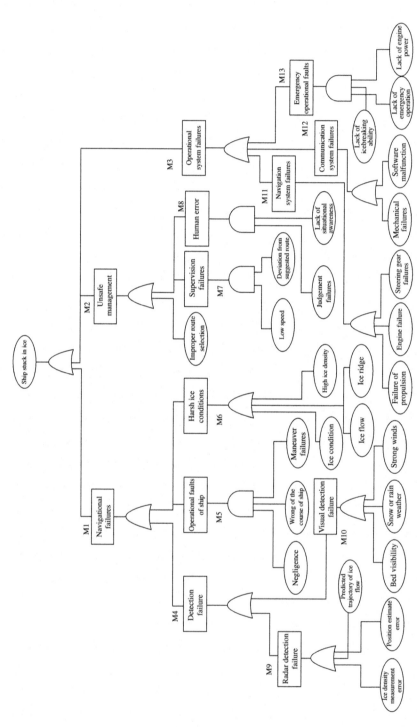

FIG. 10 FT of risk of ship being stuck in ice during independent ship navigation.

The minimum cut sets were obtained based on the dual law of Boolean algebra, which is a reflection of the system safety of navigation in ice. When there is no navigational risk factor in a minimum cut set, the accident does not occur as the top event. Therefore, by obtaining the minimum cut sets, it can be seen which basic events require RCOs in order to avoid collision accidents.

(2) The structural importance degree coefficient of risk factors

The degree of structural importance denotes the importance of each basic event based on the structure of the fault tree. This study analyzed the structural importance degree of the influence of each collision risk factor on ship collision accidents, which were then sorted, as shown in Table 3. These results were also calculated using Easy Draw.

TABLE 3 Structural importance degrees of navigational risk factors.

Scenario	Risk factors	Structural importance degree coefficient
Ship collision under icebreaker	I[[L6]]; I[[L7]]; I[[L8]]; I[[L9]]	0.86
	I[[L17]]; I[[L18]]; I[[L15]]	0.72
	I[[L12]]; I[[L20]]	0.57
	I[[L23]]; I[[L14]]	0.45
	I[[L10]]; I[[L13]]; I[[L11]]	0.42
	I[[L21]]; I[[L21]]; I[[L19]]	0.37
	I[[L4]]; I[[L2]]; I[[L16]] I[[L3]]; I[[L5]]; I[[L1]]	0.33
Stuck in ice	I[C16]=I[C19]=I[C26]	0.78
	I[C12]=I[C21]=I[C9]=I[C9]=I[C5]=I[C22] =I[C23]	0.32
	I[C4]=I[C27]=I[C10]=I[C17]=I[C28]=I[C6]	0.22
	I[C26]=I[C30]	0.17
	I[C25]=I[C18]=I[C20]	0.12
	I[C24]=I[C7]=I[C11]=I[C13]=I[C15]	0.0625
Ice-ship collision	I[C16]=I[C19]=I[C26]	0.54
	I[C14]=I[C31]	0.41
	I[C12]=I[C3]	0.32
	I[C25]=I[C18]=I[C30]=I[C29]	0.23
	I[C2]=I[C27]=I[C10]=I[C17]=I[C28]=I[C6]	0.12
	I[C24]=I[C1]=I[C11]=I[C32]=I[C13]=I[C15	0.031

Based on the proposed FT model, with the focus on accident scenarios in Arctic ice-covered waters, harsh ice conditions and unsafe supervision were found to have a higher index than others in terms of the degree of structural importance, as can be seen in Table 3. These results show that the latter factors exerted the greatest accident contribution. The results also indicate that environmental factors, such as *ice conditions, ice ridge, poor visibility, snow or rain, ice flow, and icebergs,* contribute to ship accidents in Arctic ice-covered waters, with highest of the structural importance degree coefficient. These are followed by the risk factors of unsafe supervision, such as *over safety speed, drifting with ice, and improper route selections,* with higher structural importance degree coefficient. The structural importance degree coefficients of the above navigational risk factors are higher in typical Arctic accident scenarios, indicating that these navigational risk factors have a great impact on the safety of navigation in these waters. Meanwhile, the structural importance degree coefficients of the other factors were found to be lower, indicating that they, in turn, exert a lower impact on safe navigation, but that they also contribute to shipping accidents in Arctic ice-covered waters.

4.4 Relationship analysis of different accident scenarios

In this chapter, various approaches were used to identify the navigational contributing factors in Arctic ice-covered waters, with the focus on ships under icebreaker assistance and ships traveling independently, and with reference to accident reports, expert knowledge, and scientific literature. The analysis of navigational risk factors was carried out using a fault tree analysis based on the accident procedure. Following this, the navigational risk factors were qualitatively analyzed using the structural importance degree coefficients and the minimum cut sets based on the accident scenarios. Finally, the structural importance degree coefficients of the navigational risk factors in different accident scenarios and the analysis results mentioned in the accident reports were compared, as shown in Fig. 11. This comparison indicates that the structural importance degree coefficients are closely related to real-life situations. The analysis found some differences and commonalities among these navigational risk factors contributing to three typical accident scenarios. The comparative analysis of these risk factors is shown in Fig. 11, which could provide theoretical guidance for RCOs focusing on typical accident scenarios.

5 Discussion and conclusion

The analysis in this chapter provides key ship performance and management indicators for ice navigation. The comparative analysis of navigational risk factors is shown in Fig. 11, with the focus on three typical accident scenarios. This research resulted in the identification of 37 navigational risk factors of three typical accident scenarios, using HFACS and the text mining approach based

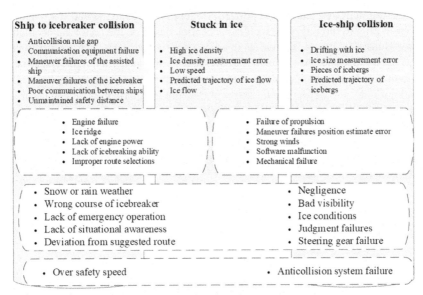

Ship to icebreaker collision

- Anticollision rule gap
- Communication equipment failure
- Maneuver failures of the assisted ship
- Maneuver failures of the icebreaker
- Poor communication between ships
- Unmaintained safety distance

Stuck in ice

- High ice density
- Ice density measurement error
- Low speed
- Predicted trajectory of ice flow
- Ice flow

Ice-ship collision

- Drifting with ice
- Ice size measurement error
- Pieces of icebergs
- Predicted trajectory of icebergs

- Engine failure
- Ice ridge
- Lack of engine power
- Lack of icebreaking ability
- Improper route selections

- Failure of propulsion
- Maneuver failures position estimate error
- Strong winds
- Software malfunction
- Mechanical failure

- Snow or rain weather
- Wrong course of icebreaker
- Lack of emergency operation
- Lack of situational awareness
- Deviation from suggested route

- Negligence
- Bad visibility
- Ice conditions
- Judgment failures
- Steering gear failure

- Over safety speed

- Anticollision system failure

FIG. 11   Comparative analysis of navigational risk factors in different scenarios.

on accident reports and the scientific literature. FT models were established for different accident scenarios based on the historical accident procedure. In addition, a qualitative analysis of typical accident risk factors was carried out, which contained a minimum cut set analysis and the calculation of structural importance degree coefficients. The results of this research capture the actual aim accurately, and the corresponding RCOs are proposed to prevent accident occurrence in Arctic ice-covered waters.

The uncertainties in the conclusions will be taken into consideration as follows: Inaccuracies in data, assumptions made about the model, and modeling procedures are conditions or choices that may affect the conclusions. The navigational risk factors in this study were identified using a reliable approach based on accident reports and scientific literature. While the number of accident reports was insufficient, the FT modeling and the qualitative risk analysis were further complemented with experts' knowledge, and the results show concordance with real-life situations. Thus, the uncertainty assessment of approach and model is high, the data of this research is medium, and the assumption is low. Above all, the FT model relies on historical accident procedures that are formulated in a specific format; therefore, additional accident reports could reduce the uncertainty of FT modeling, which could be a point of improvement in future research.

This chapter provides a comprehensive analysis of navigational risk factors, with a focus on three typical accident scenarios. First, the 37 navigational risk factors are identified using the HFACS-based ship collision risk analysis model and test mining approach based on accident reports and scientific literature.

Then, the fault tree analysis is utilized to identify the fundamental risk factors contributing to three typical accident scenarios based on the historical accident procedure. Finally, qualitative analysis is carried out to analyze navigational risk factors, where the compared analysis is formulated for three typical accident scenarios, which could provide theoretical guidance for RCOs. Overall, and notwithstanding the model assumptions, the comparison of the qualitative analysis results and historical accidents shows that the results obtained by the qualitative analysis based on the FTA are in agreement with real-life situations. Thus, the results obtained can be seen as promising, and as potentially able to aid further understanding of key navigational factors in Arctic ice-covered waters.

Appendix

● Cut sets for typical accident scenarios are shown as follow:

Minimum cut sets of the FT for collision accidents under icebreaker assistance

{L17,L19,L21,L22}; {L1,L3,L5,L13}; {L1,L3,L5,L11}; {L10,L12,L20}; {L2,L4,L10,L16}; {L12,L13,L20}; {L2,L4,L13,L16}; {L11,L12,L20}; {L2,L4,L11,L16}; {L6,L8,L14,L23}; {L6,L8,L15}; {L6,L8,L17}; {L7,L8,L14,L23}; {L7,L8,L15}; {L7,L8,L17}; {L6,L9,L14,L23}; {L6,L9,L15}; {L6,L9,L17}; {L7,L9,L14,L23}; {L7,L9,L15}; {L7,L9,L17}; {L18,L19,L21,L22}; {L6,L8,L18}; {L7,L8,L18}; {L6,L9,L18}; {L7,L9,L18}; {L14,L19,L21,L22,L23}; {L1,L3,L5,L10}; {L19,L21,L22,L15}

Minimum cut sets of the FT for ship stuck in ice

{C4,C16,C19,C12,C26};{C24C7,C11,C13,C15};{C25,C29,C14,C22,C23};{C10,C16, C19,C12,C26};{C27,C16,C19,C12,C26};{C16,C19,C12,C26,C6};{C17,C16,C19,C12 ,C26};{C28,C16,C19,C12,C26};{C4,C16,C19,C21,C26};{C4,C16,C19,C26,C24};{C4,C1 6,C19,C26,C5};{C10,C16,C19,C21,C26};{C10,C16,C19,C26,C24};{C10,C16,C19,C26 ,C5};{C27,C16,C19,C21,C26};{C27,C16,C19,C26,C24};{C27,C16,C19,C26,C5};{C16,C 19,C21,C26,C6};{C16,C19,C26,C6,C24};{C16,C19,C26,C6,C5};{C17,C16,C19,C21,C2 6};{C17,C16,C19,C26,C24};{C17,C16,C19,C26,C5};{C28,C16,C19,C21,C26};{C28,C1 6,C19,C26,C24};{C28,C16,C19,C26,C5};{C18,C29,C14,C22,C23};{C20,C29,C14,C22 ,C23};{C30,C25,C14,C22,C23};{C30,C18,C14,C22,C23};{C30,C20,C14,C22,C23}

Minimum cut sets of the FT for ice-ship collision in ice

{C2,C16,C19,C26,C12};{C24,C1,C11,C32,C13,C15};{C25,C30,C14,C31};{C10,C16, C19,C26,C12};{C27,C16,C19,C26,C12};{C16,C19,C26,C12,C6};{C17,C16,C19,C26 ,C12};{C28,C16,C19,C26,C12};{C2,C16,C19,C26,C3};{C10,C16,C19,C26,C3};{C27,C1 6,C19,C26,C3};{C16,C19,C26,C3,C6};{C17,C16,C19,C26,C3};{C28,C16,C19,C26,C3}; {C18,C30,C14,C31};{C25,C29,C14,C31};{C18,C29,C14,C31}

Note: The cut set in a fault tree is a set of basic events whose (simultaneous) occurrence ensures that the TOP event occurs, and the minimum cut set denotes that a cut set is said to be minimal if the set cannot be reduced without losing its status as a cut set.

Acknowledgment

This study was supported by the National Science Foundation of China (NSFC) under Grant Nos. 51579203 and 51711530033.

References

Afenyo, M., Khan, F., Veitch, B., Yang, M., 2017. Arctic shipping accident scenario analysis using Bayesian Network approach. Ocean Eng. 133, 224–230.

Bergström, M., Erikstad, S.O., Ehlers, S., 2016. Assessment of the applicability of goal- and risk-based design on Arctic sea transport systems. Ocean Eng. 128, 183–198.

Beveridge, L., Fournier, M., Lasserre, F., Huang, L., Têtu, P.-L., 2016. Interest of Asian shipping companies in navigating the Arctic. Polar Sci. 10 (3), 404–414.

Chauvin, C., Lardjane, S., Morel, G., 2013. Human and organizational factors in maritime accidents: analysis of collisions at sea using the HFACS. Accid. Anal. Prev. 59 (5), 26–37.

Chen, S.T., Wall, A., Davies, P., Yang, Z., Wang, J., Chou, Y.H., 2013. A human and organisational factors (HOFs) analysis method for marine casualties using HFACS-maritime accidents (HFACS-MA). Saf. Sci. 60 (12), 105–114.

Ehlers, S., Cheng, F., Jordaan, I., Kuehnlein, W., Kujala, P., Luo, Y., Freeman, R., Riska, K., Sirkar, J., Oh, Y.-T., Terai, K., Valkonen, J., 2017. Towards mission-based structural design for arctic regions. Ship Technol. Res. 64 (3), 115–128.

Franck, M., Holm Roos, M., 2013. Collisions in Ice: A Study of Collisions Involving Swedish Icebreakers in the Baltic Sea. Linnaeus University, Swedish.

Fu, S., Yan, X., Zhang, D., Shi, J., Xu, L., 2015. Risk factors analysis of Arctic maritime Transportation system using structural equation modelling[C]. In: Proceedings of the International Conference on Port and Ocean Engineering Under Arctic Conditions. Trondheim, Norway.

Fu, S., Zhang, D., Montewka, J., Yan, X., Zio, E., 2016. Towards a probabilistic model for predicting ship besetting in ice in Arctic ice-covered waters. Reliab. Eng. Syst. Saf. 155, 124–136.

Fu, S., Zhang, D., Montewka, J., Zio, E., Yan, X., 2017. A quantitative approach for risk assessment of a ship stuck in ice in Arctic ice-covered waters. Saf. Sci. https://doi.org/10.1016/j.ssci.2017.07.001.

Goerlandt, F., Montewka, J., Zhang, W., Kujala, P., 2016. An analysis of ship escort and convoy operations in ice conditions. Saf. Sci. 95, 198–209.

Goerlandt, F., Goite, H., Banda, O.A.V., Höglund, A., Ahonen-Rainio, P., Lensu, M., 2017. An analysis of wintertime navigational accidents in the northern Baltic sea. Saf. Sci. 92, 66–84.

Goncharov, V., Klementieva, N., Sazonov, K., 2011. Russian bring experience to winter navigation safety. The Naval Architect 34, 56–58.

Gupta, P.P., Agarwal, S.C., 1983. A Boolean algebra method for reliability calculations. Microelectron. Reliab. 23 (5), 863–865.

Khan, B., Khan, F., Veitch, B., Yang, M., Khan, B., Khan, F., et al., 2018. An operational risk analysis tool to analyze marine transportation in arctic waters. Reliab. Eng. Syst. Saf. 169, 485–502.

Kum, S., Sahin, B., 2015. A root cause analysis for Arctic Marine accidents from 1993 to 2011. Saf. Sci. 74, 206–220.

Lee, W.S., Grosh, D.L., Tillman, F.A., Lie, C.H., 1985. Fault tree analysis, methods, and applications: a review. IEEE Trans. Reliab. 34 (3), 194–203.

Lobanov, V., 2013. Ice performance and ice accidents fleet inland and river-sea navigation. Int. Joul. Nau. 67 (4), 12–16.

Mazaheri, A., Montewka, J., Nisula, J., Kujala, P., 2015. Usability of accident and incident reports for evidence-based risk modeling—a case study on ship grounding reports. Saf. Sci. 76, 202–214.

Montewka, J., Goerlandt, F., Kujala, P., Lensu, M., 2015. Towards probabilistic models for the prediction of a ship performance in dynamic ice. Cold Reg. Sci. Technol. 112, 14–28.

Ostreng, W., Eger, K.M., Floistad, B., Jorgensen-Dahl, A., Lothe, L., et al., 2013. Shipping in Arctic Waters: A Comparison of the Northeast, Northwest and Trans Polar Passages. Springer, Heidelberg.

Rausand, M., 2013. Risk Assessment: Theory, Methods, and Applications. vol. 115. John Wiley & Sons, Hoboken, NJ.

Schøyen, H., Bråthen, S., 2011. The Northern Sea Route versus the Suez Canal: cases from bulk shipping[J]. Journal of Transport Geography 19 (4), 977–983.

Stoddard, M.A., Etienne, L., Fournier, M., Pelot, R., Beveridge, L., 2015. Making sense of Arctic maritime traffic using the Polar Operational Limits Assessment Risk Indexing System (POLARIS). In: IOP Conference Series: Earth and Environmental Science. vol. 34. . (1). Article number 012034.

Sulistiyono, H., Khan, F., Lye, L., Yang, M., 2015. A risk-based approach to developing design temperatures for vessels operating in low temperature environments. Ocean Eng. 108, 813–819.

Valdez Banda, O.A., 2017. Maritime Risk and Safety Management With Focus on Winter Navigation. Aalto University, Finland.

Valdez Banda, O.A., Goerlandt, F., Montewka, J., Kujala, P., 2015. A risk analysis of winter navigation in Finnish sea areas. Accid. Anal. Prev. 79, 100.

Valdez Banda, O.A., Goerlandt, F., Kuzmin, V., Kujala, P., Montewka, J., 2016. Risk management model of winter navigation operations. Mar. Pollut. Bull. 108 (1–2), 242.

Wang, D., Zhang, P., Chen, L., 2013. Fuzzy fault tree analysis for fire and explosion of crude oil tanks. J. Loss Prev. Process Ind. 26 (6), 1390–1398.

Wiegmann, D.A., Shappell, S.A., 2003. A human error approach to aviation accident analysis: the human factors analysis and classification system. Reference & Research Book News.

Zhang, D., Yan, X., Yang, Z., et al., 2014. An accident data based approach for congestion risk assessment of inland waterways: a Yangtze River case. J. Risk Reliab. 228 (2), 176–188.

Zhang, D., Yan, X., Zhang, J., Yang, Z., Wang, J., 2016. Use of fuzzy rule-based evidential reasoning approach in the navigational risk assessment of inland waterway transportation systems. Saf. Sci. 82, 352–360.

Zhang, M., Zhang, D., Fu, S., Yan, X., 2017a. A method for Arctic sea route planning under multiconstraint conditions. In: Proceedings of International Conference on Port and Ocean Engineering Under Arctic Conditions, June 10-16, 2017, Busan, Korea.

Zhang, M., Zhang, D., Fu, S., Yan, X., Goncharov, V., 2017b. Safety distance modeling for ship escort operations in Arctic ice-covered waters. Ocean Eng. 146, 202–216.

Zhang, M., Zhang, D., Fu, S., Yan, X., Luo, J., Goncharov, V., 2017c. Safety distance modelling for vessels under the icebreaker assistance: taking "Yong sheng" and "50 let pobedy" as an example. In: Transportation Research Board 96th Annual Meeting, January 8-12, 2017. Washington, DC, USA.

Zhang, M.Y., Zhang, D., Yan, X.P., et al., 2018. Collision risk factors analysis model for icebreaker assistance in ice-covered waters[M]. In: Marine Design XIII. vol. 2. CRC Press, pp. 659–668.

Zhang, M., Zhang, D., Goerlandt, F., Yan, X., Kujala, P., 2019. Use of HFACS and fault tree model for collision risk factors analysis of icebreaker assistance in ice-covered waters. Saf. Sci. 111, 128–143.

Zhou, T., Wu, C., Zhang, J., Zhang, D., 2017. Incorporating CREAM and MCS into Fault Tree analysis of LNG carrier spill accidents. Saf. Sci. 96, 183–191.

Further reading

Baysari, M.T., Mcintosh, A.S., Wilson, J.R., 2008. Understanding the human factors contribution to railway accidents and incidents in Australia. Accid. Anal. Prev. 40 (5), 1750.

Chen, S.T., Chou, Y.H., 2013. Examining human factors for marine casualties using HFACS-maritime accidents (HFACS-MA). In: International Conference on ITS Telecommunications, pp. 391–396.

Ergai, A., Cohen, T., Sharp, J., Wiegmann, D., Gramopadhye, A., Shappell, S., 2016. Assessment of the human factors analysis and classification system (HFACS): intra-rater and inter-rater reliability. Saf. Sci. 82, 393–398.

Flage, R., Aven, T., 2009. Expressing and communicating uncertainty in relation to quantitative risk analysis. Reliab. Theor. Appl. 4 (2), 1–13.

Goerlandt, F., Reniers, G., 2016. On the assessment of uncertainty in risk diagrams. Saf. Sci. 84, 67–77.

Guinness, R.E., Saarimäki, J., Ruotsalainen, L., Kuusniemi, H., Goerlandt, F., Montewka, J., Verglund, R., Kotovirta, V., 2014. A method for ice-aware route optimization. IEEE PLANS, Position. In: Location and Navigation Symposium, Monterey, CA, USA, 5.5.2014-8.5.2014, Article number 6851512, pp. 1371–1378.

International Maritime Organization, 1999. Resolution A.884(21): Amendments to the Code for the Investigation of Marine Casualties and Incidents (Resolution A.849(20)). International Maritime Organization (IMO), London.

Klanac, A., Duletic, T., Erceg, S., Ehlers, S., Goerlandt, F., Frank, D., 2010. Environmental risk of collisions in the enclosed European waters: Gulf of Finland, Northern Adriatic and the implications for tanker design. In: Presented at the International Conference on Collision and Grounding of Ships. Aalto University, Espoo, Finland, pp. 55–65.

Kotovirta, V., Jalonen, R., Axell, L., Riska, K., Berglund, R., 2009. A system for route optimization in ice-covered waters. Cold Reg. Sci. Technol. 55 (1), 52–62.

Maritime Safety Administration of the People's Republic of China, 2014. Guidance on Arctic Navigation in the Northeast Route. China Communications Press, Beijing.

Montewka, J., Hinz, T., Kujala, P., Matusiak, J., 2010. Probability modelling of vessel collisions. Reliab. Eng. Syst. Saf. 95 (5), 573–589.

Patterson, J.M., Shappell, S.A., 2010. Operator error and system deficiencies: analysis of 508 mining incidents and accidents from Queensland, Australia using HFACS. Accid. Anal. Prev. 42 (4), 1379.

Qu, X., Meng, Q., Suyi, L., 2011. Ship collision risk assessment for the Singapore Strait. Accid. Anal. Prev. 43 (6), 2030–2036.

Reason, J., 1990. Human Error. Cambridge University Press, New York.

Reinach, S., Viale, A., 2006. Application of a human error framework to conduct train accident/incident investigations. Accid. Anal. Prev. 38 (2), 396–406.

Swedish Accident Investigation Board (SHK). Retrieved from: http://www.havkom.se/en/ 9 May, 2016.

Transport Canada. Arctic Ice Regime Shipping System, 1988. User Assistance Package for the Implementation of Canada's Arctic Ice Regime Shipping System (AIRSS). Transport Canada Press, Winnipeg, Manitoba.

Wickens, C., Flach, J., 1988. Information processing. In: Wiener, E.L., Nagel, D.C. (Eds.), Human Factors in Aviation. Academic, San Diego, CA, pp. 111–155.

Wróbel, K., Montewka, J., Kujala, P., 2017. Towards the assessment of potential impact of unmanned vessels on maritime transportation safety. Reliab. Eng. Syst. Saf. 165, 155–169.

Zhan, Q., Zheng, W., Zhao, B., 2017. A hybrid human and organizational analysis method for railway accidents based on HFACS-railway accidents (HFACS-RAS). Saf. Sci. 91, 232–250.

Chapter 10

Assessing the risk of potential oil spills in the Arctic due to shipping

Mawuli Afenyo[a], Faisal Khan[b], Adolf K.Y. Ng[a,c]
[a]Department of Supply Chain Management, Asper School of Business, University of Manitoba, Winnipeg, MB, Canada, [b]Memorial University of Newfoundland, St. John's, NL, Canada, [c]St. John's College, University of Manitoba, Winnipeg, MB, Canada

1 Introduction

The effects of climate change can be positive and negative. The negative side includes flooding, heat waves, extreme temperatures in some jurisdictions, wild fires, draughts, and tropical storms (Ng et al., 2018). The potential positive side, however, is the melting of ice in the Arctic at an unprecedented rate. This means that the Arctic is gradually becoming ice-free in the summer season and so the likelihood of increased activities of maritime transportation and resource exploration is high (Afenyo et al., 2017a). These activities translate to economic and social improvement in the lives of the populace living in this area. The shipping industry, for example, has already recorded an increase in the number of ships being built to take advantage of this opportunity. Recent industry news shows that many companies have signed in an order for vessels to be built for Arctic voyages. However, these opportunities, along with the economic implications, also bring with them the risk of a potential oil spill (Afenyo et al., 2017a). An oil spill has environmental and socio-economic implications for the affected area and people. For the local communities, it may even go as far as affecting their culture. The oil spill affects the reproductive cycle of species and disrupts the social make-up of the affected communities (Afenyo et al., 2016a). It is therefore important to assess the risk to make decisions for resource allocation, contingency planning, and response (Lee et al., 2015). In fact, some researchers have even advocated the idea of no shipping in the Arctic. This view is refuted by another group that sees the opportunity as a way to open up the remote areas that are linked to the Arctic for economic benefits. Furthermore, for security and strategic planning for Arctic countries, it would be unwise to lag behind while other countries advance in this regard (Ng et al., 2018).

Maritime Transport and Regional Sustainability. https://doi.org/10.1016/B978-0-12-819134-7.00011-3

With the above opportunities and challenges there is need to assess the risk of a potential oil spill in the Arctic. However, there are some challenges that include: (1) spill is a rare event and so difficult to track; (2) uncertainty and variability in climate conditions affect how the fate and transport of the potential spill is predicted; (3) there is limited data on terrestrial conditions which is critical for fate and transport modeling; (4) there is variability and uncertainty on receptor (ecological or human) data which is critical for exposure modeling; and (5) there is limited (no) data on the toxicity of ecological (aquatic and terrestrial species) for different contaminants, which is critical for risk estimation.

The issues related to the effect of climate change in the Arctic has prompted policymakers, industrial practitioners, and academic researchers to work together to generate knowledge on such. The activities carried out by some of the authors of this book include the International Workshop on Climate Change Adaptation Planning for Ports, Transportation Infrastructures, and the Arctic (CCAPPTIA) workshop held in Winnipeg, MB, Canada, in May 2018. The workshop was a key gathering forum for top academic and industrial players. Furthermore, some oil and gas companies and academia have worked together on the Arctic Response JIP (Camus and Smith, 2019). This project has generated significant knowledge on the subject matter. This is a follow-up to the pioneering work of SINTEF and other oil and gas companies (Faksness et al., 2011). Such collaboration only goes to show how urgent the issue of oil spill has become to the Arctic community. Further, the Microbial Genomics for Oil Spill Preparedness in the Canadian Arctic (GENICE)[a] project, which aims to address risk factors related to oil spill in the Arctic, is another flagship project by researchers in Canada. Emphasis is placed on shipping in the Arctic. The consortium is made up of the University of Manitoba, University of Calgary, McGill University, and the University of Ottawa (see https://www.genice.ca/).

Dealing with oil spill is a problem that often involves a lot of resources and personnel. It becomes even more challenging when there is the presence of ice where the oil is spilled (Lee et al., 2015). This is the case because in ice-covered waters, oil behaves differently. Generally, the processes of weathering and transport that occur after the release of oil are slow compared to a scenario where ice is not present. Releases are mostly from pipeline rupture, blowouts from oil and gas production, and exploration and shipping (Afenyo et al., 2016b).

The rest of the chapter is structured as follows: Section 2 describes the concept of risk; Section 3 presents the potential ways to conduct source, fate, transport, and exposure modeling of hydrocarbons from oil spills; Section 4 addresses the uncertainties and variabilities. To demonstrate the tools for oil spill modeling a scenario is presented for such purpose; this is followed by Section 5 where the scenario is analyzed with the description of tools. Discussions and conclusions are presented in Section 7 while future works are discussed in Section 8.

---

a. A Genome Canada sponsored project which is aimed at using genomics for oil spill mitigation in the Arctic. The project is worth $10.4 M and is for a period of 4 years.

2 The concept of risk

In order to assess the risk of oil spills, a number of tools are employed which represent the focus of this chapter. At each critical stage of the oil spill phenomenon, the tools available are contextualized and demonstrated. It is noted that the focus of this chapter is the release from Arctic shipping activities. Pipeline and blowouts are not addressed, although the principles and tools described here can also be applied to pipelines and blowouts with slight modifications. While risk can be static and dynamic in nature, the key determinants are almost the same, except that in describing dynamic risk, its evolution with respect to time and space is taken into consideration. Static risk is defined in Eq. (1):

$$\text{Risk} = F\{s(c,f)\} \tag{1}$$

Eq. (1) shows that it is a function of the particular scenario under study, a consequence of the event, and the frequency or probability of that particular event. In Eq. (1), s represents the scenario; c, the consequence; and f, the frequency or probability. Fig. 1 shows graphs of dynamic and static risks.

Ideally, risk is dynamic in nature and so should be treated as such. Risk can be incorporated in every stage of the engineering process. This includes conceptual design, detailed design, installation, and operation. The dynamic risk is described by Eq. (2) and it contains an element of time as illustrated in Fig. 1B.

$$\text{Dynamic risk} = F\{s(c,f),t\} \tag{2}$$

The process of risk modeling for an oil spill from shipping involves source modeling, fate and transport modeling, and exposure modeling.

3 Source, fate and transport, partition, and exposure modeling

The process of source modeling can be challenging as the release changes with time. Fig. 2 shows some factors to be considered for the source modeling process. Source modeling here entails the release and subsequently leads up to dispersion.

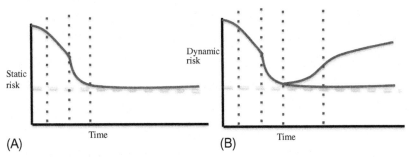

FIG. 1   (A) Static risk shows risk remains unchanged with time and (B) dynamic risk demonstration. Evolution of risk with time.

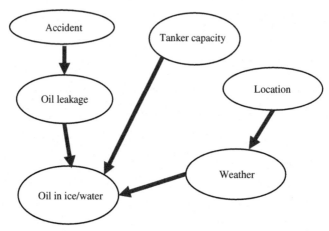

FIG. 2 Factors to consider when undertaking source modeling.

Following the source modeling is the dispersion modeling. Dispersion modeling describes how much distance the pollutant (oil) could travel in a given time after the release of hydrocarbons. Fig. 3 is taken from Afenyo et al. (2017a) and shows the dispersion of oil from a vessel. The dimensions of the plume of the pollutant are very important. In order to describe this mathematically, Eq. (3) is used.

$$C(r,t,x,y,z) = Q \frac{\exp\left[\dfrac{-(x-Wt)^2}{4D_x t}\right]}{\sqrt{4\pi D_x t}} \frac{\exp\left[\dfrac{-(y)^2}{4D_y t}\right]}{\sqrt{4\pi D_y t}} \frac{\exp\left[\dfrac{-(z)^2}{4D_z t}\right]}{\sqrt{4\pi D_z t}} \tag{3}$$

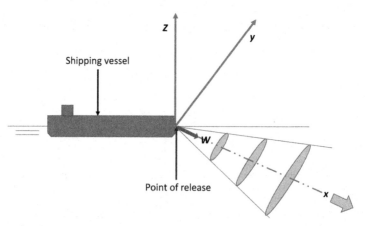

FIG. 3 The release and subsequent dispersion of oil from a vessel. *(Courtesy of Pixabay.)*

where Q is the quantity of the oil released per area, W is the wind speed, t is the time, and D is the dispersion coefficient. Once the oil is dispersed, it begins to partition into different media and considering the scenario we are looking at, it would involve four media: air, ice, water, and sediment. This is shown in Fig. 4.

The partitioning of the oil in different media can be modeled using different methods. One of the more modern and the most efficient approaches is the use of Computational Fluid Dynamics (CFD). The CFD involves the use of equations in specific software to model fluid flow. This however requires thorough verification and precise equations. Some of the most popular software for CFD include ANSYS fluent, Autodesk CFD, and SolidWorks Flow Simulation. Another method is the fugacity approach. The approach is based on the fugacity concept. Fugacity is basically described as the escaping tendency of a chemical (Afenyo et al., 2016a). It is analogous to partial pressure. The relationship describing the concentration of the released oil and the fugacity is described in Eq. (4) (Mackay, 2001). The equation to describe fugacity can be found in Eq. (4).

$$C = Z \times f \tag{4}$$

where C is the concentration (mol/L), f is the fugacity (Pa), and Z is the fugacity capacity (mol/L Pa). Here, it is important to note that a medium with a higher

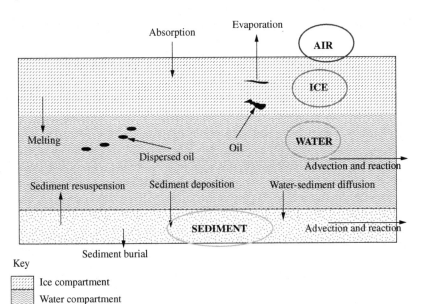

FIG. 4 Different processes and media involved when oil spills in ice-covered waters. *(From Afenyo, M., Khan, F., Veitch, B., Yang, M., 2016a. Dynamic fugacity model for accidental oil release during Arctic shipping. Mar. Pollut. Bull. 111(1–2), 347–353.)*

fugacity capacity has a high tendency to absorb more chemicals and vice versa. The fugacity-based models are divided into four distinct types namely levels 1, 2, 3, and 4 (Afenyo et al., 2016a). The focus of this chapter is the Tier IV fugacity model. This approach gives the modeler the opportunity to calculate the concentration of the pollutant with time. The result of the partition modeling is the concentration of the oil in different media. It is termed as the Predicted Exposure Concentration (PEC). This is subsequently used to estimate the level of risk in the various media described earlier.

Exposure modeling involves the assessment of oil concentration and its existence in the media of contact. In marine species, pollutant existence could be due to inhalation, ingestion of contaminated water and food, and absorption of hydrocarbon. In dynamic risk assessment, the study of temporal variability of the exposure and fate of the spilled oil is important. Season and temperature variations are temporal variables in such environments. Although the concentration in species is not the subject of this chapter, the information from such a study is used to determine the Predicted No Effect Concentration (PNEC). An oil spill in the ocean can be lethal to birds, fishes, mammals, and other marine organisms due to toxic components present in oil. An oil toxicity data and a robust toxicity model can support ecological risk assessments of spilled oil and environmental impact assessment of spills. The toxicity of oil to marine species depends on species presence and the extent of exposure of the toxic oil component. Species presence and toxicity is a function of time, space, and concentration.

The problem is that there is very limited data of oil toxicity of aquatic species in cold regions. To evaluate the risk to marine species, exposure concentration may not be sufficient. There is a need to evaluate the concentration of oil in the body of species exposed to oil. This concentration is responsible for deaths of species under study.

Dynamic risk helps to assess and manage risk in evolving conditions. It is quantitative in nature and enables the capturing of uncertainty. Also, it assists in modeling scenarios with limited data. Once the level of concentration of oil in the different media is determined, it is used to estimate the level of risk in each medium. This is achieved by comparing the estimated concentration to a standard concentration for that oil type in the medium. However, since oil is made up of different types of hydrocarbons, a surrogate may be used for the purpose of estimating the risk profile in different media. Naphthalene is a good example for a surrogate. The risk in this case is described through the risk quotient and is calculated using Eq. (5).

$$RQ = PEC / PNEC \qquad (5)$$

where Predicted Exposure Concentration (PEC) is measured through fugacity modeling and the Predicted No Effect Concentration (PNEC) is obtained from ecotoxicological studies. PNEC represents the ecosystem response. A value of RQ > 1 shows condition requiring attention. The dynamic risk concept could be extended to socio-economic risk modeling (Afenyo and Chaming, 2019). This is discussed further in the following section.

4   Addressing uncertainty and variability

While the approach described so far is one of the most adopted, there is a growing literature on the use of Bayesian Network (BN) to address the same problem. This approach helps to address uncertainty and variability. In the sections that follow, we will discuss how the BN could be used in this regard. Further, an advanced form of the BN described as the OOBN is presented as well.

BN is a graphical probabilistic-based model that is used to describe different engineering problems (Afenyo et al., 2017b) including oil spills. In the context of the oil spill, the BN is used to describe the release, fate and transport, and the socio-economic impact of an oil spill in the Arctic from shipping in this chapter. The latter is achieved by the use of Influence Diagram (ID). The ID is an extension of the BN with utility and decision nodes (Davies and Hope, 2015). Some of the many uses of the ID include the evaluation of response measures for oil spill and the evaluation of the socio-economic impact of same (Davies and Hope, 2015). To illustrate the use of the equations and tools described earlier, a scenario is presented for such purpose. The set-up is potentially an oil spill scenario that we anticipate. The sections that follow seek to illustrate the use of the tools described.

5   The scenario

The Arctic remains a place where an oil spill incident has not been recorded and so we can only rely on potential scenarios to demonstrate the tools described earlier. The intention of the scenario is not depicting the accuracy of the result but to present potentially what the outcome could be when using such tools. The scenario is taken from (Afenyo et al., 2017a) with some modifications.

The scenario involves the collision of an oil tanker in the West Siberian lowlands. The collision resulted in the release of approximately 115 kg of oil. The approximate area of the region affected by the oil spill is 1000 m^2 while the entire body of water is $300 \times 10,000$ m^2. The depth is however approximately 200 m. The ice covers the surface of the water during most part of the year. In summer, the average temperature is 0°C to 9°C and in winter it is −1.8°C to −1.2°C. The average wind speed of this area is $7\frac{m}{s}$ and the longitudinal diffusion coefficient is $5,400,000\frac{m^2}{s}$. Further, assuming that the area affected has a population of about 1000 people and depends on fish which is worth USD 400 per ton the tools are used to estimate different parameters for decision-making.

6   The analysis

Eq. (3) is applied to model the release and dispersion of the oil and the results are shown in Fig. 5. It shows the evolution of the oil in space and time. From

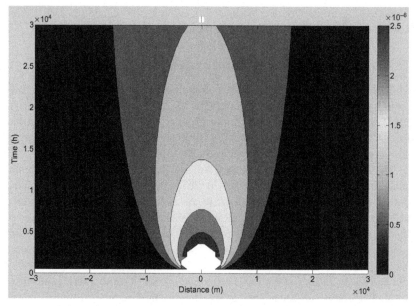

FIG. 5 Concentration profile of spilled oil in time and space. *(From Afenyo, M., Khan, F., Veitch, B., Yang, M., 2017a. A probabilistic ecological risk model for Arctic marine oil spills. J. Environ. Chem. Eng. 5(2), 1494–1503.)*

the diagram it can be inferred that the concentration of the oil is highest at the source but reduces as it moves away from it.

To obtain the concentrations in the different media under consideration, the concentration of the dispersed oil is used in the fugacity equation to estimate the concentration in air, water, ice, and sediment. The approach adopted is that the mass balance equations are derived for each of the media considering the various processes involved and then solved simultaneously using the 4th-order Runge-Kutta method. The outcome is the fugacity for each of the media and this is multiplied by the fugacity capacity of the corresponding media. The fugacity capacities (Z) are media dependent. Fig. 6A and B show samples of the concentration profile for the oil in air and in sediments. This outcome gives the level of pollution of the media.

From the graphs, it can be inferred that the concentration in the sediment is high compared to that of air. Also, the initial concentration is generally high but reduces with time. This result is subsequently compared to the PNEC for each media and the result is shown as the risk profile through the Risk Quotient (RQ). The result is obtained in a probabilistic mode by implementing a Monte-Carlo Simulation technique, details of which can be obtained from Afenyo et al. (2017a) (Fig. 7).

The risk profile shows that the quotient does not exceed 1 even at the highest probability and so the level of risk is acceptable.

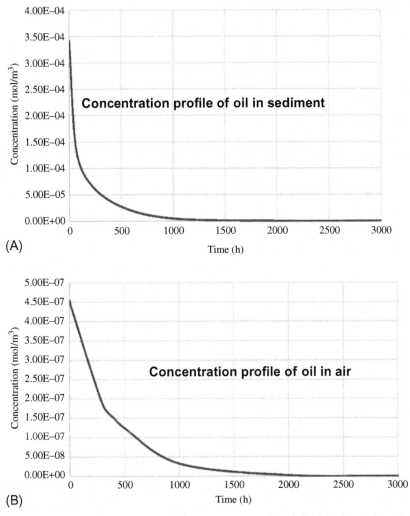

FIG. 6 (A) Concentration of oil in sediment and (B) concentration of oil in air. *(From Afenyo, M., Khan, F., Veitch, B., Yang, M., 2017a. A probabilistic ecological risk model for Arctic marine oil spills. J. Environ. Chem. Eng. 5(2), 1494–1503.)*

To evaluate the potential response measures that could be used to address the oil spill, the Object-Oriented Bayesian Network (OOBN) is used and extended to an Influence Diagram (ID). Fig. 8 is the OOBN that captures the various stages of the oil spill and Fig. 9 is the ID. A criterion called the cost-effectiveness is used for this purpose. This is the ratio of the cost to the effectiveness of each response method.

The results of the simulation are shown in Fig. 10. Here A is the in situ burning, B is the use of dispersants, C is the mechanical recovery, and D is the manual recovery.

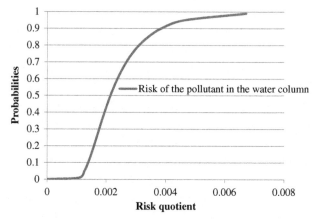

FIG. 7 Risk profile of oil in water column. *(From Afenyo, M., Khan, F., Veitch, B., Yang, M., 2017a. A probabilistic ecological risk model for Arctic marine oil spills. J. Environ. Chem. Eng. 5(2), 1494–1503.)*

Fig. 10 shows that the best combination for this scenario is the use of dispersant and in situ burning. It should be noted that no two scenarios are the same and so, the ranking may differ for other conditions.

Furthermore, to evaluate the socio-economic impact in dollar terms the ID shown in Fig. 11 is implemented.

The simulation produces the impact in dollar (USD) terms. Using the illustrated model, it is possible to simulate different scenarios. Using the information given, the socio-economic impact simulated is approximately USD101, 242,480.

7 Discussion and conclusions

The results of the simulations illustrate the flexibility that the various tools offer. That is to say that different tools can be used to evaluate the same problem, depending on what the end user really wants. There are still challenges to assessing the risk of oil spills in ice-covered waters. The first is the absence of algorithms to describe some of the processes that are very dominant when it comes to oil spill in ice. One of such processes is oil encapsulation and decapsulation in ice-covered waters.

Furthermore, there is a level of restriction on conducting outdoor experiment because of the environmental implications of oil in ice. The disposal of the oil is difficult to deal with as environmental fines for violating such laws can be consequential. However, this is changing considering the recent Arctic Oil Spill Response JIP, which produced substantial knowledge on oil-ice interaction and response of oil spill in the Arctic. In this case, the BN models can be improved tremendously with an extensive data collection regime. This will further create confidence in the method and subsequently the results. In countries,

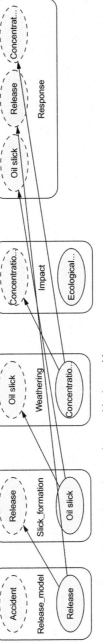

FIG. 8 The OOBN for oil release, impact due to a shipping accident.

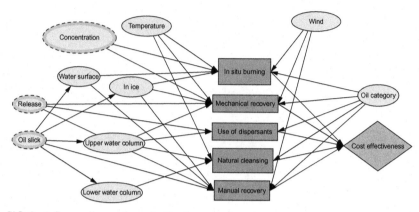

FIG. 9 Influence diagram for evaluating the response measures.

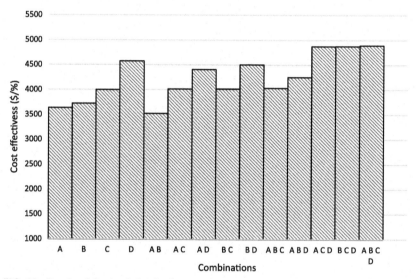

FIG. 10 Results of the simulation for the scenario.

such as Canada, where the bulk of the Arctic falls in the jurisdiction of the first nations, the involvement of these groups in developing and creating inputs for the models would go a long way to make the models acceptable. Moreover, it is noted that risk as a concept is evolving and so the methods that are required to evaluate these also need to be adapted appropriately. So, while climate change may be positive for Arctic shipping, it is also important to adequately prepare for the implications of the activities that come with it. The tools presented in this chapter are important for contingency planning, resource allocation, and emergency response.

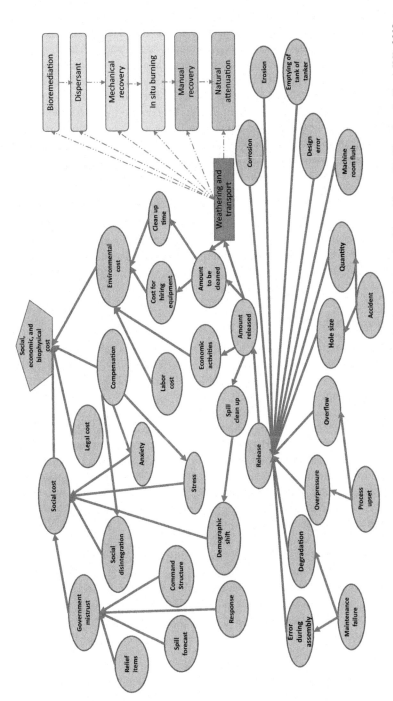

FIG. 11 An influence diagram for evaluating the socio-economic impact of oil spill given the scenario. (*From Afenyo, M., Chaming, J., Ng, A.K.Y., 2019. Climate change and Arctic shipping: a method for assessing the impacts of oil spills in the Arctic. Transp. Res. D (in press.)*)

Finally, the tools introduced in this chapter can inform regulatory framework for environmental risk assessment for oil spill in the Arctic. For example, Canada is in the process of developing a national Arctic policy. The illustrated tools can be used to determine the most vulnerable areas that require more attention. This information will enable the federal government, as well as the provincial governments, to enact laws appropriately. The indigenous communities have advocated for more involvement in issues related to the Arctic. The tools presented give the flexibility to achieve this. Thus, while collecting data from these regions, the inputs would form part of the modeling work. Also, the output would be useful to the communities as they can determine how much impact an oil spill of a particular magnitude will have on the communities.

However, the question on how much political will exists to address Arctic oil spill issues remains. In this regard, Arctic countries like Norway have already taken the lead, while Russia has also demonstrated its readiness to harness the full potential of the Arctic. Hence, it is critical for the Arctic Council to take into consideration a standard for conducting an impact risk assessment in the Arctic with regard to oil spills. Much progress has taken place especially with regard to the Polar Code. More collaborations by Arctic and non-Arctic countries are needed.

8   Future work

Despite the progress with regard to risk assessment tools for the Arctic, a lot still needs to be done. For example, there is a need to develop an innovative way to collect data for the Conditional Probability Tables (CPT) for the BN and ID diagrams. A comprehensive impact assessment needs to be performed for the Arctic shipping routes to determine the most vulnerable areas. This can be done by incorporating the models presented with real-time oceanographic database. In this way, the environmental, social, and economic impacts of an oil spill can be determined in real time as an oil spill incident happens. Also, a simulation can be made to determine where the oil would go and the overall impacts on the routes throughout the Arctic region.

Finally, other areas that need to be addressed include the evaluation of oil spills on the cruise industry, considering the significant increase of such vessels. What are the regulations in place to enable the local communities to fully benefit from this surge? Other questions that urgently need answering are: How to make the Arctic economy sustainable, considering the small number of people in this area despite the huge deposit of natural resources? How much investment should go into such areas and how would government be able to attract people to these areas to justify the investment? We hope that this chapter offers some valuable contributions to answer these questions.

References

Afenyo, M., Chaming, J., Ng, A.K.Y., 2019. Climate change and Arctic shipping: a method for assessing the impacts of oil spills in the Arctic. Transp. Res. D (in press).

Afenyo, M., Khan, F., Veitch, B., Yang, M., 2016a. Dynamic fugacity model for accidental oil release during Arctic shipping. Mar. Pollut. Bull. 111 (1–2), 347–353.

Afenyo, M., Khan, F., Veitch, B., Yang, M., 2016b. Modeling oil weathering and transport in sea ice. Mar. Pollut. Bull. 107 (1), 206–215.

Afenyo, M., Khan, F., Veitch, B., Yang, M., 2017a. A probabilistic ecological risk model for Arctic marine oil spills. J. Environ. Chem. Eng. 5 (2), 1494–1503.

Afenyo, M., Khan, F., Veitch, B., Yang, M., 2017b. Arctic shipping accident scenario analysis using Bayesian Network approach. Ocean Eng. 133, 224–230.

Camus, L., Smith, M.G.D., 2019. Environmental effects of Arctic oil spills and spill response technologies, introduction to a 5 year joint industry effort. Mar. Environ. Res. 144, 250–254.

Davies, A.J., Hope, M.J., 2015. Bayesian inference-based environmental decision support systems for oil spill response strategy selection. Mar. Pollut. Bull. 96 (1–2), 87–102.

Faksness, L., Brandvik, P., Daae, R., Leirvik, F., Børseth, J., 2011. Large-scale oil-in-ice experiment in the Barents Sea: monitoring of oil in water and MetOcean interactions. Mar. Pollut. Bull. 62 (5), 976–984.

Lee, K., Boufadel, M., Chen, B., Foght, J., Hodson, P., Swanson, S., Venosa, A., 2015. The Behaviour and Environmental Impacts of Crude Oil Released Into Aqueous Environments. The Royal Society of Canada Expert Panel, Ottawa.

Mackay, D., 2001. Multimedia Environmental Models: The Fugacity Approach, second ed. CRC Press LLC, New York.

Ng, A.K.Y., Zhang, H., Afenyo, M., Becker, A., Cahoon, S., Chen, S.L., Esteben, M., Ferrari, C., Lau, Y.Y., Lee, P.T.W., Monios, J., Tei, A., Yang, Z., Acciaro, M., 2018. Port decision maker perceptions on the effectiveness of climate adaptation actions. Coast. Manag. 46 (3), 148–175.

Chapter 11

Future Arctic shipping, black carbon emissions, and climate change

Steven Messner

e360 LLC, Sonoma, CA, United States

1 Black carbon

Black carbon is in a group of air emissions that are short-lived climate pollutants (SLCPs). SLCPs are emissions that trap heat in the atmosphere and remain in the atmosphere for a much shorter period of time than longer lived climate pollutants, such as carbon dioxide (CO_2). SLCPs include methane, fluorinated gases including hydrofluorocarbons (HFCs), and black carbon. The potency of SLCPs in terms Global Warming Potential (GWP) can be tens, hundreds, or even thousands of times greater than that of CO_2. BC has a particularly high GWP of 900 over a 100-year time frame, and 3200 over a 20-year time frame (IPCC, 2013).

BC emissions in the Arctic are a major contributor to ice melt as well as to local public health impacts. BC absorbs heat, typically falls out of sky in a matter of days or weeks, but then can settle on snow and ice and heat up the ice surface. As a result of this localized deposition and heat absorption from BC particles, Sand et al. (2013) estimated that 1 kg of BC emitted in the Arctic causes five times higher Arctic surface temperature change than 1 kg emitted at midlatitudes. The GWP impacts of BC emissions in the Arctic are then approximately five times more potent than the GWPs noted previously—equivalent to 16,000 over a short-term (20-year) time frame (Fig. 1).

2 BC emissions and maritime shipping

BC emissions are caused by the incomplete combustion of fossil fuels and other organic matter. Because heavy fuel oil (HFO) is still in common use in global shipping and since HFO combustion creates a relatively high amount of BC emissions compared to other fuels, increased shipping through the Arctic as ice

Maritime Transport and Regional Sustainability. https://doi.org/10.1016/B978-0-12-819134-7.00012-5
195

FIG. 1 BC and particulate deposits in the Arctic. *(Photo ©Henrik Egede Lassen/Alpha Film, from the Snow, Water, Ice, and Permafrost in the Arctic report from the UN Arctic Monitoring and Assessment Programme.)*

FIG. 2 Arctic shipping with visible plume of black carbon and particulates. *(Courtesy of David Mark, Pixabay.)*

melts earlier will create impacts of concern—i.e., accelerated ice melting and increased atmospheric haze (Fig. 2).

Comer et al. (ICCT 2017) reported that HFO was the most commonly used marine fuel in the IMO Arctic in 2015 as well as in the "Geographic" Arctic. There is no standardized definition of the "Geographic" Arctic boundaries but the ICCT 2017 study set it as everything above 58.95°N for comparison to prior shipping specific emission inventories. The IMO Arctic is generally set as waters above 60°N, but with a large area of the North Atlantic between Norway and Greenland removed. Although the IMO Arctic boundary is useful for issues like oil spill safety, for BC emissions, it is relevant to consider all high latitude BC emissions since they have the potential for atmospheric transport and deposition within the Arctic.

Complete emission inventories have not typically been conducted specific to high latitudes to date. For example, the 8 nation Arctic Council,[a] those countries with territory above the Arctic Circle at 66.5°N, have to date only developed inventories at the national level and not tailored to high-latitude areas—e.g., above 60°N. The Arctic Council's Arctic Monitoring and Assessment Programme (AMAP) assessment (2015) did some statistical evaluation of global emission inventories above 40°N and developed ratios of emissions that could be emitted above 60°N but concluded that there was a considerable variability in emission estimates and further work is needed to improve their accuracy. The AMAP assessment also recommended additional work on identifying the location of high-latitude sources and on improving the accuracy of the spatial distribution of emissions in the Arctic.

The ICCT 2017 study was sectoral specific to the shipping industry. The study reported that in 2015, 59% of the 4.37 million tons of fuel combusted by ships in the Geographic Arctic area was HFO, followed by distillate fuel at 38% and liquefied natural gas (LNG) at 3%. Ro-pax vessels (primarily used for cars, trucks, and other wheeled cargo) consumed the most HFO in the Geographic Arctic, followed by oil tankers, and cruise ships. Distillate fuel is lighter than the HFO and mainly used by the largest fleet operating in Arctic waters—fishing vessels which account for 13% of the BC emissions in the Geographic Arctic.

Overall BC emissions in 2015 from shipping in the Geographic Arctic were then calculated to be 1453 tons, 67% from HFO vessels, 33% from distillate fuel, and 2% from LNG. The overall results were compared to previous BC inventory work performed using Danish Maritime Authority vessel activity data in the Geographic Arctic (Winther, 2014). The Winther study estimated 1584 tons BC emitted from shipping in this area. There were a number of differences in the two studies' methodologies, notably fishing vessel emission factors and assumptions on ship engine load factors. Even with the methodology differences, the results were very comparable between the two studies.

The ICCT 2017 study also presented a forecast of shipping emissions using diversion factors estimating the amount of traffic diverted from the Panama and Suez canals through the Arctic. The diversion factors were assumed to be 1% in 2020, 3% in 2030, and 5% in 2050 which was the approach used in previous studies (Corbett et al., 2010; Winther et al., 2014). The forecast took into account the mandatory use of lower sulfur HFO in 2020, but BC emission factors were assumed to be the same.

---

[a] United States, Canada, Russian Federation, Norway, Sweden, Iceland, Finland, Denmark (Greenland).

The forecast showed a 48% increase in BC emissions by 2025 from shipping in the Geographic Arctic to 2144 tons. CO_2 emissions in 2025 from shipping were projected to be 20,063,600 tons. When multiplying the BC emissions by its GWP of 3200, the result is 6,860,800 tons of CO_2e emissions. And if these emissions are deposited in the Arctic, the CO_2e emissions become 34,304,000 tons—a greater impact to climate change than the CO_2 emissions from the fuel combustion. As will be discussed in the following section, future IMO targets to reduce carbon emissions should then take into account the BC emissions as well.

The ICCT 2017 study also noted that BC emissions from shipping are likely transported some distance before they are deposited on the surface but recommended additional research to better understand the transport of BC emissions within the Arctic.

Other high-latitude emission sources besides shipping that contribute to Arctic BC deposition. The shipping specific emission inventory can be compared to another sectoral inventory at high latitudes for context, namely oil and gas flaring. Stohl et al. (2013) used the global model ECLIPSE (Evaluating the CLimate and Air Quality ImPacts of ShortlivEd Pollutants) to estimate BC emissions at high latitudes focusing on oil and gas flaring. Using the ECLIPSE model, the Stohl study also developed estimates for other sectors including residential combustion and biomass burning in forests. Table 1 summarizes the emission estimates. The emission figures are in k tons.

The 2015 shipping BC inventory from the ICCT 2017 study was 1.45 k tons, roughly 1% of the modeled inventory data from the Stohl study (2013). The ICCT 2017 study forecasted an increase of 2.14 k tons, still under 2% of the overall total BC emissions above 60°N.

The Stohl study also modeled BC transport and deposition and estimated that 42% of BC landing in the Arctic is from flaring. This is due to the greater

TABLE 1 High-latitude sectoral emission estimates

Emission sector	Latitude	>60°N	>66°N
Residential		6.2	0.6
Flaring		52.2	26.4
Ag waste burning		0.2	0.0
Biomass burning		92.4	12.3
Other (industry)		8.0	1.0
Total		159.0	40.3

(Modified from Stohl, A., Klimont, Z., Eckhardt, S., Kupiainen, K., Shevchenko, V.P., Kopeikin, V.M., Novigatsky, A.N., 2013. Black carbon in the Arctic: the underestimated role of gas flaring and residential combustion emissions. Atmos. Chem. Phys. 13, 8833–8855.)

prevalence of flaring emissions at very high latitude and the fact the emissions occur year round, rather than primarily in the summer months (i.e., biomass burning) when transport mechanisms of BC are not strong.

The ICCT 2017 and Stohl studies would indicate that shipping is not a major contributor to BC in the Arctic but other estimates have been made that shipping could be as much as 5% of overall BC emissions in the Arctic (Arctic Council, 2017). In either case, because shipping lanes are typically located close to coastal villages, it still has localized health impacts that can be eliminated or reduced through fuel shifting and clean technologies. Also, increased shipping and commerce at high latitudes are already inducing industrial and residential growth which could have a much larger impact than increased shipping alone. For example, cargo volumes in the Russian Arctic ports were up almost 25% in 2018.[b]

3   Localized health impacts

A comprehensive study of the health impacts of BC and related particulate matter (PM) (includes both PM2.5 and PM10) was conducted by the World Health Organization (WHO) in 2012 (Janssen et al., 2012). The WHO study concluded that there was insufficient information to distinguish the health effects of BC from the known effects of PM, which include cardiopulmonary disease and lung cancer. However, the study noted that BC may "operate as a universal carrier of a wide variety of chemical constituents of varying toxicity, such as semi-volatile organics, to sensitive pulmonary and cardiovascular targets. Because of this role, EC may very well act as a good indicator for combustion-derived and potentially very harmful parts of PM."

Aliabadi et al. (2014) studied localized emissions during the 2013 shipping season (June 1–November 1) near two Arctic communities in Nunavut—Cape Dorset and Resolute. Cape Dorset is near the northern entrance to Hudson Bay and Resolute is further north and west near the Barrow Strait. The study found consistently degraded air quality during the shipping season. The estimated range for percent ship contribution to local particulate emissions measured as PM2.5 was 19.5%–31.7% for Cape Dorset and 6.5%–7.2% for Resolute. Additional measurements in Resolute indicated that percent ship contribution to local black carbon was 4.3%–9.8% and that black carbon constituted 1.3%–9.7% of total PM2.5. Percent ship contribution to local sulfur dioxide (SO_2) emissions from the use of high sulfur HFO was 16.9%–18.3% for Cape Dorset and 5.5%–10.0% for Resolute. SO_2 particles can directly affect respiratory health but can also react with other emission particles to form PM.

---

[b] Boom times for Russia's Arctic ports http://www.rcinet.ca/eye-on-the-arctic/2019/01/25/russia-ports-arctic-shipping-murmansk-sabetta-yamal-arkhangelsk-lng-gas-oil-coal-varandey/ (accessed May 7, 2019).

FIG. 3 Brooks Range in Alaska showing BC and particulate haze layer. *(Reproduced with permission from Lack, D.A., 2016. The Impacts of an Arctic Shipping HFO Ban on Emissions of Black Carbon. A report to the European Climate Foundation.)*

The Aliabadi study (2014) indicates that shipping emissions can indeed have profound localized impacts on health in the Arctic even if the overall contribution to regional warming is relatively low (Fig. 3).

4 International Maritime Organization (IMO) regulatory activities

The IMO is an agency of the United Nations established in 1948 which is responsible for measures to improve the safety and security of international shipping and to prevent pollution from ships. It is also involved in legal matters, including liability and compensation issues and the facilitation of international maritime traffic. It currently has 174 Member States.

Some IMO nations are calling for a ban on the use of HFO in the Arctic. A heavy fuel oil ban in the Arctic is supported by Finland, Sweden, Norway, Denmark, Iceland, and the United States, which comprise six of the eight member nations of the Arctic Council, which are nations with geographic territory in the Arctic. Although the Arctic Council is not a UN agency, their full consensus is typically necessary to move forward Arctic specific shipping issues. NGOs working on the HFO issue at IMO meetings are advocating for a ban in the Arctic to be adopted in 2021 and phased in by 2023.

The use of HFO by vessels in Arctic waters has been debated at the IMO for nearly a decade, and its use in the Antarctic area has been banned since August 2011. Russia and Canada, the two remaining Arctic Council nations whose ships also use the most fuel oil, are not currently supporting the Arctic

HFO ban. Canada is concerned about economic effects to its indigenous Arctic populations. Russia is only recommending additional safety precautions to reduce the risk of HFO spills in the Arctic.

In October 2018, the IMO enacted a new requirement effective January 1, 2020 that lowers the sulfur content in international shipping fuels to a maximum of 0.5%, 7 times lower than the current limit of 3.5%. This will have positive effects on local health in many parts of the world but the impacts to climate change will vary. According to the AMAP assessment of SLCFs (AMAP Assessment, 2015), reduced sulfur and aerosol clouds in the Arctic can have a net cooling effect in the winter (dark most hours) and neutral to slight heating effect in the summer due to less radiation being absorbed in the aerosol clouds.

Fuglestvedt et al. (2014) investigated the overall competing effects of projected international shipping emissions on global climate change and concluded that a shipping shift to shorter Arctic transit will incur a net overall climate penalty over the first one and a half centuries as emissions of SLCPs like BC increase in the Arctic in the near term. The initial net warming for the first one and a half centuries from the SLCPs gradually declines and transitions to net cooling as the effects of CO_2 reductions from using less overall fuel become dominant.

To comply with the IMO sulfur rule in 2020, ICCT 2017 estimated that 12% of ship fuel consumption would still be HFO with a greater than 0.5% sulfur content and that 88% of fuel consumption will be residual fuel that is less than 0.5% sulfur. The underlying assumption here is that 12% of ships will continue to operate on high-sulfur HFO but install scrubbers to comply with the IMO sulfur content regulations.

In April 2018, The IMO adopted a long-term goal of achieving 50% reductions in overall GHG emissions by 2050 compared to 2008 emissions. The IMO goals also call for reducing CO_2 emissions per transport work (tons∗miles) by at least 40% by 2030 aiming for a 70% reduction by 2050. Their initial strategy appears to focus on CO_2 emissions, but as was noted earlier in this chapter, BC emissions from shipping can have a greater warming impact in the Arctic than the CO_2 emissions. Reductions in BC emissions could then contribute to achievement of the IMO GHG target, especially reductions of BC in the Arctic.

5   Arctic Council goals

The eight nation Arctic Council set targets in May 2017 to reduce black carbon emissions collectively between 25% and 33% below 2013 levels by 2025. Member country representatives attending the 10th Arctic Council Ministerial Meeting in Fairbanks, Alaska signed the Fairbanks Declaration. The Declaration noted that the Arctic is warming at more than twice the rate of the global average and that the pace and scale of continuing Arctic warming will depend on future emissions of greenhouse gases and short-lived climate pollutants.

The Council adopted the recommendations of their Expert Group on Black Carbon and Methane and included measures aimed at diesel-powered mobile sources (BC), oil and gas production (methane and BC), residential biomass combustion equipment (BC), and solid waste (methane). The Expert Group's report (Arctic Council, 2017) also recommended incentivizing the uptake of emission abatement technologies, electrification of ports, fuel efficiency improvements, or use of alternative fuels and engaging in ongoing work within the IMO's Sub-Committee on Pollution Prevention and Response to identify appropriate methods for measuring black carbon emissions from international shipping and to consider control measures.

6 Future Arctic shipping emission control measures and cleaner technologies

Before developing a comprehensive program to reduce BC emissions from shipping, it is important to understand BC emission factors from the current fuels primarily used for international shipping—HFO, distillate, and blends of these fuels to achieve the IMO limit of 0.5% sulfur content in fuels. The relationship between overall PM emissions and fuel type is reasonably well understood, with distillate combustion from ships emitting approximately 5× less PM than HFO (Sax and Alexis, 2007).

The relationship between BC and PM emissions can vary with fuel and engine loads and engine type. HFO is not a uniform fuel—the sulfur and ash content can vary considerably which in turn will affect the BC emissions. Lack (2016) notes that the HFO fuel variability affects ship engine efficiency which in turn leads to variable emissions of BC. Lack and Corbett (2012) conducted 19 experiments where the same ship engines were used to produce emissions from HFO and higher quality fuels such as distillate. The study concluded that improvements in fuel quality reduce BC emissions by an average of 50% (range of 30% to 80%).

A recent study led by UC Riverside which was conducted for the ICCT in 2016 (Johnson et al., 2016) measured BC emissions from a newer (post 2010) Tier 2 ship engine designed to meet lower NOx emission requirements set by the IMO. The study concluded that BC emission factors (EFs) were extremely low from the large Tier II slow speed diesel (SSD) engine, suggesting that electronic controls and in-cylinder approaches to reduce NOx may also reduce BC and PM emissions. BC EFs in prior inventory studies have ranged between 0.1 and 1.0 g/kg of fuel. The Tier II engine tests showed much lower emissions at all loads.

Johnson et al. (2016) recommended that more research be conducted to validate the low BC EFs from Tier II engines. Table 2 also shows that BC emissions tend to decrease as engine load increases, consistent with the understanding that improved combustion efficiency occurs at higher loads.

The UC Riverside 2016 study also tested the effectiveness of a scrubber designed to reduce sulfur emissions to meet new IMO sulfur emission standards on BC emissions. The study found that the scrubber reduced BC emissions by approximately 30%.

TABLE 2 Tier II SSD engine BC emission factors

Load (%)	57%	41%	28%	Vessel speed reduction (VSR) (9%)
Emission factor (g/kg fuel)	0.002	0.009	0.051	0.019

(Data from Johnson, K., Miller, W., Durbin, T., Jiang, Y., Yang, J., Karavalakis, G., Cocker, D., 2016. Black Carbon Measurement Methods and Emission Factors From Ships. University of California, Riverside.)

Other key findings from the study were

- Slow-steaming/vessel speed reduction (VSR): On a per unit distance basis, VSR reduced BC emissions compared to higher speed operations.
- The use of scrubbers to meet global or regional fuel sulfur limits may have BC benefits. Scrubbers appear to allow compliance with regional gaseous phase sulfur limits but they do not appear to control sulfur particulates. This finding has implications for local public health and points to the need for better data on scrubber performance.
- Overall, distillate fuels had the lowest BC emissions, followed by conventional HFO. The low sulfur residual fuel tested, however, had the highest BC EF of the fuels tested. This raises concerns about the potential impact of IMO's 2020 global sulfur limit of 0.5% for marine fuels in 2020 on BC emissions if met primarily through the use of blended fuels (mixtures of lower sulfur residual crude oils and distillates).

As was noted earlier in this chapter, the ICCT 2017 study determined that in 2015, 59% of fuel used by ships in the Geographic Arctic area was HFO, followed by distillate fuel at 38% and LNG at 3%. Blended fuels are not currently in common use in the Arctic but are expected to be by 2020, estimated at 88% by ICCT 2017, as an option to meet the lower IMO sulfur requirements. So the UC Riverside 2016 finding on blended fuels and high BC emissions is significant and needs to be investigated further. LNG use in ship engines is currently low, but is expected to grow as more ship operators try to comply with IMO emission requirements as well having favorable costs compared to other fuels. BC EFs from LNG ship engines are typically 20–40× lower than BC EFs from distillates and HFO (Comer et al., 2017).

Which fuels and clean technologies are eventually used in Arctic shipping will be driven by economics, regulation, and incentives. Existing and future IMO regulations, including the potential of a HFO ban, have been discussed in this chapter. The Arctic Council's goal of 25%–33% reduction of BC emissions by 2025 will also be a driver of actions in the near term.

Regarding the economics of shipping fuels and engine technologies, Comer (2019) considered five Arctic shipping routes with ships that can use HFO or an alternate fuel. For each case, Comer (2019) compared the costs of using HFO, 0.5% S-compliant fuel, distillate fuel, LNG, electricity from batteries, or liquid hydrogen (H_2) in fuel cells by the year 2023. The five routes considered were (1) a tanker carrying LNG from Norway to South Korea, (2) a cargo ship carrying equipment from Shanghai to the Netherlands, (3) a small container ship servicing western Greenland, (4) a bulk carrier transporting nickel ore from Canada to the Netherlands, and (5) a 20-night northern Europe and Arctic cruise originating from Amsterdam. Five specific ships currently performing these duties were analyzed.

One of the Comer (2019) study's conclusions is that in all cases, the ships can stop using HFO and avoid the use of 0.5% S-compliant residual fuels by 2020. All five ships considered could immediately use distillate fuels and the tanker in Case (1) could immediately use LNG. The study estimated that the cost of distillate fuels to be slightly more expensive than 0.5% S-compliant fuels and 26% more expensive than HFO in 2023. This cost differential can rapidly change—a check of current HFO and Marine Gas Oil (MGO) (mainly distillates) prices shows that MGO is 56% more costly than HFO.[c] The study also concluded that distillate spills are 30% less expensive than 0.5% S-compliant fuel spills and 70% less expensive than HFO spills, which needs to be taken into account over the long term as spills will inevitably occur.

Comer (2019) predicted that LNG will be less expensive to use than any other fossil fuel in the Arctic in 2023, as it is currently in 2019. If this continues to be the case, this will be a strong incentive for many ships to make the costly engine conversions needed to run on LNG. Using LNG also contributes to local health improvements since particulate, SO_2, and BC emissions from LNG are so much lower than from HFO and distillate. The impacts of a potential oil spill would be minor, since LNG evaporates very quickly, even at low Arctic surface temperatures. A shift to LNG in Arctic shipping would also contribute to the Arctic Council's 2025 BC reduction goal.

In the longer term, the use of LNG may not be helpful to deal with climate change and Arctic warming. First, LNG is composed primarily of methane which, like BC, is another SLCP. Methane is 40–50× less potent than BC from a GWP perspective, but has a tendency to leak when stored or transported. The climate-related tradeoffs of using LNG versus other alternatives need to be evaluated in more detail before being used on a widespread basis. In addition, recent research on the US supply chain for methane indicates a leak rate during production and transportation of 2.3% which is 60% higher than previously estimated by the US EPA (Alvarez et al., 2018). This is significant because it would make LNG a higher contributor to climate change emissions on a unit (per-mile) basis than diesel and gasoline in the US road transportation sector. The overall climate impacts of methane and LNG should be carefully evaluated for the shipping

---

[c] https://shipandbunker.com/prices (accessed May 14, 2019).

sector as the IMO will need better information before developing more refined strategies to meet their 2050 carbon reduction requirements.

The Comer (2019) study also considers two emerging technologies—all electric ships driven by battery power and ships with engines than run on hydrogen. Regarding battery power, the study notes that there is limited space for batteries on board to power long trips as their energy densities are quite low compared to liquid fuels. Still, they could be used for vessels on short routes in the Arctic such as the small container ship servicing western Greenland evaluated in Comer (2019)—route 3 noted above.

The Comer (2019) study concluded that hydrogen appears to be the most promising solution for zero-carbon, long-range Arctic shipping. However, the projected fuel costs of hydrogen are much higher than fossil fuels—estimated to be 259% more than HFO in 2023. As hydrogen becomes more prevalent as a fuel in the decades ahead, this cost difference will likely be reduced. Also, sufficient hydrogen to power a long trip from China to Europe—route 2 noted above—would need an estimated two refueling stops versus none for the fossil fuels. A hydrogen refueling infrastructure would need to emerge to support this kind of trip.

These zero-emission emerging technologies (electric battery, hydrogen) would also have the cobenefits of no damages from fuel spills as well as local reductions in BC emissions.

Still, the IMO's ambitious carbon reduction target of 50% by 2050 will need all potential fuels and solutions to be supported as much as possible. Some of the reductions can and will certainly come from conventional improvements—ship engine efficiency improvements and reducing speeds. Some will also need to come from emerging technologies and cleaner fuels.

Direct subsidies from governments and public entities will help move forward the technologies that will be needed to achieve the 2050 IMO carbon target, as well as the 2025 Arctic Council BC targets. However, subsidies typically cannot sustain emerging technologies over a long period of time when the market prices of fuel (e.g., HFO versus distillate versus hydrogen) are significantly different. Carbon trading has been suggested by some experts as a way to support low carbon technologies in the shipping sector over the long term.

The EU has recommended in the past that shipping be included in their carbon trading program—the EU Emission Trading Scheme. However, the IMO raised implementation concerns about such a program only applying in one part of the world and not in others.[d] If the IMO administered a carbon trading program on an international level, it could support their goal of 50% reduction by 2050 and give a long-term signal to the markets that a fuel or technology shift will be supported over a long period of time.

For example, a ship that could convert to hydrogen, electricity, or high-efficiency engine systems could sell their carbon emissions allowances for

---

[d] http://www.imo.org/en/MediaCentre/PressBriefings/Pages/3-SG-emissions.aspx.

"over-compliance" to a ship owner that may not be able to afford modifications. Given the technological challenges ahead for shipping to achieve the 2050 target, an IMO carbon trading program would seem likely and could definitely help reduce absolute emissions in addition to ongoing technological improvement efforts.

7 Conclusions

BC emissions are a growing problem in the Arctic leading to increased warming, ice melting, as well as measured health impacts to local communities.

Recent studies such as ICCT 2017 and Winther (2014) indicate that shipping is currently not a major contributor to BC in the Arctic (1%–2%) but other estimates have been made that shipping could be as much as 5% of overall BC emissions in the Arctic (Arctic Council, 2017). This amount of difference points to the need for more thorough inventories being performed specific to high-latitude emission sources.

Because shipping lanes are typically located close to coastal villages and because HFO is still the primary shipping fuel used in the Arctic, it contributes to localized health impacts that can be eliminated or reduced through fuel shifting and clean technologies. Also, increased shipping and commerce at high latitudes are already inducing industrial and residential growth which would have a much larger impact than increased shipping alone. As was noted earlier in this chapter, cargo volumes in the Russian Arctic ports were up almost 25% in 2018.

The use of HFO has negative health impacts, higher emissions, as well as higher oil spill risks and risks of ecological damage. A ban on HFO use in the Arctic would reduce environmental impacts, but the next lowest cost shipping fuels—residual crude/distillate blends—may have similar impacts on air pollution and BC emissions. While additional research on the impacts of blended shipping fuels is warranted, an immediate shift to only distillates would reduce air emissions and potential oil spill impacts simultaneously in the near term.

Due to favorable economics, LNG is likely to be more commonly used in Arctic shipping in the near term. Using LNG also contributes to local health improvements since particulate, SO_2, and BC emissions from LNG are so much lower than from HFO and distillate. A shift to LNG in Arctic shipping would also contribute to the Arctic Council's 2025 BC reduction goal. In the longer term, the use of LNG may not be helpful to deal with climate change and Arctic warming due to its potential for leaking and its relatively high GWP. The climate-related tradeoffs of using LNG for shipping fuel versus other alternatives need to be evaluated in more detail before it is used on a widespread basis.

BC emissions from Arctic shipping appear to have a greater warming impact in the Arctic than the CO_2 emissions from shipping. The IMO 50% reduction target by 2050 appears to focus on CO_2 emission reductions at this time. It should be expanded to include other emissions, especially SLCPs that contribute to an increase in Arctic warming like BC. Reductions in BC emissions could then contribute to achievement of the IMO 2050 target as well as supporting the 2025 Arctic Council goal of 25%–33% reductions in BC.

In the longer term, economic incentives to shift entirely away from less expensive crude oil-derived fuels are needed. Achieving a 50% carbon reduction target by 2050 will likely not be achieved without ambitious government support programs as well as a trading market that could help sustain technological innovations and emission reductions over a long period of time.

References

Aliabadi, A.A., Staebler, R.M., Sharma, S., 2014. Air quality monitoring in communities of the Canadian Arctic during the high shipping season with a focus on local and marine pollution. Atmos. Chem. Phys. Discuss. https://doi.org/10.5194/acpd-14-29547-2014.

Alvarez, R.A., Zavala-Araiza, D., et al., 2018. Assessment of methane emissions from the U.S. oil and gas supply chain. Science https://doi.org/10.1126/science.aar7204.

AMAP Assessment, 2015. Black Carbon and Ozone as Arctic Climate Forcers. Arctic Monitoring and Assessment Programme (AMAP), Oslo, Norway. ISBN: 978-82-7971-092-9.

Arctic Council, 2017. Expert Group on Black Carbon and Methane: Summary of Progress and Recommendations 2017. Arctic Monitoring and Assessment Programme(AMAP), Oslo, Norway. ISBN: 978-82-93600-24-4.

Comer, B., 2019. Transitioning Away From Heavy Fuel Oil in Arctic Shipping. Working Paper 2019-03, International Council on Clean Transportation, Washington, DC.

Comer, B., Olmer, N., Mao, X., Roy, B., Rutherford, D., 2017. Prevalence of Heavy Fuel Oil and Black Carbon in Arctic Shipping, 2015 to 2025. International Council on Clean Transportation.

Corbett, J.J., Lack, D.A., Winebrake, J.J., Harder, S., Silberman, J.A., Gold, M., 2010. Arctic shipping emissions inventories and future scenarios. Atmos. Chem. Phys. 10 (19), 9689–9704. https://doi.org/10.5194/acp-10-9689-2010.

Fuglestvedt, J.S., Dalsøren, S.B., Samset, B.H., Berntsen, T., Myhre, G., Hodnebrog, Ø., Eide, M.S., Bergh, T.F., 2014. Climate penalty for shifting shipping to the Arctic. Environ. Sci. Technol. 48 (22), 13273–13279.

IPCC, 2013. Climate Change 2013: The Physical Science Basis. Contribution of Working Group I to the Fifth Assessment Report of the Intergovernmental Panel on Climate Change. C. U. Press, Cambridge, United Kingdom and New York.

Janssen, N.A., Gerlofs-Nijland, M.E., Lanki, T., Salonen, R.O., Cassee, F., Hoek, G., Fischer, P., Brunekreef, B., Krzyzanowski, M., 2012. Health Effects of Black Carbon. World Health Organization, ISBN: 978-92-890-0265-3.

Johnson, K., Miller, W., Durbin, T., Jiang, Y., Yang, J., Karavalakis, G., Cocker, D., 2016. Black Carbon Measurement Methods and Emission Factors From Ships. University of California, Riverside.

Lack, D.A., 2016. The Impacts of an Arctic Shipping HFO Ban on Emissions of Black Carbon. A report to the European Climate Foundation.

Lack, D.A., Corbett, J.J., 2012. Black carbon from ships: a review of the effects of ship speed, fuel quality and exhaust gas scrubbing. Atmos. Chem. Phys. 12, 3985–4000.

Sand, M., Berntsen, T.K., Kay, J.E., Lamarque, J.F., Seland, Ø., Kirkevåg, A., 2013. The Arctic response to remote and local forcing of black carbon. Atmos. Chem. Phys. 13 (1), 211–224. https://doi.org/10.5194/acp-13-211-2013.

Sax, T., Alexis, A., 2007. A Critical Review of Ocean-Going Vessel Particulate Matter Emission Factors. California Air Resources Board, Sacramento, CA.

Stohl, A., Klimont, Z., Eckhardt, S., Kupiainen, K., Shevchenko, V.P., Kopeikin, V.M., Novigatsky, A.N., 2013. Black carbon in the Arctic: the underestimated role of gas flaring and residential combustion emissions. Atmos. Chem. Phys. 13, 8833–8855.

Winther, M., Christensen, J.H., Plejdrup, M.S., Ravn, E.S., Eriksson, O.F., Kristensen, H.O., 2014. Emission inventories for ships in the Arctic based on satellite sampled AIS data. Atmos. Environ. 91, 1–14. http://dx.doi.org/10.1016122014.03.006.

Chapter 12

Opportunities and challenges of the opening of the Arctic Ocean for Norway

Naima Saeed[a], Adolf K.Y. Ng[b]

[a]*Department of Working Life and Innovation, School of Business and Law, University of Agder, Grimstad, Norway, [b]Department of Supply Chain Management, Asper School of Business, University of Manitoba, Winnipeg, MB, Canada*

1 Introduction

The Arctic Ocean is considered a semienclosed sea that is surrounded by the following five coastal states: Canada, Denmark (Greenland), Norway (Svalbard), Russia, and the United States. In addition, Finland, Iceland, and Sweden are also generally considered to be Arctic states (Potts and Schofield, 2008). There has been a drastic change in the Arctic environment, as evidenced by the fact that, over the past 50 years, the Arctic has warmed more than twice as quickly as the rest of the world. It is expected that, by the late 2030s, the Arctic Ocean could be largely free of sea ice during summers (AMAP, 2017). As a result, navigating the Arctic is becoming commercially viable during part of the year (Zhang et al., 2019), especially during September when the level of shipping activity is highest because of the reduced ice coverage (Eguíluz et al., 2016).

There has been a tremendous increase in the number of vessels navigating the Arctic during the summer since 2005 (Lasserre and Alexeeva, 2015; Lasserre et al., 2016). The three main trans-Arctic sea routes are the Arctic Northwest Passage, the Northern Sea Route, and the Transpolar Sea Route (Stevenson et al., 2019).

It is expected that shipping through the Arctic will divert global shipping traffic (Eguíluz et al., 2016). For instance, new seaports along the Arctic coastline will handle more traffic due to increased Arctic shipping. On the other hand, transit seaports (such as Singapore) along the traditional southern routes through the Suez Canal or around the Cape of Good Hope will lose traffic, especially during the northern hemispheric summer season (Zhang et al., 2019). This trend will also increase shipping traffic in Norway, especially Northern Norway, which would become a focal point for port- and shipping-related activities.

Maritime Transport and Regional Sustainability. https://doi.org/10.1016/B978-0-12-819134-7.00013-7

In addition to the increase in shipping traffic, the opening up of Arctic sea routes also brings a number of opportunities in areas such as oil and gas activities, mining, tourism, fisheries, and economic development (AMAP, 2017). However, there are also certain challenges, which cover aspects such as environmental pollution, ecological damage due to oil and gas transportation/extraction, and geopolitical risks associated with new resources and trade opportunities. To fully utilize the benefits of the opening of the Arctic Ocean, it is essential that the various stakeholders involved are able to manage the challenges associated with the opening of the Arctic Ocean.

This chapter analyzes the potential benefits and challenges associated with the opening of the Arctic Ocean and discusses policies that could be considered to minimize the risks. We have taken Norway as a case study. The rest of the chapter is organized as follows: the second section describes a general overview of benefits and risks associated with the opening of the Arctic Ocean, the third section discusses benefits and risks with Norway as a case study, and the final section concludes this chapter.

2　Benefits and risks of the opening of the Arctic Ocean

Based on a literature review, Tables 1 and 2 provide an overview of benefits and risks associated with the opening of the Arctic Ocean. The benefits are broadly divided into the following four categories: reduction in shipping distance, economic benefits, changes in wildlife and ecosystems, and effects on the environment. The risks are generally divided into the following nine categories: access and transportation within the Arctic region; Arctic communities; wildlife, sea species, and ecosystems; environmental effects; vessels/ships; Arctic ice and weather; infrastructure; political disputes; and impacts on regions outside the Arctic.

It is not currently possible to conduct a cost-benefit analysis by simultaneously considering all the factors listed in Tables 1 and 2. A detailed risk assessment of the Arctic passage is at an early stage (Zhang et al., 2019) and the analysis of some aspects of benefits and risks of the Arctic requires either a very large amount of data or data that is scarce (see, for example, Schøyen and Bråthen, 2011; Zhang et al., 2019). However, based on the factors listed in Tables 1 and 2, we might conclude that the number of factors reflecting the risks is more than factors showing benefits. Measuring the strength of these factors to decide whether the benefits outweigh the risks or vice versa is complicated and beyond the scope of this study.

In this chapter, we analyze the benefits and risks of the functioning of the Arctic Ocean by focusing only on Norway. Norway is a part of the European Arctic, which Hønneland (2003, p. 141) defined as "the parts of Norway, Sweden, Finland, and European Russia that are located north of the Arctic Circle, plus the Barents Sea, the Svalbard Archipelago and the Russian archipelagos of Novaya Zemlya and Franz Josef Land". Åtland (2007) identified the

TABLE 1 Benefits of the opening of the Arctic Ocean.

Category	Description
Reduction in shipping distance	The use of the Arctic Northeast Passage instead of the present route via the Suez Canal reduces navigational distance between East Asia and Europe by 30%–40%; similarly, it reduces the distance by 40%–50% compared to the Panama Canal route and by 50%–60% compared to the route around the Cape of Good Hope (Lasserre and Pelletier, 2011; Zhang et al., 2019)
Economic benefits	The Arctic Ocean is creating more opportunities for marine shipping and tourism. The Arctic could become a potential future source of freshwater and hydropower for southern areas. It may facilitate access to oil, valuable minerals (like nickel, cobalt, palladium, and platinum are found in Russia, Alaska, Canada, and Greenland), gas, and other resources (Lindholt, 2006; Potts and Schofield, 2008; Rhéaume and Caron-Vuatari, 2013; AMAP, 2017) There is also a trend to shifting of some fish species (for example, mackerel recently migrated into waters around Svalbard and Greenland) and there is a potential for 17 species to migrate into the Arctic (AMAP, 2017; The Norwegian Polar Institute, 2015), which will promote fisheries. The shippers and carriers can achieve benefits of economies of scale by utilizing mega vessels because, as Zhang et al. (2019) explained, the mega vessels are unable to pass through the Suez Canal and must navigate around the Cape of Good Hope. There are no canal fees for Arctic navigation routes and, by avoiding politically unstable regions and the piracy affected regions in traditional routes, the cost of insurance for transportation can be reduced (Zhang et al., 2019). Changing climate and rising temperature may benefit agriculture and aquaculture in Northern Norway (The Norwegian Polar Institute, 2015)
Changes in wildlife and ecosystems	The growth of marine phytoplankton and creation of more habitats for open-water species (AMAP, 2017)
Effects on the environment	Ships navigating via the Arctic Northeast Passage could reduce carbon dioxide emissions by 49%–78% compared to traditional southern routes (Schøyen and Bråthen, 2011)

following five attributes of the European Arctic: it is a region (1) of peripheries, (2) that is rich in natural resources, (3) that has unresolved legal issues, (4) that is strategically significant, and (5) of transnational cooperation.

In the next section, we discuss some of the benefits and risks listed in Tables 1 and 2 that are most relevant to the Norwegian context.

TABLE 2 Risks of the opening of the Arctic Ocean.

Category	Description
Access and transportation within the Arctic region	Ice roads are affected because of the decrease in the thickness of lake and river ice and changes in permafrost. Because of the shorter snow cover season, it has become difficult for some northern communities to obtain wild sources of food and access to resources. There are also risks to food and water security (AMAP, 2017)
Arctic communities	The safety and, in some cases, the very existence of coastal communities are under threat because of coastal erosion and flooding resulting from melting of coastal sea ice. There is an increase in severe wildfires in the Arctic areas of North America and Eurasia. Thawing permafrost is affecting the communities and infrastructure built on frozen soils, especially in Siberia. There is also a probability of health risks to people such as increased incidence of West Nile fever. Climate change will also negatively affect Sámi culture (AMAP, 2017; The Norwegian Polar Institute, 2015)
Wildlife, sea species, and ecosystems	There is a change in the availability of habitat for microorganisms, plants, animals, and birds. Vegetation can be damaged, which affects the conditions for grazing animals such as caribou, reindeer, and musk ox. Loss of ice-associated algal species disturbing the feeding platforms and life cycles of seals, polar bears, and, in some areas, walrus and finally affecting the food web (AMAP, 2017; The Norwegian Polar Institute, 2015). There is a possibility that Arctic species like North East Arctic Cod, which are generally slow-growing due to their cold environment, will be especially vulnerable to overfishing (Barber et al., 2005; Potts and Schofield, 2008)
Environmental effects	More greenhouse gas emissions because of extraction of oil and gas and acidification of the ocean by carbon dioxide (AMAP, 2017; The Norwegian Polar Institute, 2015). Ship accidents may increase oil spills and other types of contamination of the Arctic environment (Zhang et al., 2019). Pollution threats from economic activities such as mining, heavy industry, tourism, mineral resource development, and military activities (Potts and Schofield, 2008)
Vessels' operation	Possibility of the vessels being trapped in the Arctic ice and ship damage and accidents (Liu and Kronbak, 2010; Lasserre and Pelletier, 2011; Kum and Sahin, 2015; Goerlandt et al., 2016; Zhang et al., 2017, 2019), higher building costs for ice-classed ships, ice-breaking fees, and other additional costs such as route recommendation, communication service, special vessel steerage, bunker-filling fee, and supply of fresh water (Liu and Kronbak, 2010; Aksenov et al., 2017); the increased consumption when plowing ice packs; no precise opening and closing date of Arctic sea routes and the cost of operating an ice-class vessel in non-Arctic waters (Lasserre et al., 2016)

TABLE 2 Risks of the opening of the Arctic Ocean.—cont'd

Category	Description
Arctic ice and weather	Arctic ice specified by shipping lines in various forms such as drifting ice, growlers, icebergs, ice ridges, and multiyear ice; and the Arctic weather (coldness, icing, and fog) (Lasserre et al., 2016)
Infrastructure	Poor infrastructure of existing Arctic ports and inadequate support facilities for commercial shipping such as deep-water access, places of refuge, marine salvage, port reception facilities for ship-generated waste, and towing services.[a] Existing transport infrastructure and municipal utilities like water mains and drains, and buildings are exposed to floods (The Norwegian Polar Institute, 2015)
Political dispute	The possibility of the opening up of Northwest Passage led to the re-emergence of the dispute between Canada and the US over the legal status of the waterway (Byers and Lalonde, 2006; Kraska, 2007). Conflict between Norway and Russia in the Barents Sea (Potts and Schofield, 2008)
Impacts on outside regions	A potential rise in sea level by 0.6 m by 2099 can have a negative effect on low-lying areas and islands of countries such as Bangladesh, the Netherlands, and the Maldives (Miller, 2007)

[a] See http://www.chnl.no/publish_files/Future_of_Arctic_Shipping_Routes.pdf (Accessed 29 April 2019).

3   Benefits and risks of the opening of the Arctic Ocean related to Norway

In this section we discuss certain risks and benefits by focusing only on Norway. We divide these risks and benefits into two broad groups. The first group includes the benefits of "reduction in shipping distance" and the risks related to "vessels' operation" in the Arctic. There are two reasons for addressing these two themes together: first, both themes are related to shipping; second, the risks associated with vessel operation can outweigh the benefits achieved from the reduction in shipping distance. The second group consists of "economic benefits" (see Table 1) and specific risks that can outweigh benefits such as poor infrastructure, overfishing, and political disputes (see Table 2).

3.1   Reduction in shipping distance and vessels' operation

A significant decrease in shipping distance can be utilized by selecting the Arctic routes instead of traditional current sea routes. For instance, selecting the Arctic Northeast Passage instead of the present route via the Suez Canal reduces navigational distance between East Asia and Europe by 30%–40%; similarly, it reduces the distance by 40%–50% compared to the Panama Canal route and by 50%–60% compared to the route around the Cape of Good Hope (Lasserre and Pelletier, 2011; Zhang et al., 2019). This reduction in sea distance means it is expected that more vessels will use the Arctic Ocean routes. However, as mentioned by Liu and Kronbak (2010), a reduction in distance does not necessarily mean a corresponding decrease in cost. This is because there are number of other cost-related factors, such as higher building costs for ice-classed ships, ice-breaking fees, additional charges such as route recommendation, communication service, specialized vessel steerage, bunker-filling fee, supply of fresh water (Liu and Kronbak, 2010; Aksenov et al., 2017) and the increased consumption when plowing ice packs, and the cost of operating an ice-class vessel in non-Arctic waters (Lasserre et al., 2016).

Arctic shipping can be broadly classified into two categories. The first is Intra-Arctic transport, which includes a voyage or marine activity that remains within the general Arctic region and links two or more Arctic states. Examples of Intra-Arctic transport include marine route between the port of Churchill, Manitoba, Canada on Hudson Bay and Murmansk, Russia, known as an "Arctic-bridge" between the two continents; and an Icelandic fishing vessel working in Greenlandic waters, and tug-barge traffic operating between Canada's Northwest Territories and the US Beaufort Sea off the Alaskan coast. The second category is trans-Arctic transport, which is taken across the Arctic Ocean from the Pacific to the Atlantic Oceans, or vice versa. These are full voyages that connect the North Pacific and North Atlantic Oceans through the Arctic (AMSA, 2009). Three main trans-Arctic routes are the Arctic Northwest Passage, the Northern Sea Route, and the Transpolar Sea Route (Stevenson et al., 2019).

Regarding the trans-Arctic shipping, Norway's Statoil does not consider it very attractive, after sending several tankers, including cargoes of naphtha and LNG, to Japan in previous years (Lasserre et al., 2016).

"Statoil has not used the Northern route since 2013 and we currently have no plans to use it." "The attractiveness of a route depends on direct costs, and sailing time as well as the market characteristics of the respective commodities at the time of sailing."

(Statements by a company spokeswoman, quoted in Pettersen, 2016)

In order to acquire updated information, we conducted a telephonic interview in June 2019 with a concerned person in Equinor[a] (Statoil), who confirmed that Statoil still does not use the Northern Sea Route to deliver cargo to countries outside the Arctic region.

Similarly, according to the Norwegian Shipowners' Association (2014, p. 3):

---

a. In May 2018 Statoil changed its name to Equinor. See https://www.gasworld.com/statoil-changes-name-to-equinor/2014750 (Accessed 19 June 2019).

Transit through the northern sea routes will gain in importance but will remain limited in volume in the next few years.

As in the previous case, we conducted a telephonic interview in June 2019 with a concerned person at the Norwegian Shipowners' Association, who mentioned that the situation is the same as stated in 2014 report that the transit through the Northern Sea Route will remain limited in volume in the next few years. This is mainly because of the shallow water near Russia, the expense of building specialized ships, and the regulations set by the Russian government that shipping lines must adhere to.

These findings are consistent with those obtained by Lasserre and Alexeeva (2015) and Lasserre et al. (2016), which explained that trans-Arctic (transit) traffic in the Northwest Passage (NWP) and the Northeast Passage (NEP) remains low. For their study, the transit data for the NEP was provided by the center for High North Logistics (CHNL) located in Kirkenes, Norway, with a branch in Murmansk, Russia. Thus, we may conclude that, even for Norway, as explained by Lasserre et al. (2016), the benefits realized from a reduction in costs because of shorter shipping distance and the technical feasibility are not currently sufficient to select the Arctic Sea routes for transit. In this situation, environmental benefits—reducing carbon dioxide emissions by 49%–78% by choosing the Arctic Northeast Passage instead of traditional southern routes, as described by Schøyen and Bråthen (2011)—are not currently being realized (see Table 1).

Norway has taken several measures to overcome the risks associated with vessel operation in the Arctic Ocean, including the provision of specialized vessels, as explained by the Norwegian Shipowners' Association (2014, p. 5):

A number of shipping companies have under construction, or already in service, specialised ships and rigs adapted to working in extreme climatic conditions.

To provide adequate protection in the case of accidents at sea, Norway has joined the international regulatory framework that consists of the International Convention on Oil Pollution Preparedness, Response and Cooperation, and also the International Oil Pollution Compensation Funds, including the Supplementary Fund. From these frameworks, Norway has total available compensation equal to 750 million Special Drawing Rights (SDR) (approx. NOK 7.5 billion). There is a very low possibility that a claim for compensation following an accident off the Norwegian coast would exceed this amount. By introducing an excellent compensation scheme, Norway has protected itself from compensation issues being put forward as an argument against increased oil transport along the Norwegian coast (Norwegian Shipowners' Association, 2014).

In addition to compensation for accidents at sea, there is compensation claim for the discharge of bunker oil from ships not transporting oil as cargo. This scheme was set by the 2001 International Convention on Civil Liability for Bunker Oil Pollution Damage, which entered into force on November 21, 2008. The convention was accompanied by a resolution encouraging all states to ratify or accede to the 1996 Protocol to the 1976 Convention on Limitation of

Liability for Maritime Claims (LLMC, 1976). Consequently, Member States are given the option to set higher national rates instead of the standard international rates. Among the Arctic states, the Russian Federation, Denmark, Finland, and Norway follow LLMC 1976 (AMSA, 2009). Norway is one of the countries that has selected this option and has recently adopted a further increase to ensure that the compensation amount is available to cover most accidents, including the largest scale ones (Norwegian Shipowners' Association, 2014).

3.2 Economic benefits, sea species, infrastructure, and political dispute

In terms of economic benefits, Norway has the potential to utilize several resources from the Arctic region. Norway's sea area is six times larger than its land area and most of the sea area lies in the High North[b] (Norwegian Shipowners' Association, 2014). The Barents Sea off the coast of Norway and Russia in the Arctic region is recognized as the area with the highest potential in both fishing and petroleum. Svalbard in Norway and Franz Josef Land in Russia are considered significant protected land and sea areas in the region (Eliasson et al., 2017).

From the perspective of Norway's maritime industry, the three main areas of interest in the Arctic region are offshore energy extraction, intraregional transport, and trans-Arctic transport. The most important area is offshore energy extraction, and intraregional transport provides a support activity for the offshore energy extraction (Norwegian Shipowners' Association, 2014), while Arctic transit transport is limited, as discussed in the previous section.

Oil and gas extractions have already started in High North Norway. Norway has one of the most advanced maritime industries and, as mentioned by the Norwegian Shipowners' Association (2014, p. 16):

> *Offshore activities are not new to the High North and, with the world's most modern fleet, we have demonstrated that we can solve the challenges within the defined frameworks.*

Statoil made significant investments in North Norway in 2018. As a part of this investment, Aker Solutions will connect a total of 30 wells on the oil field in the Barents Sea. Six of the subsea templates will be delivered in 2019, and four in 2020. In addition, Aker Solutions, Sandnessjøen has a contract for developing the subsea template and suction anchor for Snefrid Nord, a gas discovery 12 kilometers from the Aasta Hansteen field in the north of the Norwegian Sea. Snefrid Nord is expected to be in operation in the fourth quarter of 2019[c] and will generate employment opportunities in the region, as described by the managing director of the supplier network for petroleum activities in the North, Petro Arctic:[c]

---

b. The Norwegian Shipowners' Association defines the High North as the entire circumpolar Arctic, including the Barents region and the Barents Sea area.

c. See https://www.equinor.com/en/news/13apr2018-investments-north.html (Accessed 1 May 2019).

This will generate employment and add value at Helgeland for many years to come.

In addition to the two above-mentioned projects, Statoil and its partners also invested about NOK five billion in Askeladd, which is the second part of the phased development plan of the Snøhvit field in the Barents Sea. It is expected that, as an outcome of this project, Askeladd will deliver 21 billion cubic meters of gas and two million cubic meters of condensate to Hammerfest LNG.[c]

The Norwegian Arctic has some of the largest and most valuable fishing stocks in the world, such as the northeast Arctic cod and the Norwegian spring spawning herring (DNV GL, 2019). There is also a trend in shifting of some fish species; for example, mackerel recently migrated into waters around Svalbard and Greenland (AMAP, 2017; The Norwegian Polar Institute, 2015). Southern and pseudo-oceanic temperate fish species stocks are migrating toward North Norway (Barents and Bering Seas), which could result in unprecedented harvest levels and, thus, promote commercial fisheries (Hunt Jr. et al., 2013; Christiansen et al., 2014; Falk-Petersen et al., 2015).

According to the report by the Ministry of Local Government and Modernisation: Regionale utviklingstrekk (2018), fisheries and aquaculture contributed to the economy in 2015 as follows: The two industries combined represented 7.0% of the gross domestic product of Nordland (county in Norway) and 6.1% in Tromso (a municipality in Norway) and Finnmark (county in Norway), contributing to annual growth of 2.8% and 2.6% in the period 2010–15. The fishing industry has traditionally been labor-intensive and because of technology development and economies of scale, efficiency in the industry has increased significantly (DNV GL, 2019).

The opening of the Arctic Ocean has generated significant maritime traffic in the Norwegian Arctic. The large tankers that are used for oil and gas extraction in Russia and Finnmark are sailing outside the coast in traffic separation lanes; bulkers and deep-sea vessels are calling at Norwegian ports such as Narvik and Mo i Rana; coastal traffic including cruise and passenger ships are operating throughout the fairways; fishing vessels are operating on the fishing grounds; and the offshore service vessels are serving the oil fields at Haltenbanken and the Barents Sea. Fishing vessels and cruise traffic are common in the sea area between mainland and Svalbard (DNV GL, 2019). The majority of this maritime traffic falls into the category of intra-Arctic transport. Finally, changing climate and rising temperature may benefit agriculture and aquaculture in Northern Norway (The Norwegian Polar Institute, 2015).

In addition to economic benefits there are some challenges, and one of the main challenges in North Norway is the poor infrastructure, which includes roads and streets, airports, harbors (lighthouses, navigation aids, etc.), railways, energy supply, telecommunications (including broadband), buildings, water and sewage, and waste management. Norwegian officials have already started to work on this issue. For instance, the Norwegian Government has set out infrastructure in the High North as one of the five priority areas in its Arctic policy.

In the National Transport Plan 2018–29, priority is given to several major infrastructure investment projects in the North, which will shorten travel times significantly. The Barents Euro-Arctic Transport Plan, which focuses on better connectivity between Sweden, Finland, Russia, and Norway, is of great benefit to national and regional transport planning of Norway. Norway carried out the follow-up work on the Joint Barents Transport Plan at the national level through planning processes and budget priorities. To improve the port facilities and to handle the increasing number of cruise ships, as well as other types of vessels, the Norwegian Coastal Administration recommends building a new floating dock with a terminal building in Svalbard. The government has allocated 300 million NOK to finance this project.[d]

Regarding fisheries, the biggest challenge is the fact that Arctic species such as North East Arctic cod, which are generally slow-growing due to their cold environment, are probably especially vulnerable to overfishing (Barber et al., 2005; Potts and Schofield, 2008). To overcome this challenge and to achieve economic benefits from fisheries, overfishing should be avoided under yet insufficient Arctic fisheries biological data (Christiansen et al., 2014).

Another challenge is the dispute between Norway and Russia in the Barents Sea, mainly over oil and gas and fishing resources. To overcome this dispute Norway and Russia signed a maritime boundary agreement on 11 July 2007, which was intended to clarify, update, and reconfirm an agreement dating from 1957 and extend it into the southern Barents Sea (Potts and Schofield, 2008). Despite this agreement, the tension remains because of the two countries' overlapping claims further north in the Barents Sea and their conflict over Norway's maritime claims from Spitsbergen (Svalbard).[e]

However, the Norwegian Shipowners' Association (2014) considers the agreement between the two countries to be a positive step in the political stability of the region:

> It is important to point out that a stable region characterised by low levels of tension and by international cooperation would form the basis for all maritime activity in the High North. In this context, the Norwegian Shipowners' Association finds the Norwegian authorities' foreign and security policies in the High North to be stable and sound. The five Arctic coastal states share the view that maritime law must govern the resolution of outstanding issues concerning control of sea areas, and that generally accepted principles for resource management shall also apply in the Arctic. We believe that such an approach is beneficial for all stakeholders in the High North and the Arctic.

(Norwegian Shipowners' Association, 2014, p. 7)

---

d. See https://www.regjeringen.no/contentassets/7c52fd2938ca42209e4286fe86bb28bd/en-gb/pdfs/stm201620170033000engpdfs.pdf (Accessed 15 June 2019).

e. See http://www.regjeringen.no/en/dep/ud/Press-Contacts/News/2007/Agreement-signed-between-Norway-and-Russ.html?id=476347 (Accessed 15 May 2019).

4   Conclusion

It is generally expected, and also highlighted in the media, that the opening of the Arctic Ocean will bring a lot of opportunities, not only for the Arctic states but also for the rest of the world. However, there are several risks that also emerge with the melting of polar ice. In this chapter, based on a literature review, we have identified potential benefits and risks associated with the warming of the climate in the Arctic region. It is difficult to conduct a cost-benefit analysis by considering all the factors related to benefits and risks because this requires data that is either scarce or required on a large scale. However, based on the listed factors, we can conclude that there are more risk factors than benefit factors.

We conducted a detailed analysis of certain benefits and risks that are relevant to Norway, with the help of the literature review and case studies. We divided the relevant factors into two broad groups. The first group considers the benefits of reduction in shipping distance by selecting the Arctic route for Norway. In the same group, we also analyzed certain risks factors that are related to vessels' operation in the Arctic Ocean and can challenge or offset the benefits of reduction in shipping distance. The second category includes the economic benefits and specific risks that can outweigh the economic benefits.

Our analysis shows that, for Norway, trans-Arctic (transit) shipping is currently not attractive. As a result, it is not desirable for the cargo owners and shipping companies to gain the benefits of reduction in cost by selecting Arctic routes instead of traditional routes. The analysis also revealed that the strong and advanced maritime industry of Norway is capable of overcoming the risks related to vessels operation by investing in specialized vessels that can handle the harsh weather and ice. Moreover, the Norwegian Government has developed a substantial compensation scheme for ship accidents and the discharge of bunker oil in the Arctic Ocean. In its Arctic policy and its National Transport Plan 2018–29, the Norwegian government has prioritized the development of infrastructure in High North. The government has allocated NOK 300 million to improve the port facilities in Svalbard.

This study shows that Norway has the potential to utilize the abundant opportunities emerging from climate change, such as oil and gas extraction and commercial fishing in the Barents Sea. The analysis indicates that despite the reduction in oil prices, investment in oil and gas sector by Statoil is increasing and will also generate employment opportunities in Northern Norway. There is also a potential for growth in other sectors such as agriculture, aquaculture, and fishing in High North, although fishing is also subject to the risks of overfishing. The best strategy to overcome this risk is to avoid overfishing, which is challenging and difficult because it requires data related to the amount of fishing.

Finally, with the emergence of abundant natural resources in the Barents Sea, the tension between Norway and Russia has increased over the sea boundary. However, both countries' governments have attempted to overcome this dispute by signing a maritime boundary agreement. The Norwegian Shipowners' Association considers this as a good step for regional stability and maritime-related activities in the Barents Sea.

The study shows that although Norway is a small and open economy compared to the other Arctic states, its strong maritime industry, along with support from the government, is well prepared to achieve the benefits and to overcome certain risks related to the melting of ice in Arctic Ocean.

References

Aksenov, Y., Popova, E.E., Yool, A., George Nurser, A.J., Williams, T.D., Bertino, L., Bergh, J., 2017. On the future navigability of Arctic sea routes: high-resolution projections of the Arctic Ocean and sea ice. Mar. Policy 75, 300–317.

AMAP, 2017. Snow, Water, Ice and Permafrost in the Arctic (SWIPA). Arctic Monitoring and Assessment Programme (AMAP), Oslo, Norway. xiv + 269 pp. Available at: https://www.amap.no/documents/download/2987.

AMSA, 2009. Arctic Marine Shipping Assessment 2009 Report. https://www.pmel.noaa.gov/arctic-zone/detect/documents/AMSA_2009_Report_2nd_print.pdf.

Åtland, K., 2007. The European Arctic After the Cold War: How Can We Analyze It in Terms of Security? . Rapport for Norwegian Defense Research Establishment.

Barber, D., Fortier, L., Byers, M., 2005. The incredible shrinking sea ice. Policy Options, 66–71. (December 2005–January 2006).

Byers, M., Lalonde, S., 2006. Who controls the Northwest Passage? Vanderbilt J. Transl. Law 43, 1133–1210.

Christiansen, J.S., Mecklenburg, C.W., Karamushko, O.V., 2014. Arctic marine fishes and their fisheries in light of global change. Glob. Chang. Biol. 20, 352–359.

DNV GL, 2019. Sustainable blue economy in the Norwegian Arctic. Centre for the Ocean and the Arctic , pp.1–103.

Eguíluz, V.M., Fernandez-Gracia, J., Irigoien, X., Duarte, C.M., 2016. A quantitative assessment of Arctic shipping in 2010–2014. Sci. Rep. 6 (30682), 1–6.

Eliasson, K., Ulfarsson, G.F., Valsson, T., Gardarsson, S.M., 2017. Identification of development areas in a warming Arctic with respect to natural resources, transportation, protected areas, and geography. Futures 85, 14–29.

Falk-Petersen, S., Pavlov, V., Berge, J., Cottier, F., Kovacs, K., Lydersen, C., 2015. At the rainbow's end: high productivity fueled by winter upwelling along an Arctic shelf. Polar Biol. 38, 5–11.

Goerlandt, F., Montewka, J., Zhang, W., Kujala, P., 2016. An analysis of ship escort and convoy operations in ice conditions. Saf. Sci. 95, 198–209.

Hønneland, G., 2003. Russia and the West: Environmental Co-Operation and Conflict, first ed. Routledge, London.

Hunt Jr., G.L., Blanchard, A.L., Boveng, P., Dalpadado, P., Drinkwater, K.F., Eisner, L., Hopcroft, R.R., Kovacs, K.M., Norcross, B.L., Renaud, P., Reigstad, M., Renner, M., Skjoldal, H.R., Whitehouse, A., Woodgate, R.A., 2013. The Barents and Chukchi Seas: comparison of two Arctic shelf ecosystems. J. Mar. Syst. 109–110, 43–68.

Kraska, J., 2007. The law of the sea convention and the Northwest Passage. Int. J. Mar. Coast. Law 22 (2), 257–282.

Kum, S., Sahin, B., 2015. A root cause analysis for Arctic Marine accidents from 1993 to 2011. Saf. Sci. 74, 206–220.

Lasserre, F., Alexeeva, O., 2015. Analysis of maritime transit trends in the Arctic passages. In: Lalonde, S., McDorman, T. (Eds.), International Law and Politics of the Arctic Ocean: Essays in Honour of Donat Pharand. Brill Academic Publishing, Leiden, pp. 180–193.

Lasserre, F., Pelletier, S., 2011. Polar super seaways? Maritime transport in the Arctic: an analysis of shipowners' intentions. J. Transp. Geogr. 19 (6), 1465–1473.

Lasserre, F., Beveridge, L., Fournier, M., Têtu, P.-L., Huang, L., 2016. Polar seaways? Maritime transport in the Arctic: an analysis of shipowners' intentions II. J. Transp. Geogr. 57, 105–114.

Lindholt, L., 2006. The Economy of the North: 3: Arctic Natural Resources in a Global Perspective. Statistics Norway, Oslo.

Liu, M., Kronbak, J., 2010. The potential economic viability of using the Northern Sea Route (NSR) as an alternative route between Asia and Europe. J. Transp. Geogr. 18, 434–444.

LLMC, 1996. Protocol of 1996 to Amend the Convention on Limitation of Liability for Maritime Claims. 1976, http://folk.uio.no/erikro/WWW/LLMC-96.html.

Miller, H.L. (Ed.), 2007. Summary for Policymakers. Contribution of Working Group I to the Fourth Assessment Report of the Intergovernmental Panel on Climate Change. Climate Change 2007: The Physical Science Basis, Cambridge University Press, Cambridge, United Kingdom and New York, USA.

Ministry of Local Government and Modernisation: Regionale utviklingstrekk. 2018. Report derived from: https://www.regjeringen.no/contentassets/59c12ad48f604486926d605f1680871e/no/pdfs/regionale-utviklingstrekk-2018.pdf.

Norwegian Shipowners' Association, 2014. Maritime Opportunities in the Arctic. Available at: https://rederi.no/en/rapporter/.

Pettersen, T., 2016. Declining interest in use of Northern Sea Route. The Independent Barents Observer (March 18). https://thebarentsobserver.com/en/industry/2016/03/declining-interest-use-northern-sea-route.

Potts, T., Schofield, C., 2008. An Arctic scramble? Opportunities and threats in the (formerly) frozen north. Int. J. Mar. Coast. Law 23, 151–176.

Rhéaume, G., Caron-Vuatari, M., 2013. The Future of Mining in Canada's North. The Conference Board of Canada, Canada.

Schøyen, H., Bråthen, S., 2011. The Northern Sea route versus the Suez Canal: cases from bulk shipping. J. Transp. Geogr. 19 (4), 977–983.

Stevenson, T.C., Davies, J., Huntington, H.P., Sheard, W., 2019. An examination of trans-Arctic vessel routing in the Central Arctic Ocean. Mar. Policy 100, 83–89.

The Norwegian Polar Institute, 2015. The Arctic Region. Available at: https://www.environment.no/topics/the-polar-regions/the-arctic-region/.

Zhang, Z., Huisingh, D., Song, M., 2019. Exploitation of trans-Arctic maritime transportation. J. Clean. Prod. 212, 960–973.

Chapter 13

Climate change, a double-edged sword: The case of Churchill on the Northwest Passage

Yufeng Lin[a,b], Adolf K.Y. Ng[a,b], Mawuli Afenyo[a]
[a]Transport Institute, University of Manitoba, Winnipeg, MB, Canada, [b]Department of Supply Chain Management, Asper School of Business, University of Manitoba, Winnipeg, MB, Canada

1 Introduction

Shipping activity in the Arctic has increased lately (Ng et al., 2018; Østreng et al., 2013; Smith and Stephenson, 2013; Afenyo et al., 2016). This is made possible by the rapid melting of the ice at an unprecedented level (Ng et al., 2018; Østreng et al., 2013; Lasserre and Pelletier, 2011). The phenomenon has prompted governments of Arctic countries to develop the shipping routes further in order to make them safe and viable commercially (Østreng et al., 2013). Many studies have reviewed the feasibility of the new Arctic sea route: the Northern Sea Route and the Northwest Passage (Østreng et al., 2013; Ng et al., 2018). The authors concluded that they are viable alternatives to conventional shipping routes like the Suez Canal. Most works have gone further to compare these new routes and the old ones and agreed that, in terms of distance, the new ones are shorter (Liu and Kronbak, 2010; Schøyen and Bråthen, 2011; Østreng et al., 2013; Lasserre and Pelletier, 2011). However, scholars have not reached a consensus on the cost implications in terms of insurance, escorts, emergency response, and implication of potential oil spill; when these are considered, the comparison may not be straight forward (Østreng et al., 2013).

Of the new sea routes, the Northern Sea Route has experienced the most traffic. This is due to several reasons; these include: (i) the rate of ice melt and, for that matter, the quantity of ice along the routes are less compared to the Northwest Passage; (ii) Russia, the country sovereign over this area, has invested a lot in terms of icebreakers, infrastructure, and other necessary logistics to make it safe and navigable, which, therefore, creates confidence for the ship owners and operators to use this route; (iii) the routes are better defined compared to those on the Canadian side (Northwest Passage) (Østreng et al., 2013).

Maritime Transport and Regional Sustainability. https://doi.org/10.1016/B978-0-12-819134-7.00014-9
223

These reasons mean that the Northwest Passage needs a lot of investment before it is ready to take advantage of the potential positive effect of climate change. For an area that contains approximately 30% of the world's hydrocarbon deposits, as well as other precious minerals (Lee and Kim, 2015; Lasserre and Pelletier, 2011), it is also very strategic for the future energy security and wealth of the individual countries (Østreng et al., 2013).

Knowing how important the Arctic and, for that matter, the Northwest Passage is likely to become, the Canadian government has stepped up work in this region. Though the sovereignty of the Northwest Passage is still a matter of discussion between Canada and the United States (Østreng et al., 2013), a well-developed route with good infrastructure will not only be beneficial to both countries but to the world at large. This is because shipping will become faster, jobs will be created, and scientific research will also be enhanced (Østreng et al., 2013; Lasserre and Pelletier, 2011).

One of the key benefits of developing such infrastructure, as pointed out earlier, is the benefit the northern communities of Canada are likely to receive. Unlike the Russian side, which has quite a high population, this is not the case on the Canadian side. The increased activities in the Northwest Passage will therefore open the northern communities up and make movement of goods easier. Due to the nature of the Canadian Arctic, communities are mostly accessible by air; this is normally expensive. A special case is a town called Churchill, which has a railway, a port, and an airport. This makes a town such as Churchill, located in the Canadian Arctic, very important to opening up the north. This critical infrastructure can be so important that damage to or nonfunction of any create a huge impact in the community (Ng et al., 2018).

This chapter examines the role of the Churchill port as an example of how climate change and its impact on shipping are likely to bring wealth to an otherwise remote community and the consequences as well when these facilities do not function or are underutilized. The latter examines some of the challenges that these communities are facing and what can be done to improve the situation. The case of Churchill may be peculiar in terms of how the effects of climate change have both positive and negative impact on one community. On one hand, the effect of climate change is likely to open up Churchill to businesses and increase revenue for the port and the government. On the other hand, there is the challenge of the huge impact on infrastructure, such as the erosion of the railway through flood.

The European Union (EU), in their report on the effect of climate change on transport infrastructure, outlined how bad weather conditions can cause flooding and so interrupt essential transportation infrastructure like bridges and railways (EEA, 2014). Further, Schweikert et al. (2014) presented a study that is representative of the worldwide impact of climate change on roads. The study concluded that there is a tremendous repercussion of climate change on the cost and maintenance of roads as well as the ability to connect communities.

In addition, Nemry and Demirel (2012), in their report for the EU on the impact of climate change on transport, noted that the impact of damaged infrastructure has a cascading effect for economic activities. That is to say that damage to a bridge or, for example, railway will affect the economic activities of the areas where the railway is linked. Further, Baker et al. (2010) examined the effect of climate change on railways. They identified, among other things, the effect that extreme weather conditions may have on infrastructure such as the railway.

The case presented is very special, in that it involves two key infrastructures located in an Arctic setting affected by climate change. The case study presented shows some of the many challenges the town of Churchill faces and this may be true for other such locations in the Arctic in the different countries. Churchill port, which is the only deep water port in the Canadian Arctic, will be very strategic to successful shipping business in the future should the Northwest Passage begin to experience increased traffic. The port infrastructure is, however, still not very developed to cope with a potential increase in business. This has been due mainly to less shipping activity and the high cost involved in operating such a facility. Currently, there is no incentive to continue to ship items through Churchill port, since ports like Vancouver and Montreal offer better services to China and Europe.

However, looking ahead into the future, the Churchill Township and its infrastructure will play a key role in world shipping business. Churchill is key to delivering goods to the northern communities. The port is very strategic to achieving this. The nonfunctioning of the Churchill port would create a very difficult life for the populace. Should the Northwest Passage reach the level of the Northeast Passage, transport of goods to Churchill and other communities in the north will speed up. It is also likely to create jobs for the local people. Further, people would be willing to move to the community to work knowing they are assured of supplies.

Cost of living is very high in these areas and so this discourages people from taking up employment. An improved and fast means of delivering cargo means that employees are assured of food and other essentials. The uncertainty about food and travel would be laid to rest.

Tourism would improve tremendously. Like many other Arctic cities or towns, accessibility in terms of cruise ships would improve the tourist potential of these places. Many want to know how the Arctic looks and the natural fauna and flora in this area, but the difficulty in traveling through ice has prevented such voyages. The ice melt therefore presents an unprecedented opportunity for tourism in the Arctic areas.

The rest of the chapter is organized as follows. Section 2 gives a brief description of Churchill, Section 3 describes the problem of the railways caused by climate change, Section 4 is the chapter that describes the problems currently facing the people residing in Churchill, and Section 5 provides the conclusion.

2   Case study: The Churchill example

The town of Churchill is located in the province of Manitoba in Canada, on the shores of Hudson Bay. Two of its key infrastructures are the railway and the port. The Churchill railway is part of the Hudson Railway. On May 23, 2017, Omnitrax claimed that because of severe floods, the railway between Amery and Churchill was closed indefinitely (CNW, 2017). It was uncertain when the railway would be repaired due to a disagreement over which entity took responsibility for the repairs. Losing access to rail service influenced the tourism industry, imposing high freight cost on Churchill's already-fragile economy (Robertson, 2017). In the end, the railway was fixed in October, 2018, but the loss was substantial.

2.1   The implications

Across the north of Canada, there are remote communities like Churchill that only have access to either airline or railway. Churchill is lucky to have both. Losing its access to railway resulted in sever social-economic crisis. Due to the importance of transport, any disruption has serious consequences (Shinozuka et al., 1998). The interregional transportation network increases interregional trade and human mobility, contributing to regional economic development (Tatano and Tsuchiya, 2008). The Churchill Railway washout of 2017, therefore, has serious repercussions for the current state of the northern comminutes and the future of Arctic shipping in the area; especially at a time when there are efforts to close the gap between development along the Northern Sea Route and Northwest Passage.

Fig. 1 shows the problem of Churchill: a double impact of climate change on railway and the port. In Fig. 2, the current state of the railways as washed by flood is shown.

Literature abounds with works that have examined the vulnerability of individual company supply chains from a micro-level perspective (Juttner et al., 2003; Peck, 2005; Svensson, 2000). In the Churchill case as presented in this chapter, however, a macro-level assessment of the impact of a railway disruption of the transportation system is carried out. We explore the influence of railway disruption with specific reference to Churchill in northern Canada. We touch on a number of issues that relate to social sustainability, transportation system, community-cooperation relationship, and community empowerment. This next section as discussed and presented is informative for public, private, and people sectors to understand how critical the railway is to the survival of the Churchill township and also its role in the future of Arctic shipping.

2.2   Impacts on essential services

The sectors to be examined in this section have the following characteristics: their distribution was exclusively or predominantly by railway and they play

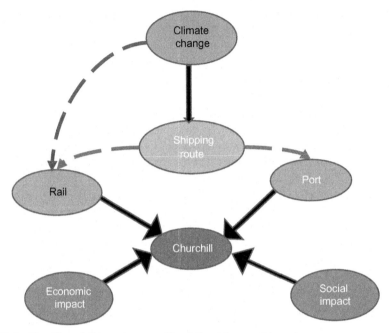

FIG. 1   The effect of climate change on Churchill and the overall impact.

FIG. 2   2017 Churchill Railway washout.

a critical role in economic activity or quality of life. These sectors include, but are not limited to, grocery retailing, fuel supply, postal services, health services, and liquor.

2.2.1 Impact on grocery

The diet in Churchill comprises a mixture of "market" and "country" foods. Country foods include plants, seafood, and animals that are harvested locally, both from domestic and wild sources. Markets foods, on the other hand, are those that are transported into the communities either on barges in the summer, by air, via railway, or in the winter, via ice road. Due to transportation costs, market foods are generally expensive.

The price of groceries in Churchill has skyrocketed as groceries now arrive by plane. For example, the price of goods at the local Northern Store can be two to four times higher than provincial average price. The new subsidy rate of $1.60 per/kg offered to Churchill is the highest rate offered to Manitoba communities in both the Affordable Food in Remote Manitoba (AFFIRM) and Nutrition North Canada (NNC) programs. Even with such programs, residents can barely afford the basics. For example, a $4.97 jar of 2.63 liter Orange Juice is going for $14.29 in Churchill. Other items like detergent are three times more expensive than in south Manitoba. People used to take the overnight train to a nearby town, Thompson, a few times a year for cheaper meat and suppliers, but they cannot do that any longer as the railway is destroyed. Table 1 is a snapshot of the comparison of the cost of some of the items in Churchill and the provincial capital, Winnipeg.

2.2.2 Impact on fuel supply

As of June 2017, the average price of gasoline in Winnipeg is $1.17 for a liter of regular but selling for $2.17 per liter in Churchill's only filling station. The gasoline was transported by railway but it now relies on shipment arranged by the Manitoba government and tanks normally reserved for jet fuel by Calm Air. Also, the provincial government agreed to subsidize the cost of the shipment to ease the financial burden on the local people.

2.2.3 Impact on postal services

Cars or furniture used to be transported by train, but currently a barge is employed for this purpose. Canada Post may raise their rate because of this challenge; therefore, businesses would be further affected by the disruption of parcel deliveries. When shipping rates increase, cost of goods and services is passed to customers. For the Northern Study center, a research center located in Churchill, they order things like chemical and oil products they need for six to eight months and put them in a barge in Montreal. Researchers who come up and are willing to get several barrels of fuel for a project now need to plan well ahead due to the disruption of the rail.

TABLE 1 Price index in Churchill and Winnipeg.

Content	Volume	Price in Churchill (northern)	Price in Winnipeg (superstore)	Pc/Pw (%)
Grocery				
Tropicana orange juice	2.63 L	14.29	4.97	287.53%
Danone activia yougurt	650 g	6.09	3.87	157.36%
Beatrice milk	4 L	10.89	4.59	237.25%
Beatrice milk	2 L	5.95	3.2	185.94%
Lays potato chips	255 g	5.85	2.47	236.84%
Wonder bread white	570 g	5.25	2.48	211.69%
Produce				
Apples gala bag	3 LB	4.99	3.49	142.98%
Broccoli iceless	Per each	10.79	2.97	363.30%
Blade steak, boneless	Per kg	25.79	13.48	191.32%
Houseware				
Tide liquid high efficiency, original	1.47 L	15.89	5	317.80%

2.2.4 Impact on health care

The most hit by this disruption are women. This is mainly due to the inability of these women to participate in the Breast Check program. The program helps women in Manitoba over the age of 50 check for early signs of breast cancer. The mammograms machine used to be accessible via train every two years. Letters are sent to qualifying women to make an appointment. However, since there are no more trains coming to Churchill until the railway is fixed, the women have no choice but to remain in Churchill. Visiting Winnipeg for such an endeavor would cost these women not less than 1400 Canadian dollars.

2.2.5 Impact on liquor

The Liquor Mart in Churchill itself lost over 50% of sales. Without the train, the liquor store does not have people coming from other communities. In addition, it lost the far north customers as Churchill can't supply them at the same price, with the freight cost going up to 95 cents a pound. The store is not as profitable as it used to be. This has resulted in the sales going back to the distribution center in Winnipeg.

2.3   Other impacts

2.3.1   Tourism and commuting cost

Churchill heavily relies on tourism, especially in the summer season, when Beluga whales show up in the Churchill River and polar bear also come out. This period typically starts from mid-October and ends in late November. Due to the disruption of the railway, hotels and tour companies cancel most of the tours. Independent travelers are expected to fill the gaps between the busy seasons or even the hours between tour groups but they are more sensitive to the fluctuation of commuting cost than family or tour travelers. Without the train, there is a sharp decrease in the number of independent tourists this summer. People need to think twice before they head to Churchill or leave Churchill for a trip as the commuting cost has doubled. Box 1 is an extract from an interview of a tourist to Churchill. The names of the individuals are indicated with letters for privacy reasons.

2.3.2   Employment

According to what our participants shared with us, there are only 20 people currently working in the port at this moment, as many people did not get a callback from the company. The subsequent closure of the Hudson Bay Railway resulted in a spinoff effect on the tourism and transshipment industry. Due to the decline in customers and increasing commuting cost, employers have cut down on the number of employees. The closure of the railway resulted in layoffs at transshipment firms as well as hotels in Churchill. A local hardware store used to have ten employees but has laid off eight. Meanwhile most laid-off Churchill workers left the community to find jobs or remain unemployed. Without an

---

Box 1

Since early 2017, H has been organizing a four-member family trip to Churchill. The family intended to fly from New Brunswick to Winnipeg, take the train all the way to Thompson and then Churchill, and finally, fly back home. However, the flood washed out the railway in May 2017. Alternative plan sought from Calm Air shows a cost of $4700 dollars to get a car to Thompson and fly from there. H eventually got on the flight with her family and a friend from Australia. They made it to Churchill but at an exorbitant fee. Churchill may not be able to welcome such tourists if the trend continues. Definitely not so many people can afford the high price of travel.

> **Box 2**
>
> E accepted a job offer to work in Churchill in September 2016. Her employer offered her a free train ticket from Thompson to Churchill, so she drove to Thompson to take the train. Her friends are unable to visit her because of the disruption of the railway. Further, she also notices how this disruption influences her local friends' attitude towards livelihood. People are sad and always talk about prices and how government has abandoned them. Lots of people lose their jobs and they can't even commute to the next city to find one as there is no means to do that. Trying to support the local community, she mainly buys groceries in the only two grocery stores, even though it means that she needs to spend much more money. Luckily, she makes enough money to support herself but she is not sure about the future as prices are still going up.

adjacent community, people either leave their family in Churchill or move south together. Box 2 is an extract of somebody's experience in terms of employment in Churchill.

2.3.3   Psychology effect

The washout and uncertainty of the Churchill railway is a concern to the local community. More than half of our participants have mentioned the words "frustrated" and "uncertain." They are frustrated about the disagreement and conflict between the government and Omnitrax, leaving them without a rail connection. One of our participants said:

> And then psychologically, if you continue seeing how people are talking and are frustrated by the situation, it creates a lack of optimism everywhere. If a community lacks optimism, then you are not going to have entrepreneurs. You are not going to have someone come here, thinking they should take a chance and spend some money to start a business when you have that uncertainty.

It also creates a mist to the public. One of the participants commented on the situation of Churchill this way:

> There was a mist like. Don't go to Churchill, they have a state of emergency. They have no food. The media advertisement goes like please come to our community and support our economy. Some people are coming because they want to help and some people are not coming because they want to help us as well.

3   Opportunities for substitutions

In terms of model substitution, there is little incentive for switching within the proposed time-scale. Very few warehouses or shops have direct access to a barge. Even if the premises were air-connected, it would take weeks rather than days to start an alternative air or barge freight service, given constraints on infrastructure, rolling stock and operating procedures (McKinnon, 2006).

Although there is a port and an airport in Churchill, the freight is delivered via rail. If rail was required for the supply chain, the cost and time of the product delivery would critically increases or the flow of product would be interrupted. For rail substitution, an ice road to the remote Manitoba town, about 1000 km north of Winnipeg, has been completed and is ready for goods to be transported before Christmas. The journey is through the woods, across the tundra, and could take between 30 and 36 h.

In terms of locational substitution, which in this case involves replacing distant suppliers with local suppliers, not much can be done. People used to order groceries online or go south to Thompson or Winnipeg to load their cars with groceries or do it via train. If the absence of railway were to last for weeks or even months, one would anticipate some return to localized sourcing.

4   The development of Northwest Passage as a potential solution

Following the development in the Arctic, it is likely that government and other organizations will come to the aid of Churchill. A massive investment in the infrastructure in this area will mean the avoidance of some of these scenarios. Climate change may just be bringing relief to the people of the north. It is expected that the Canadian government will invest massively in the Arctic. Already, a flagship project by the University of Manitoba dubbed the Churchill Marine Observatory (CMO) is a huge step. The CMO is the first of its kind in the world. When completed it is likely to attract more researchers and other projects to the region. This is a big indication of how seriously the Canadian government views the Arctic. After all, climate change may just be a solution to another climate change issue.

5   Conclusions

Climate change has two sides. There is a good side with opportunity and the bad side with destruction of key infrastructure through flooding and other natural phenomena. The positive side has been identified as potential hydrocarbon exploration and very critical is shipping. The Northwest Passage does not have the same advantages as the Northeast Passage, even though both are alternatives to conventional routes like the Suez Canal. Using the Arctic routes may also help avoid pirates and reduce cost and distance of transporting goods and people.

Despite this positive future outlook, current preparation for intensified shipping in the Northwest Passage is far behind. This is exemplified by the case study presented on the town of Churchill. The case study shows the serious challenges a disrupted infrastructure can have on a typical town in the Arctic. Considering the nature of the Northwest Passage, even more resources are required relative to the Northeast Passage in order to ensure safety and reap the full economic benefit of this opportunity.

Following the disruption of the railway, the level of economic activity dropped sharply. In a short time period, the community is plunged into a deep economic and social crisis like unemployment and food scarcity. Without help from the government, many businesses would not survive. Most local businesses have fragile trading and cash flow. They would not be able to withstand even a temporary loss of business. One participant who has worked for the port and lived in Churchill over 10 years sums up the importance of the port and the role in the current predicament as well as what the future will look like without intensified development of the Northwest Passage. The participant said:

I think the potential for Churchill is bright but the opportunity is slowly fading away because some of what is happening here. No freight coming through so no one can do business. They have to look elsewhere to get their products or material. So a lot of the shipping that was done through here to the North now goes through Montreal. Once this becomes the norm, even if the rail is fixed, there may not be any incentive to revert back, bearing in mind the high level of uncertainty. They may say it is easier to get stuff in Montreal, I just need to make a phone call. Customers may not come back here. They may say there are too many problems with Churchill, I don't want my goods to get stuck there again. Things are just moving slowly away from Churchill.

Further, the current problem of Churchill brings into light the challenges indigenous communities face. It should be noted that the Churchill railway is not merely lines that run from one place of origin taking tourists or cargoes to another. It is a component of vast, complex technological and economic systems that connect many indigenous and local communities in North Manitoba. The debate of repairing the railway is significant in that it also highlights issues of social and cultural diversity, which are also often neglected in the discussion of human-cooperation relations in the North in the context of plans for transportation infrastructure. As the railway passes through a number of communities, especially communities having no access to airline, it cannot be seen as a project that affects solely the Churchill Township.

With increased voyages through the Northwest Passage and research activities on the rise, it is clear the Canadian government has identified the significance of the Arctic to the future of the country. Flagship projects like the Churchill Marine Observatory (CMO) and the Microbial Genomics for Oil Spill Preparedness in Canada's Arctic Marine Environment (GENICE) have already resulted in the first research vessel, the William Kennedy, heading to the Hudson Bay. It should be noted that this vessel is solely dedicated for research in this area. In addition, the CCGS Amundsen ice-breaker is going to Churchill for scientific studies; on board are many scientists and some politicians. Canada is highly committed to see the North open up through the opportunity of Arctic shipping.

References

Afenyo, M., Veitch, B., Khan, F., 2016. A state-of-the-art review of fate and transport of oil spills in open and ice-covered water. Ocean Eng. 119, 233–248.

Baker, C.J., Chapman, L., Quinn, A., Dobney, K., 2010. Climate change and railway industry: a review. Proc. Inst. Mech. Eng. C J. Mech. Eng. Sci. 224 (C3), 519–528.

CNW, 2017. OmniTRAX Announces Indefinite Closure of the Hudson Bay Railway From Amery to Churchill. Retrieved from: https://www.newswire.ca/news-releases/omnitrax-announces-indefinite-closure-of-the-hudson-bay-railway-from-amery-to-churchill-627505613.html.

European Environment Agency, 2014. Europe's transport network vulnerable to climate change. Retrieved from https://www.eea.europa.eu/highlights/europe2019s-transport-network-vulnerable-to.

Juttner, U., Peck, H., Christopher, M.G., 2003. Supply chain risk management: outlining an agenda for future research. Int. J. Log. Res. Appl. 6 (4), 197–210.

Lasserre, F., Pelletier, S., 2011. Polar super seaways? Maritime transport in the Arctic: an analysis of shipowners' intentions. J. Trans. Geogr. 19 (6), 1465–1473.

Lee, T., Kim, H., 2015. Barriers of voyaging on the Northern Sea Route: a perspective from shipping Companies. Mar. Policy 62, 264–270.

Liu, M., Kronbak, J., 2010. The potential economic viability of using the Northern Sea Route (NSR) as an alternative route between Asia and Europe. J. Trans. Geogr. 18 (3), 434–444.

McKinnon, A., 2006. Life without trucks: the impact of a temporary disruption of road freight transport on a national economy. J. Bus. Logist. 27 (2), 227–250.

Nemry, F., Demirel, H., 2012. Impacts of Climate Change on Transport: A focus on road and rail transport infrastructures. European Commission, Joint Research Centre (JRC). Institute for Prospective Technological Studies (IPTS), Chicago.

Ng, A.K., Andrews, J., Babb, D., Lin, Y., Becker, A., 2018. Implications of climate change for shipping: opening the Arctic seas. Wiley Interdiscip. Rev. Clim. Change 9 (2), E507.

Østreng, W., Eger, K.M., Fløistad, B., Jørgensen-Dahl, A., Lothe, L., Mejlænder-Larsen, M., Wergeland, T., 2013. Shipping in Arctic Waters: A Comparison of the Northeast, Northwest and Trans Polar Passages. Springer Science & Business Media, Berlin, Heidelberg.

Peck, H., 2005. The drivers of supply chain vulnerability: an integrated framework. Int. J. Phys. Distrib. Logist. Manag. 35 (4), 210–232.

Robertson, D., 2017. "Churchill dying," business leader says, vowing blockade if via tries to ship rail cars south by barge. Winnipeg Free Press.

Schøyen, H., Bråthen, S., 2011. The Northern Sea Route versus the Suez Canal: cases from bulk shipping. J. Trans. Geogr. 19 (4), 977–983.

Schweikert, A., Chinowsky, P., Kwiatkowski, K., Espinet, X., 2014. The infrastructure planning support system: analyzing the impact of climate change on road infrastructure and development. Transport Policy. 35, 146–153.

Shinozuka, M., Rose, A., Eguchi, R.T., 1998. Engineering and socioeconomic impacts of earthquakes: An analysis of electricity lifeline disruptions in the new Madrid area (No. PB-99-130635/XAB; MCEER-MONO-2). Multidisciplinary Center for Earthquake Engineering Research, Buffalo, NY (United States); National Science Foundation, Arlington, VA (United States); New York State Government, Albany, NY (United States); Federal Highway Administration, Washington, DC (United States); Federal Emergency Management Agency, Washington, DC (United States). Chicago.

Smith, L.C., Stephenson, S., 2013. New Trans-Arctic shipping routes navigable by midcentury. Proc. Natl. Acad. Sci. 110 (13), E1191–E1195.

Svensson, G., 2000. Conceptual framework for the analysis of vulnerability in supply chains. Int. J. Phys. Distrib. Logist. Manag. 30 (9), 731–749.

Tatano, H., Tsuchiya, S., 2008. A framework for economic loss estimation due to seismic transportation network disruption: a spatial computable general equilibrium approach. Nat. Hazards 44 (2), 253–265. https://doi.org/10.1007/s11069-007-9151-0.

Part V

Other key issues

Chapter 14

An investigation into the responsibility of cruise tourism in China

Yui-Yip Lau[a], Xiaodong Sun[b]
[a]Division of Business and Hospitality Management, College of Professional and Continuing Education, The Hong Kong Polytechnic University, Kowloon, Hong Kong, [b]School of Business Administration, East China Normal University, Shanghai, China

1 Introduction

In the context of passenger transport, shipping is designed for the movement of people from one place to another place; in the maritime regime, however, cruise ships perform more than a basic type of passenger transport. A cruise ship provides a service to serve the passenger who inclines toward going for relaxation, interest, and pleasure. Cruise ships generally make occasional calls in a specified region of a geographical continent. Kendail (1986, p. 360) defined a cruise as the "transportation of pleasure-seeking travelers on ocean voyages offering one or more glamorous ports of calls." Also, Wild and Dearing (2000, pp. 319–320) described a cruise as "any fare paying voyage for leisure onboard a vessel whose primary purpose is the accommodation of guests and not freight normally to visit a variety of destinations rather than to operate on a set route." In the past few decades, we identify that the cruise ship experience appears to be a recreational experience (Mileski et al., 2014) and in a more relaxed atmosphere (Lau et al., 2014).

Cruise tourism can be traced back to the 1840s (Sun et al., 2011). Cruising was the preferred travel mode for the world's social elite during the 1920s. After the Second World War (1939–45), the decline of the cruising industry led to losing trade to passenger aircraft (Johnson, 2002). However, the turning point appeared in the 1960s. The blooming of the aviation industry had caused the cruise industry to undergo a revolution. The cruise ship was transformed from its original use as a postal or passenger transport in the historical era to a luxury cruise ship purely used for tourism, holiday, and leisure purposes in the contemporary era (Sun et al., 2014). Thanks to technological and scientific advancement in the

Maritime Transport and Regional Sustainability. https://doi.org/10.1016/B978-0-12-819134-7.00015-0

239

past 30 years, it has created significant improvements in the power supply, design, catering facilities, and accommodation of cruise ships. The modern cruise ships generate stiff competition with the land-based holidays, including hotels (Lois et al., 2004). In doing so, cruise tourism has been considered an area of tremendous growth in the global hospitality and tourism industry (Marti, 2004). From 1990 to 2015, the number of cruise passengers increased at an average of 7% per year, up to over 23 million cruise passengers in 2015 (CLIA, 2016). More than 154 million passengers have taken a 2+ day cruise. Over 68% of the total passengers have been generated in the past 10 years (Rodrigue and Notteboom, 2012). In accordance with *2018 Cruise Industry Outlook,* Table 1 summarizes which geographical regions cruise passengers mainly came from during 2011 to 2016. In general, cruise tourism generates substantial economic benefits, including coastal city transformation, regional economic cooperation, and port economic development (Sun et al., 2014). In the long term, cruise tourism can create a significant contribution to a destination's economy, creating jobs and generating revenue. Also, cruise tourism can avoid urban relocation by generating local jobs (CBI Ministry of Foreign Affairs, 2016).

Starting from 2000, various frequent cruisers are looking for attractive destinations, diverse cultures, and wonderful experiences which they would find in Asian regions, notably in China (Lau et al., 2014). The influx of a large number of cruisers within a limited time and an increasing arrival number of

TABLE 1 Demand for cruising from different regions (2011–16)

Regions	Number of cruise passengers (millions)
United States	11.5
China	2.1
Germany	2.0
United Kingdom	1.9
Australia	1.3
Canada	0.8
Italy	0.8
France	0.6
Brazil	0.5
Spain	0.5

(Data from Cruise Lines International Association, Inc., 2017. 2018 Cruise Industry Outlook. Cruise Lines International Association, Washington, United States.)

cruise ships will aggravate the problems. Since the Chinese cruise industry is definitely gaining much attention and rapid development in recent years, there is an urgent demand for exploring the negative externalities behind the cruise tourism and investigating the association between cruise industry economic development and environmental impacts. Reducing the negative impact of cruise tourism means that we can decrease the potential risks of cruise industry development. In other words, it is necessary to promote the cruise industry in a reasonable and moderate approach under the framework of responsibility and sustainability, so as to ensure the healthy development of the cruise industry in the forthcoming years.

Due to cruise industry being a resource-dependent industry, it brings various negative impacts, consisting of environmental degradation, resource squeeze, marine pollution, to name but a few. Thus, how to reduce the negative impacts of cruise tourism and generate responsibility in cruise tourism is now highly addressed in the cruise industry. The accountability and sustainability of cruise tourism will be a key issue that the stakeholders cannot avoid in the development process (Carić, 2016).

In this chapter, we have conducted a literature review about the research trend in cruise shipping. Traditionally, the researchers (e.g., Rodrigue and Notteboom, 2012; Jia et al., 2013; Sun et al., 2015; Lau et al., 2014; Lau and Yip, 2018) have mainly highlighted that the development of the cruise industry has generated a significant economic and social development. That the number of tourists and cruise liners creates negative impacts on local communities seem seriously overlooked. In terms of Chinese literature, we manually searched the key words from the CSSCI or CSSCD journals containing "cruise + impact," "cruise + environment," "cruise + pollution," "cruise + community," "cruise + society," "cruise + residents," "cruise + sustainability," "cruise + responsibility," "cruise + employee," and "cruise + conflict," to name but a few. All these key words are highly relevant with cruise tourism studies. Although the research on cruise tourism has emerged dramatically in the last decade, only 11 articles were relevant with the sustainability, accountability, and the negative impacts of cruise tourism. For instance, Xie et al. (2010) investigated the different aspects of environmental impact pertaining to cruise construction, facility operation, transportation and distribution, consumption utilization, and waste disposal by using a life cycle theory so as to generate the ideas of environmental pollution control. Ji (2015) used the stakeholder theory to give a comprehensive discussion of the potential negative impacts of cruise tourism. Du et al. (2016) investigated the green shore power experience in Hamburg Ferry Terminal, including engineering design, regulations and policies, financing channels, and social participation. The study has given a valuable insight on China's cruise port shore power projects implementation. In terms of English literature, we also manually searched keywords closely associated with cruise tourism studies, including "cruise + environment," "cruise + polluting or pollution," "cruise + community," "cruise + society or social," "cruise + resident," "cruise + sustainable or sustainability," "cruise + responsible or responsibility," "cruise + crew or worker," and "cruise + impacts or negative impacts," to name but a few. We found that different scholars began to explore the negative impacts, the sustainability, and liability issues of cruise tourism, which has created a research hotspot since 2010. The negative impacts of the cruise industry has been a hot topic in the international academic community and provided significant research results. However, the analysis and discussion provided small numbers of case studies in China.

2   Negative effects of cruise tourism

The booming of cruise tourism stimulates negative impact on environmental resources, marine ecosystems, and marine environment (Tang et al., 2013; Carić et al., 2016; Xu and He, 2016). The different stakeholders design the cross-regional cruise routes together with the trend of larger sizes of cruise ships speeds up the negative environmental impacts of cruise tourism. To this end, depletion of natural resources, excessive demand for energy and water in destinations, and climate change will occur (Kaldy, 2011). In addition, cruise ships not only generate air pollutants, but also release large amounts of waste and waste water that damages the marine environment, including miscellaneous drainage (i.e., gray water), sewage, hazardous wastes, oily sewage (i.e., oily bilge water), solid waste, and ballast water (Lester et al., 2016). The details are given in Table 2.

TABLE 2   Key sources of waste generated by cruise ships

Source of pollution	Description
Black water	Refers to domestic sewage. Each passenger produces 10 gallons of sewage per day
Gray water	Contains sinks and shower waste. Each passenger produces 90 gallons of gray water per day
Garbage and solid waste	A cruise ship produces 3.5 kg of solid waste per day
Hazardous waste	Generated from dry cleaning materials, the printing shop, chemical cleaning, and batteries. A cruise ship produces 15 gallons of toxic waste per day
Oily bilge water	The contaminants collected from the ship's hull, including fuel, oil, and wastewater. A cruise ship produces 7000 gallons of toxic waste per day
Ballast water	The exchange of ballast water causes local species exchange
Diesel exhaust emissions	A cruise ship generates emissions equivalent to 12,000 vehicles

(*Based on* Lester, S.E., White, C., Mayall, K., 2016. Environmental and economic implications of alternative cruise ship pathways in Bermuda. Ocean Coast. Manag. 132, 70–79.)

In addition, cruise ships create bacteria and toxic substances that affect the ecology and the health of residents. For example, in response to residents' concerns about the ecological and health impacts of cruise tourism, in 2000, the US Department of Environmental Conservation carried out a research study of the cruise ship activity area. It found that the waste water did not meet the sanitary conditions. The black and gray water contained toxic levels also exceeding national standards (Loehr et al., 2006). Kaldy (2011) established an initial dilution equation model including key elements like the ship width, cruise draft, navigation speed, and sewage discharge rate. The study pointed out that adjusting the size of the cruise ship and the speed of navigation, the effect of decreasing the discharge of cruise wastewater can be achieved. When the speed of the cruise ship is greater than or equal to 6 nautical miles/hour and the offshore is greater than or equal to 1 nautical mile, the sewage discharge from the cruise will not create water pollution.

In addition, the researchers explored that the environmental impact of cruise tourism is much larger than general types of tourism. Toh et al. (2005) addressed that the energy demand for hotel functions by a cruise ship is five times higher than that of ordinary high-end hotels. Moreover, Klein's (2009) study indicated that cruise ships produced an average carbon footprint over three times that of trains, airplanes, and ferries. Furthermore, Howitt et al. (2010) assessed the carbon dioxide emissions from cruise ships after considering the key factors of cruise ship size, engine life, engine size, onboard facilities, crew number, and cruise ship frequency. The carbon emissions of cruise ships were significantly higher than those of passenger aircraft.

In addition, Zhong et al. (2016) identified that providing ecotourism products to cruisers at cruise destinations will further reduce the negative environmental impacts of cruise tourism. The two main features of ecotourism are to protect nature and sustain community benefit. Most onshore activities still failed to perform with ecotourism characteristics. The common onshore activities like forest trips, all-terrain vehicle (ATV) expeditions, jet boats, and a lack of guided diving leads to physical damage to the natural environment (Johnson, 2006). Up to now, the international organizations, cruise associations, government bodies, policymakers, cruise operators, cruise terminals, and industry practitioners have coordinated together in order to strive to respond to the sustainability and responsibility of cruise tourism through setting up a unified management organization and drafting waste management plans and site policies. Nevertheless, the notion of responsibility and sustainability of cruise tourism remains unpopular, notably in China. In the next section, we adopt the representative example on how the shore power can be applied in China cruise ports in order to foster cruise industry sustainability.

3   The use of shore power in China's cruise ports

With the rapid development of the cruise industry in China, the cruise ports induce various environmental problems. When a cruise ship is berthed, it uses a large amount of heavy oil and diesel oil to generate a large electricity demand. Heavy oil and diesel oil not only produce a large amount of toxic gasses like carbon dioxide, sulfide, and nitride, but also create noise pollution. This affects a quality of life and atmosphere environment adversely. Furthermore, cruise tourism brings other aspects of environmental problems, for examples, climate change, depletion of natural resources, and excessive demand for water and energy.

The cruise tourism effect on the port environment has received notable attention. Establishing a green cruise port is a "must" to become an inevitable trend and common pursuit in the cruise industry. The use of shore power for cruise ships has become a common practice in different cruise home ports worldwide. As an alternative power source, the adoption of shore power technology has a remarkable effect on reducing the emission of harmful substances like NOX, SOX, and PM10. Shore-based power supply refers to cruise ships using shore-based power supply instead of using marine generators during berthing period. Shore-based power is also known as shore power or alternative maritime power (AMP). Shore power can significantly reduce the emission of toxic substances and air pollution to meet environmental protection requirements (Shanghai Observer, 2016).

In accordance with the *Enhancement of China's Cruise Industry Sustainable Development* in March 2014, enterprises should give priority to using advanced technologies, low energy consumption, and safety environment facilities. In addition, the Chinese government encourages using cleaner fuels such as gas, liquefied petroleum gas, and natural gas. The new established cruise terminals are preferred to employ shore power systems. In order to promote the construction of a green cruise port, Shanghai has implemented the *Three Years Action Plan for Green Cruise Port (2015–2017)*. This clearly defines the overall objectives, key tasks, and protection measures. In order to build the concept of green shipping process, the idea of shore power has been launched in Shanghai. Wusongkou International Cruise Terminal—Berth One was considered as one of seven shore power demonstration projects in 2016.

China has actively promoted the construction of green cruise ports. More and more cruise ports have launched a series of green port development plans including the enlargement of cruise shore power projects, as cruise green energy is an important aspect of a green cruise port concept. The cruise fuel produces large emissions of pollutants and thus, we need actively to promote the installation of cruise shore power equipment to align with creating green cruise ports.

Shanghai Wusongkou International Cruise Port not only received the largest number of cruisers in China, but also provided the earliest start of shore power facilities for cruise ships. The first phase of shore-based power supply project was officially put into operation on July 13, 2016. Until now, this has been the world's largest cruise frequency conversion shore power system as well as the first set of cruise shore power system in Asia (Shanghai Observer, 2016).

In general, the construction of the shore-based power supply project in Shanghai Wusong International Cruise Port is divided into two main phases. It will cover four berths after completion. Currently, the first phase of the project can provide a total capacity of 16,000 kVA with the coverage of two berths. SkySea Cruise Company has taken the lead in using shore power facilities. In addition, the cruise ship "Majestic Princess" successfully completed adoption of shore power on November 20, 2017. In 2016, Shanghai Wusong International Cruise Port planned to arrange 488 voyages to dock at the port; the docking cruise would use 36.6 million kWh of alternative electricity to achieve zero emissions. In 2017, the annual cruise berthing number in Shanghai Wusong International Cruise Port would exceed 1200. We expected that the emission reduction effect was to be more obvious when the replacement power was 87.8 million kWh. Based on the average 150,000 cruise tonnage, carbon dioxide can be further reduced by 36,000 tons, sulfur dioxide by 750 tons, and nitrogen oxide by 65 tons, annually.

Shanghai has tried to establish an international shipping center and a world-famous tourist city in response to the direction of the Belt and Road Initiative. In addition, it has realized the cruise economy has been evolved from a small to large scale. In October 2018, the General Office of the Shanghai Municipal Government issued a full text of *Intensifying the Shanghai Cruise Economy Development*. It indicates Shanghai cruise ports should continue to upgrade the cruise waste disposal support capacity, improve the shore power system utilization, and improve the standard of health management of the cruise port. Eventually, all cruise ships are required to use shore power by January 1, 2021.

However, the use of shore power systems in cruise ports is now facing unfolding challenges. First of all, the cruise ports lack clear guidance, mechanism, and systems to construct the cruise port and create effective monitoring and tracing of green indicators. Also, the cruise port operators and public have demonstrated a lack of environmental protection awareness. In addition, the huge construction and associated facilities costs pose a restriction to the use of shore power. In terms of electricity use, Shanghai Coastal Power charges electricity according to the general commercial electricity use with the basic monthly electricity charges on the basis of the power access capacity. Taking Wusong cruise port as an example, the actual shore power consumption was about 470,000 kWh in 2016, generating about 350,000 yuan of electricity charge. It is equivalent to the average electricity consumption of 0.75 yuan/kWh. Wusong Cruise Port has set up a shore power technology company to provide shore power services. Based on the annual 10% utilization rate of shore power facilities, the annual expenditure on human resources costs, equipment depreciation costs, and equipment maintenance fees is about 5.75 million yuan (Shanghai China, 2018).

4 Responsible cruise tourism

With the rise of an increasing trend in scrutiny and criticism of industry practitioners and researchers on the negative impacts of cruise tourism activities, the notion of responsible cruise tourism has become a hot issue of tourism research and practice. Initially, responsible cruise tourism concentrated on participants like tourists and travel agents understand a wide range of tourist interactions can minimize negative effects of environmental and social impacts and maximize the local communities benefit (Frey and George, 2010). Now, we need to consider stakeholder theory to extend it to cruise tourism activities in different regions, different areas, different levels, and different linkages. Also, we need to create a framework of action including cruise liners, cruise ports, travel agencies, onshore service providers, organizations (i.e., government bodies, cruise associations, nonprofit organizations, and industry alliances), and individuals (i.e., local residents, crews, and cruisers). As shown in Fig. 1, all stakeholders require improved involvement through drafting a master plan, formulating activities, and providing clear guidelines so as to make cruise tourism into a sustainable and responsible industry.

Responsible cruise tourism systems include various stakeholders and effective implementation of responsible cruise tourism is largely determined by the attitudes and behaviors of government departments, organizations, enterprises, and associations toward the idea of cruise tourism. Firstly, government departments, industrial organizations, and industry associations are responsible for drafting and supervising the laws, policies, and industry standards to improve the tourists travel experience and the well-being of residents (Klein, 2009). How to align between the international standard and local government practice is of significant importance. Secondly, industry associations and industry organizations play a leading role in maintaining the industry standards, and

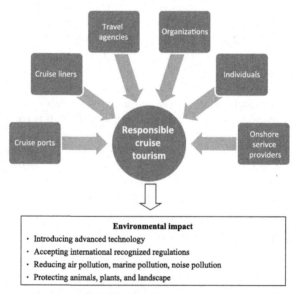

FIG. 1 A framework for responsible cruise tourism.

provide professional advice on responsible travel; for example (1) protecting the environment, including animals, plants, and landscapes; (2) minimizing pollution consisting of noise, waste disposal, and congestion (Goodwin and Francis, 2003). In addition, cruise liners, travel agencies, cruise ports, industry associations, and other organizations must jointly create a responsible cruise culture to increase the cruise visitors' sense of responsibility for the natural environment, marine ecology, and community culture through publicity and education.

5 Conclusion

This chapter provides an insight into preventing problems before they happen. This is the time for all stakeholders to take responsibility at the time of the development of cruise tourism. Although the international organizations and associations highlighted the negative effect of the cruise tourism, and various industry management practices have been raised, the local countries or cities still do not deeply recognize the negative externalities of the cruise industry in terms of environmental impacts, notably in China. To this end, we identify different aspects in terms of responsibility and sustainability of China's cruise tourism that should be investigated in the forthcoming years:

- The environmental impact of cruise tourism on China's coastal port areas is a hot topic and contains a large number of research results. With the increasing ecological impacts of tourism development, achieving the dual goals of minimizing environmental impact and maximizing economic benefits should be addressed (Yao and Chen, 2015). Thus, we can use the carbon emissions and carbon footprint during the cruise stops to explore the impact of cruise tourism on air quality and marine ecology.
- The research study explored that the influx of a large number of cruisers will occupy the living space and resources of local residents. As a result, rising local community prices, traffic jams, and even garbage mountains will reduce the residents' quality of life and cruiser travel experience. In the future research, we can carry out research focusing on local residents' attitudes toward cruise port infrastructure construction and risk perception of cruise tourism development. This can truly reflect how cruise tourism can bring substantial benefits to local residents.
- From the perspective of economic impact, the negative effects of cruise tourism are mainly reflected in the uncertainty of regional economy. Currently, China's cruise industry development is at the initial stage. The economic benefits of cruise tourism are mainly generated from the income sources such as port fees, shipping agency fees, and ticket sales commissions. The indirect economic benefits of cruise tourism are difficult to evaluate. Thus, we need to take into account cruise economic indicators including the indicator system, the cruise economy climate index, and the economic risk assessment in the future research.

References

Carić, H., 2016. Challenges and prospects of valuation–cruise ship pollution case. J. Clean. Prod. 111, 487–498.

Carić, H., Klobučar, G., Štambuk, A., 2016. Ecotoxicological risk assessment of antifouling emissions in a cruise ship port. J. Clean. Prod. 121, 159–168.

CBI Ministry of Foreign Affairs, 2016. Available at: https://www.cbi.eu/market-information/tourism/cruise-tourism. (Accessed June 26, 2019).

Cruise Lines International Association (CLIA), 2016. CLIA Cruise Industry Outlook. Available at: http://www.cruising.org. (Accessed June 1, 2019).

Du, X., Li, H.T., Wen, Y.Y., 2016. Experience and implications of the onshore power supply facility at the cruise terminal Altona in Hamburg, Germany. Environ. Sustain. Dev. 41 (4), 40–43.

Frey, N., George, R., 2010. Responsible tourism management: the missing link between business owners' attitudes and behaviour in the Cape Town tourism industry. Tour. Manag. 31 (5), 621–628.

Goodwin, H., Francis, J., 2003. Ethical and responsible tourism: consumer trends in the UK. J. Vacat. Mark. 9 (3), 271–284.

Howitt, O.J.A., Revol, V.G.N., Smith, I.J., 2010. Carbon emissions from international cruise ship passengers' travel to and from New Zealand. Energy Policy 38 (5), 2552–2560.

Ji, J., 2015. An interpretation on the sustainable development of cruise tourism based on stakeholders theory. Tour. Forum 8 (2), 68–74.

Jia, P., Liu, R.G., Sun, R.P., 2013. A prediction model for cruise tourism demand based on BP neural network. Sci. Res. Manag. 34 (6), 77–83.

Johnson, D., 2002. Environmentally sustainable cruise tourism: a reality check. Mar. Policy 26, 261–270.

Johnson, D., 2006. Providing ecotourism excursions for cruise passengers. J. Sustain. Tour. 14 (1), 43–54.

Kaldy, J., 2011. Using a macroalgal δ^{15}N bioassay to detect cruise ship waste water effluent inputs. Mar. Pollut. Bull. 62 (8), 1762–1771.

Kendail, L., 1986. The Business of Shipping. Cornell Maritime Press, Centreville, MD.

Klein, R.A., 2009. Keeping the cruise tourism responsible: The challenge for the ports to maintain high self esteem. In: International Conference for Responsible Tourism in Destination, Belmopan, Belize.

Lau, Y.Y., Yip, T.L., 2018. Location characteristics of cruise terminals in the Asian region: lessons in Hong Kong and Shanghai. In: 2018 World Transport Convention, Beijing, China.

Lau, Y.Y., Tam, K.C., Ng, A.K.Y., Pallis, A.A., 2014. Cruise terminals site selection process: an institutional analysis of the Kai Tak Cruise Terminal in Hong Kong. Res. Transp. Bus. Manag. 13, 16–23.

Lester, S.E., White, C., Mayall, K., 2016. Environmental and economic implications of alternative cruise ship pathways in Bermuda. Ocean Coast. Manag. 132, 70–79.

Loehr, L.C., Beegle-Krause, C.J., George, K., 2006. The significance of dilution in evaluating possible impacts of wastewater discharges from large cruise ships. Mar. Pollut. Bull. 52 (6), 681–688.

Lois, P., Wang, J., Ruxton, W.T., 2004. Formal safety assessment of cruise ships. Tour. Manag. 25, 93–109.

Marti, B.E., 2004. Trends in world and extended-length cruising (1985–2002). Mar. Policy 28 (3), 199–211.

Mileski, J.P., Wang, G., Beacham IV, L.L., 2014. Understanding the causes of recent cruise ship mishaps and disasters. Res. Transp. Bus. Manag. 13, 65–70.

Rodrigue, J.P., Notteboom, T., 2012. The geography of cruise shipping: itineraries, capacity deployment and ports of call. In: International Association of Maritime Economists (IAME) Conference, Taipei, Taiwan, September 5–8, 2012.

Shanghai China, 2018. Available at: www.shanghai.gov.cn/nw2/nw2314/nw2315/nw4411/u21aw1316761.html. (Accessed April 23, 2019).

Shanghai Observer, 2016. Available at: www.jfdaily.com/news/detail?id=92124. (Accessed June 26, 2019).

Sun, X.D., Jiao, Y., Tian, P., 2011. Marketing research and revenue optimization for the cruise industry: a concise review. Int. J. Hosp. Manag. 30 (3), 746–755.

Sun, X.D., Feng, X.G., Gauri, D.K., 2014. The cruise industry in China: efforts, progress and challenges. Int. J. Hosp. Manag. 42, 71–84.

Sun, X.D., Wu, X.R., Feng, X.G., 2015. Basic characteristics and key elements of cruise itinerary planning. Tour. Tribune 30 (11), 111–121.

Tang, C.C., Zhong, L.S., Cheng, S.K., 2013. A review on sustainable development for tourist destination. Prog. Geogr. 32 (6), 984–992.

Toh, R.S., Rivers, M.J., Ling, T.W., 2005. Room occupancies: cruise lines out-do the hotels. Int. J. Hosp. Manag. 24 (1), 121–135.

Wild, P., Dearing, J., 2000. Development of and prospects of cruising in Europe. Marit. Policy Manag. 27 (4), 315–333.

Xie, F., Li, H.M., Li, D., 2010. Cruise environmental pollution control mechanism and tactics based on life-cycle assessment. Mar. Sci. Bull. 29 (6), 702–706.

Xu, F.F., He, Y.M., 2016. A review of environmental ethics and sustainable tourism behavior. Prog. Geogr. 35 (6), 724–736.

Yao, Z.G., Chen, T., 2015. Review on overseas tourism eco-efficiency studies. J. Nat. Resour. 30 (7), 1222–1231.

Zhong, L.S., Ma, X.Y., Zeng, Y.X., 2016. Progresses and prospects of ecotourism research in China. Prog. Geogr. 35 (6), 679–690.

Further reading

Cruise Lines International Association, Inc, 2017. 2018 Cruise Industry Outlook. Cruise Lines International Association, Washington, United States.

Chapter 15

Government initiatives on transport and regional systems: The development and management of Chinese high-speed rail

Changmin Jiang, Adolf K.Y. Ng, Xiaoyu Li
Department of Supply Chain Management, Asper School of Business, University of Manitoba, Winnipeg, MB, Canada

1 Introduction

There have been considerable efforts made by the governments of many countries and regions to engage remote or peripheral regions within national or continental networks. Indeed, governments often play hugely significant, and sometimes decisive, roles in the evolution and development of transport and regional systems. Substantial research in transport and transport policies has supported this notion (e.g., Hou and Li, 2011; Panayides et al., 2015; Bonnafous, 2015). The introduction of new and major government initiatives often generates heated and controversial public debate, not only because they involve huge transport investments and institutional changes, but also because it may cause transformation of transport and regional systems (Daamen and Vries, 2013; Monios and Lambert, 2013; Ng et al., 2018, 2019). For instance, many European governments have encouraged and funded the expansion of the European Community to the less developed and remote regions, and most of these funds involve transport infrastructures (Dall'Erba and Le Gallo, 2008). In some cases, new circumstance has led to the redistribution of population (e.g., HSR is likely to change the spatial distribution of population along the corridors). Therefore, government initiatives and transport systems are particularly relevant.

In transport systems, the introduction of high-speed rail (HSR) is no doubt one of the most remarkable changes in the past decades. It has been more than half a century since Japan launched the first modern HSR "Shinkansen" on the

Maritime Transport and Regional Sustainability. https://doi.org/10.1016/B978-0-12-819134-7.00016-2

Tokyo-Osaka route in 1964. HSR has now been regarded as a common transport mode in countries like France, Italy, Spain, Germany, South Korea, and Taiwan and become the dominant transport mode on some corridors, especially for short-distance markets, serving millions of passengers every day (Jiang and Li, 2016). Worldwide experience has indicated that HSR is the most competitive transport mode for routes between 400 and 1000 km where airlines usually lose ground to HSR, being either pulled out of markets or into a big decline in market share (Rothengatter, 2010).

HSR is not just a faster transport mode but poses substantial impacts on regional issues and reshapes spatial structures (Ureña et al., 2009; Vickerman, 2017). Previous studies have reported that HSR networks play a critical role in spatial transformation and regional accessibility. For instance, Cao et al. (2013) indicate that the large-scale implementation of the HSR network in China generates the redistribution of demographic and economic activities. Zheng and Kahn (2013) discuss that HSR implementation triggered the development of surrounding second-tier and third-tier cities. Wang et al. (2012) report that the development of HSR will enlarge and transform tourism market space, intensify market competition on a large scale, and redistribute urban tourism centers. On the other hand, even with the massive realignments in construction and operations, the impacts of HSR and the resulting spatial patterns of transport systems are likely to be countered by certain inertia and forces. Active government initiatives (e.g., public funds) do not necessarily transform regional and transport systems (Breidenbach and Mitze, 2015; Monios, 2016). There is thus a need to have more critical analysis of the far-reaching impacts of government initiatives on HSR development and evolution to direct its future planning.

Compared with its counterparts, China developed HSR relatively late, but its expansion rate has increased dramatically over the past decade. Indeed, it was not until the construction and operation of Chinese HSR that this term has regained the world's attention. The operation of the route between Guangzhou and Wuhan on December 26, 2009, marked that China railway transport entered a new high-speed era. Travel time has been reduced from 12 to 3 h because of the launch of the network, with over 23,000 km in service, more than the rest of the world's high-speed lines (HSLs) combined. According to the National Development and Reform Commission (NDRC), by 2020, the Chinese rail network will be 150,000 km in total, including about 30,000 km HSLs. With a system comprising four horizontal and vertical lines, China's HSR served 80 million people in 2014, and is expected to serve 90% of the population by 2020. With the commitment from the Ministry of Railway, the Chinese HSR network has expanded at an unprecedented speed and scale. This development is an extraordinary achievement, especially considering the fact that it started only less than a decade ago and is still in the middle of network expansion today. More aggressive plans have been proposed; many are under construction. With a well-established HSR network, the development and management of Chinese HSR serves as an illustrative exposition in understanding the impacts of major government initiatives on the spatial transformation of transport and regional systems. By starting with a review of the Chinese HSR network development and then moving to a discussion of its future development, we investigate the spatial transformation that the Chinese HSR has generated.

The rest of the chapter is organized as follows. Section 2 extends the theoretical discussions through a relevant case investigating the development of HSR in China. Then Section 3 projects its future development by discussing four possible development trends. Finally, concluding remarks can be found in Section 4.

2   Chinese HSR development

The Chinese government played a key role in HSR development from the initial design stage to the later expansion plan. In China, the HSR network has been encouraged and financially supported by the Chinese government, and in turn, the opening of HSR has had significantly positive impacts on regional economies, which are often regarded as important factors in planning and decision-making for policy makers. A major indicator of regional economic impacts is the effect of transport-induced agglomeration on business productivity. According to the World Bank (2014a), the estimated agglomeration effects of HSR on various second-tier and third-tier cities are substantial. As shown in Table 1, they occupy 0.55% of total gross domestic product (GDP) in Jinan per year, 1.03% in Dezhou, and 0.64% in Jilin. Although economic factors are critical to the decision on whether the government should invest in HSR projects, the rationale for the investment goes far beyond them. In fact, it is the combination of political and strategic factors, normally related to regional development objectives, that determines the feasibility of building an HSR network. The development of the Chinese HSR network is an excellent example

TABLE 1 Regional economic impacts of HSR on cities.

City	Agglomeration effects	
	Increase in GDP	Benefits (RMB in billions)
Jinan (Beijing-Shanghai HSR)	0.55%	3.65
Dezhou (Beijing-Shanghai HSR)	1.03%	3.59
Jilin (Changchun-Jilin HSR)	0.64%	2.39

(Data from World Bank, 2014a. Regional Economic Impact Analysis of High Speed Rail in China. Available at: http://www.worldbank.org/content/dam/Worldbank/document/EAP/China/high_ speed-rail-%20in-china-en.pdf.)

to show the impact of the government initiative on spatial transformation, as many previous studies have shown. A list of literature on the impacts of HSR on spatial transformation is shown in Table 1. Apart from the studies mentioned in Table 1, we further discuss two examples in the following, i.e., the announcement of 10 megacity regions and the 4 trillion RMB stimulus package, to illustrate the impact of the Chinese government initiative on infrastructure development and spatial transformation.

2.1 Ten megacity regions

In 2007, NDRC announced 10 megacity regions, including seven in inland areas and three along coastal economic zones, which are expected to become engines of domestic economic development. The idea of megacity regions refers to an economic concept embodied in a geographical location. It links multiple centers and subcenters and then integrates them in a hierarchical order via essential communication and transport infrastructures (Hall and Pain, 2006). The productivity of these 10 megacity regions is more than half of the whole country. For example, in 2005, the GDP of these 10 megacity regions accounted for about 53% of the total output and their population was about 35% of the total population in China (Xiao and Yuan, 2007). In fact, the development of the Chinese HSR network has been closely related to this spatial development strategy. HSR enables the formation of polycentric agglomerations of urban areas, i.e., the megacity regions. On the one hand, the spatial development strategy facilitates the exchange of information, economic, technological, and labor flows along HSR corridors, which would have positive effects on regional development from urban centers. On the other hand, the Chinese HSR network is expected to link all megacity regions, unify the whole national economy, and eventually achieve the goal of reconfiguring both regional and national economies. The cross-boundary HSR network development is likely to stimulate the integration of the Chinese regional economy, while the development of transport infrastructure would enable local governments to break fixed administrative boundaries and hence optimize their policies. As a result, the lower value-added activities are expected to shift from coastal areas to the western parts of China, and then the booming coastal economic zones would focus on developing high value-added services. The rapid development of the Chinese HSR network in parallel with the improvement of the regional competitiveness would decrease the level of regional uneven development.

2.2 Four trillion RMB (585 billion USD) stimulus package

From 1978 to 2008, the market share of road transport rose from 29.9% to 53.8%. The market share of aviation experienced a similar trend, increasing from 1.6% to 12.4%. However, the market share of rail has dropped from

62.7% to 33.3% (Fu et al., 2012). An important reason for this situation has been the capacity constraint of the rail industry. The capacity increase had not met traffic growth over these years. However, such capacity constraint has been largely alleviated due to the fact that the Chinese government increased its investment in infrastructure development, especially in HSR infrastructure. In 2008, the Chinese government announced an economic stimulus package of 4 trillion RMB (585 billion USD), including 1.5 trillion RMB (225 billion USD) on public infrastructure (e.g., railway, road, and airport construction). Since then, the importance of infrastructure development has been further emphasized. The Chinese HSR network is no doubt one of the main beneficiaries of this stimulus package. Fig. 1 shows the progression of the Chinese HSR network, which partially indicates that the stimulus package has a positive effect on the expansion of the Chines HSR network. As shown in Fig. 1, the Chinese government built new HSR lines every year, indicating that China adopted a simultaneous instead of sequential plan for the construction of multiple HSR corridors at the same time. In addition, we can see that the growth of HSR lines was relatively slow from 2003 to 2007. However, the development of HSR lines started to become faster at the beginning of 2008, right after the announcement of the stimulus package. The speed of growth has experienced a dramatic increase since 2011, causing the total length of HSR lines to more than triple within 4 years. We suppose that this phenomenon was largely due to the financial crisis in 2008—with a weakened global demand, China had to transform itself from an export-driven economy to rely on investment for GDP growth.

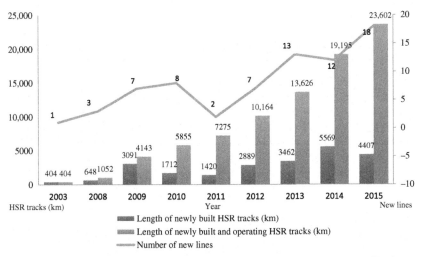

FIG. 1   The progression of the Chinese HSR network. (*Data from Chinese Ministry of Railways, 2016. Available at: http://www.china-railway.com.cn/.*)

2.3   Government support

In the period when the Chinese government began to plan the large-scaled HSR project, there was no shortage of financial resources. This is the main reason why the Chinese HSR network could be built and has achieved substantial success in such a short period of time. However, according to the China State Railway Group Co., Ltd,[a] which is a state-owned HSR operator, about two-thirds of its debt was from HSR construction, and most of the HSR corridors are not profitable. The big exception is the Beijing-Shanghai HSR corridor, which earned 6.58 billion RMB (1.01 billion USD) in 2015. Five other lines including the Ningbo-Hangzhou line also started to gain profits. However, other lines and corridors, especially those in western China, remain stuck in losses, such as the Xi'an-Zhengzhou line, operating at well below capacity (Garst, 2016). Even so, the government still invests billions of dollars every year in its development. In fact, there are a lot more indirect benefits of HSR, including spatial transformation. This is also the case in Spain, where the first HSR line began operation along the corridor between Madrid and Seville in 1992. It now has the longest HSR system (2515 km) in Europe, due to the fact that the Spanish government has regarded the HSR development as a priority in its transport policy. Spain was the only country in Europe that started to build the HSR network from a less populated city (Seville) since the main objective of HSR development is to promote economic growth in poor regions. This case also perfectly proves the impact of the government initiative on spatial transformation. Also, the Spanish HSR is not financially viable. Even in corridors with high passenger demand, the HSR system still needed to receive huge public subsidies from the government to maintain daily operations. In fact, HSR lines in Spain have received European Union subsidies for reginal development, ranging between 30 percent to 50 percent of total construction costs.

By contrast, other countries, such as the United States and Canada, are still hesitant about (or even shied away) from HSR investments (Albalate and Bel, 2012). Indeed, the US government faces far more obstacles in HSR development than the Chinese government does. In April 2009, they announced the blueprint for the construction of a national HSR system, but the debate about the costs and benefits of building an HSR system in the US still remains a controversial issue, and some policymakers did not think that it is socially profitable to do that. Moreover, labor costs are low in China, and it is not much of an issue to acquire land when the government owns all the land. However,

---

a. China State Railway Group Co., Ltd. (formerly named as China Railway Corporation) is a state-owned sole proprietorship enterprise that provides railway passenger and freight transport services via 21 subsidiaries in the People's Republic of China. It is a state-owned industrial enterprise established under the "Law of the People's Republic of China on All-Ownership Industrial Enterprises." The Ministry of Finance acts on behalf of the State Council to perform the duties of shareholders.

this is difficult to achieve in the United States. From this point of view, we can conclude that it is very difficult to develop HSR without government support, and as a result, there would be no spatial transformation caused by HSR development.

Even though the political and economic centers, i.e., the capital region, the Pearl River Delta region, and the Yangtze River Delta, did enjoy priority in the development of the Chinese HSR network, its evolution is relatively even in China, especially in eastern and southern parts of China (Chen et al., 2018). This is in stark contrast to countries where a radial network was developed and centered on particular cities (e.g., Madrid and Paris in Spain and France, respectively) (i Queralt and Falco, 2011). The cost advantage is another unique characteristic of the Chinese HSR network. According to Ollivier et al. (2014), the Chinese HSR network has been accomplished at a relatively low unit cost that is at most two-thirds of that in other countries. They further argued that the relatively low labor cost and the large scale of the HSR network planned in China are the two main reasons for this cost advantage. Besides, this cost advantage is closely related to the Chinese government initiative, i.e., the huge investment in building viaducts and bridges. In fact, the Chinese HSR network prefers to lay track on viaducts so as to mitigate harmful environmental effects and to minimize the use of fertile land and resettlement costs. The estimated cost of viaducts for a double track line ranges from RMB 57 to 73 m/km (USD 8.34 to 10.69 m/km) (World Bank, 2014b). Such costs can be kept low due to the standardization of the design and construction process for casting and laying bridge beams on viaducts. Therefore, the spatial patterns of HSR network development, as well as the government initiative, play an essential role in gaining competitive advantages of the Chinese HSR.

The Chinese government has promoted HSR as the key project due to the fact that it has not only construction and operation contributions but also positive effects on economy, society, culture, and environment. The vast land area and densely populated cities in central and eastern provinces trigger Chinese HSR development. Cities with high population densities along the corridors, suffering from road and air congestion, are more likely to ensure the financial viability of HSR operations. Some important metropolitan areas linked by HSR are also important aviation destinations, and hence these two modes are competing with each other, especially in short-haul to medium-haul markets. For example, HSR traffic volumes were significantly larger than air traffic volumes in the market between Shanghai and Beijing in 2009. This is understandable, since the majority of the Chinese population still regard air transport as a "luxury transport mode" because of the relatively low average income (Fu et al., 2012). In this case, the government is now building new lines over greater distances and expanding the HSR network to the less-developed and less-populated western regions in order to solve these problems to some extent.

3 Looking to the future: HSR and regional development

In China, HSR entries have huge and direct impacts on transport and regional systems. In terms of HSR's impact on existing transport modes, it appears to be competitive with airlines (Givoni and Dobruszkes, 2013; Bergantino et al., 2015; Jiang and Zhang, 2016). However, with its rapid development and expanded network, HSR can be a catalyst to stimulate integration with other transport modes and with the transport system. It is important to integrate HSR with other modes due to the fact that integration is able to mitigate environmental pollution and relieve airport congestion (Pfragner, 2011; Jiang and Zhang, 2014; D'Alfonso et al., 2015, 2016), and some issues related to HSR's integration need to be further discussed in the following section.

In 2013, China initiated the "One Belt One Road" (OBOR) (now the "Belt and Road Initiative" (BRI)), which involves countries and regions across Asia, Europe, and Africa and aims to promote the connectivity, and establish and strengthen partnerships among these countries. The implementation of BRI would definitely bring more opportunities and challenges to HSR development. Given the history of a strong Chinese capacity to engineer changes in past decades, including the transport systems (e.g., expansions of ports along the Chinese coastline in the 1990s and early 2000s, the HSR network in the late 2000s), it is expected that at least part of BRI will occur. Therefore, it is necessary to take BRI into consideration when discussing future issues of HSR development.

Based on this background, looking to HSR future development, we hereby make four predictions:

(1) HSR services are likely to affect the spatial distribution of employment and population.
(2) An integrated HSR network with other transport modes, especially airlines, is to be expected.
(3) HSR Freight will connect major Chinese logistics sites, and regional impacts of this new logistic mode are going to favor established and large locations.
(4) The long-distance international HSR services will begin operation and have positive effects on regional development, but the impacts may be less than we conventionally expect.

3.1 Spatial distribution of employment and population

Since the busiest HSR line, the Beijing-Shanghai line, started operation in 2011, we can conclude that HSR is helping to create a deeply connected economy. The advent of HSR has started to change life and work, especially in Beijing, Shanghai, and Guangzhou (three megacities in China located in the northern, central, and southern part of the country, respectively). Now, each of these three megacities is developing commuter corridors in order to reduce commute time. Many people have begun to live within one hour of megacities by HSR because of the high real estate prices in these cities (The Economist, 2017). In the foreseeable future, more people will enjoy the benefits of urban agglomeration while not suffering from high levels of traffic congestion, pollution, and rents, because HSR enables passengers to access megacities without living within its boundaries. In addition, the introduction of HSR facilitates firm fragmentation and firm sorting depending on their unique requirements for megacity access. HSR services provide the possibility that firms locate their headquarters in major cities and send other activities to surrounding low-cost cities. For example, after the launch of Tianjin-Beijing HSR line, which only takes half an hour in travel time, several large firms have kept their headquarters in Beijing while transferring their manufacturing factories to Tianjin. As a result, the lower value-added activities[b] are expected to shift from megacities to (nearby) low-cost cities, while the headquarters in megacities will focus on developing high (or even higher) value-added services. This trend is likely to be adopted by more and more companies with the expansion of the HSR network towards second- (*erxian*) and third-tier (*sanxian*) regions. Therefore, it is expected that HSR will have increasing impacts on the spatial distribution of employment and population.

3.2 An integrated HSR network

In China, air transport is not fully deregulated and well-developed: some remote areas cannot be reached by air. Although airport connectivity has improved in general, its density will not likely be sufficient in the coming decades, considering the country's huge population. Hitherto, almost every major Chinese airport has faced runway capacity shortages, resulting in increased delay and congestion. In this case, the government is now building new lines over greater distances and expanding the HSR network to the less-developed and less-populated western regions so as to solve these problems. Simultaneously, in some particular cases, governments encourage the cooperation between airline and HSR by providing connections between nearby cities and airports. The hub-and-spoke network adopted by most major airlines makes such airline-HSR complementarities possible. In China, there are several airline-HSR cooperation cases, mainly aimed at reducing airport runway congestion, subject to capacity constraints. For example, the Beijing-Tianjin HSR line helps transfer passengers from Beijing Capital International Airport to Tianjin Binhai International

---

b. Lower value-added activities refer to activities that consume resources but add little value to customers or companies. Examples of such activities include setting up machines, moving product parts, waiting, reworking, inspecting, and storing.

Airport, and all travel costs between Beijing and Tianjin will be subsidized by the airport. Most of these cases connect regional spaces to main transport infrastructures so as to reinforce the secondary transport networks (Jiang et al., 2017; Li et al., 2018; Xia et al., 2019). We believe that there will be more and more such cooperation cases in the future. Such an integrated transport system is likely to change the spatial structure of the whole network, and the reinforcement of secondary transport networks can modify the urban hierarchy of the spatial system. In the existing HSR network, the political and economic megacenters enjoyed priority in the development process, so there is a big city bias in impacts. However, we also perceive that a more integrated HSR network with other transport modes will help to relieve such a bias, since it enables more currently underdeveloped regions to be integrated into the network. The spatial penetration of HSR can be improved by the integration between HSR and other transport modes (Yin et al., 2015). By doing so, the HSR network grants access to regional population, and finally, decreases the level of uneven regional development.

3.3 HSR freight

At first, Chinese HSR lines were only passenger-dedicated lines, and then passenger-freight mixed lines were introduced in recent years. According to the State Post Bureau, China's express delivery volumes exceeded 30 billion in 2016, rising by 53%. Nowadays, China is the world leader in terms of express delivery volumes. In October 2016, HSR Express services launched on trial to offer customers high-end door-to-door small parcel express service. We expect the HSR Express to become a new logistics mode connecting major logistics sites in China due to the surging demand for express delivery and the rapid development of e-commerce in recent years. The infrastructure and technological improvements made by HSR freight will lead to increased freight capacity, which creates new and growing routes to trade and business. At the same time, more firms may shift from offline to online trade. This will be a stimulus to boost intraregional trade.

In recent decades, the role of transport, and the scale and location of transport activities, have been reshaped by a complex set of interfirm linkage associated with the global production network. Saxenian (2002, 2006) showed that communities play an important and complementary role in the development of global production network, and emphasized the "transnational elite community," which has the tactical knowledge and skills to create new responses within the industry and in its related infrastructure. They have proven that there are now several forces operating within and through transport infrastructure and logistics services at a global scale that reinforce the idea that regional impacts of transport systems favor established and large locations. According to this line of discussion, we believe that regional impacts of HSR freight are going to favor established and large locations.

3.4 Long-distance international HSR services

As part of the BRI initiative, China plans to connect its HSR lines to 17 countries in Asia and Eastern Europe. Additional rail lines will be built into Russia as well as South East Asia, in what will likely become the largest transport infrastructure project in history. In this case, China plans to install 81,000 km (50,000 miles) of HSR lines, involving 65 countries. With this improved connectivity between different regions, a new intercity relationship is going to be expected.

In most cases, HSR is regarded as a short- or medium-distance transport mode. However, long-distance international HSR services are gaining more attention and may even revolutionize the concept of HSR. They are likely to facilitate the exchange of information, economic, technological, and labor flows along these international HSR corridors. These new corridors may provide positive effects on regional development from urban centers and reconfigure urban economic geography both regionally and nationally, due to the fact that the primary cities in BRI are underdeveloped northwestern cities. This indicates that the Chinese government provides an opportunity for peripheral areas initially disadvantaged by distance from the core areas. The cross-nation HSR network is likely to stimulate the integration of the Chinese regional economy.

However, even though the cross-border network has been seen as a priority of the HSR future development strategy in terms of BRI, problems of jurisdictional segregation and competition between various countries may prevent the creation of new services that could transform regional performance. For example, the China-Pakistan economic corridor (CEPC) passes through some of the world's most vulnerable and conflict-ridden territories. Pakistan has been combating an Islamist insurrection for more than a decade. Beijing worries about militants from Pakistan's federally administered tribal area possibly penetrating China's western Xinjiang province, which has its own unrest. Meanwhile, some political parties in Pakistan have expressed deep reservations about the CPEC, claiming that the ruling party is deliberately trying to alter the design of the corridor to favor its own constituencies. Moreover, the relationship between central and local governments may pose threats to the construction of these international HSR lines. Although agreement and consensus have been reached between national governments, local governments, who will implement the whole project, may not be willing to cooperate with foreign investors and ignore the national government's policies. Therefore, the expansion of the Chinese HSR network worldwide in parallel with the improvement of the regional competitiveness may decrease the level of regional uneven development, but the real impacts may be less than we conventionally believe, given that international projects covering so many countries may be promising at the beginning but will be difficult to pursue.

The above suggests that HSR can be an important mode to improve equality of intercity accessibility and the external agglomeration economies. In the future, a more mobile workforce and newly accessible markets, a more integrated HSR network, and a new logistic mode will be expected. Such large-scale implementation of HSR network cannot be achieved without government support. On the other hand, new and major government initiatives (e.g., BRI) may generate fewer impacts than we have initially hoped for on the spatial transformation of transport and regional systems when other countries are involved.

4 Conclusion

The governments of many countries and regions attempt to engage remote or peripheral regions within national or continental networks, notably transport investments and institutional changes. In some cases, this has led to spatial transformation in both the transport and regional systems. However, the impacts of government initiatives on the transformation of transport and regional systems are currently underresearched, and this chapter attempts to address this deficiency. It consists of an in-depth case study on the development and management of HSR in China. By doing so, it offers important contributions through establishing a comprehensive framework on the impacts of major government initiatives on the spatial transformation of transport and regional systems, and therefore filled this gap. It illustrates the major reason why the Chinese HSR network has achieved extraordinary success, namely the Chinese government always financially supports and encourages its development and expansion. At the national level, we argue that government initiatives are very likely to have positive impacts on the transformation of transport and regional economies. However, at the international level, such impacts may be more limited. Our analysis suggests that government initiatives on transport and regional systems can be highly diversified based on physical locations, economic and social systems, institutional, and even cultural issues between various countries.

The regional focus in this study may have its own limitations and the results may not reflect the situation of transport systems in other countries. Despite this, it offers a solid theoretical foundation and practical reference for the implementation of other government initiatives in the future. Moreover, we recognize that governments and related public sectors play pivotal roles in the spatial transformation of transport infrastructure programs. Thus, our analysis contributes valuable food for thought on what the right approach should be for policymakers when initiating major investments under diversified circumstances, and how success should be assessed. The insight gained from this paper can help policymakers to develop effective approaches in transforming transport and regional systems, improving their efficiency and management, thus facilitating the development of international trade and the global economy. Otherwise, substantial money and time may be wasted during implementation processes. In terms of the environmental perspective, however, some of BRI's major corridors are known to pass through ecologically sensitive areas. Increasing interconnectivity between countries through the initiative could mean dissecting these natural environments with the construction of roads and rail, and such disruptions would threaten the surrounding environment system. It is worth considering the environmental impacts caused by the new government initiatives, to design a sustainable framework that could determine the overall environmental impact, and ensure the project participants' willingness and ability to adhere to it.

However, one thing that we need to emphasize here is the approach of how government initiatives are presented. Although new and major government initiatives are always regarded as the locomotive for transformation, the actual impacts remain a vivid debate among politicians, industrial practitioners, and scholars. Given the persistence in economic differences between regions and countries within the BRI network, there is a real possibility that some countries may reject this initiative. Moreover, the new institutional settings created by the new government initiatives may not be suitable to all the involved regions due to diversified institutional and political histories and systems. The spatial transformation of transport systems caused by the government initiatives may make some of the involved regions' transport infrastructures uncompetitive. Due to this, they may reject the initiatives in order to protect self-interests. Such initiatives might induce a competition effect among neighboring regions for potential regional investors. In this case, governments need to consider the extent of spatial competition and implement mechanisms, which help prevent such behaviors. Therefore, we strongly advise policymakers to be careful on the arts of presenting and managing new and major initiatives. If the initiatives are presented too aggressively, they may not get the desired results that they look for.

References

Albalate, D., Bel, G., 2012. High-speed rail: lessons for policy makers from experiences abroad. Public Adm. Rev. 72 (3), 336–349.

Bergantino, A.S., Capozza, C., Capurso, M., 2015. The impact of open access on intra-and inter-modal rail competition. A national level analysis in Italy. Transp. Policy 39, 77–86.

Bonnafous, A., 2015. The economic regulation of French highways: just how private did they become? Transp. Policy 41, 33–41.

Breidenbach, P., Mitze, T., 2015. The long shadow of port infrastructure in Germany: cause or consequence of regional economic prosperity? Growth Chang. 47 (3), 378–392.

Cao, J., Liu, X.C., Wang, Y., Li, Q., 2013. Accessibility impacts of China's high-speed rail network. J. Transp. Geogr. 28, 12–21.

Chen, C., D'Alfonso, T., Guo, H., Jiang, C., 2018. Graph theoretical analysis of the Chinese high-speed rail network over time. Res. Transp. Econ. 72, 3–14.

D'Alfonso, T., Jiang, C., Bracaglia, V., 2015. Would competition between air transport and high-speed rail benefit environment and social welfare? Transp. Res. B Methodol. 74, 118–137.

D'Alfonso, T., Jiang, C., Bracaglia, V., 2016. Air transport and high-speed rail competition: environmental implications and mitigation strategies. Transp. Res. A Policy Pract. 92, 261–276.

Daamen, T.A., Vries, I., 2013. Governing the European port–city interface: institutional impacts on spatial projects between city and port. J. Transp. Geogr. 27, 4–13.

Dall'Erba, S., Le Gallo, J., 2008. Regional convergence and the impact of European structural funds over 1989–1999: a spatial econometric analysis. Pap. Reg. Sci. 87 (2), 219–244.

Fu, X., Zhang, A., Lei, Z., 2012. Will China's airline industry survive the entry of high-speed rail? Res. Transp. Econ. 35 (1), 13–25.

Garst, D., 2016. China's Central Asian HSR Dreams. Available at: http://danielgarst.com/articles/1B%201R%20art,%20revised.pdf.

Givoni, M., Dobruszkes, F., 2013. A review of ex-post evidence for mode substitution and induced demand following the introduction of high-speed rail. Transp. Rev. 33 (6), 720–742.

Hall, P.G., Pain, K., 2006. The Polycentric Metropolis: Learning From Mega-City Regions in Europe. Routledge.

Hou, Q., Li, S.-M., 2011. Transport infrastructure development and changing spatial accessibility in the Greater Pearl River Delta, China, 1990–2020. J. Transp. Geogr. 19 (6), 1350–1360.

i Queralt, G.B., Falco, I., 2011. Espanya, capital París: tots els camins porten a Madrid: Edicions La Campana.

Jiang, C., Li, X., 2016. Low cost carrier and high-speed rail: a macroeconomic comparison between Japan and Western Europe. Res. Transp. Bus. Manag. 21, 3–10.

Jiang, C., Zhang, A., 2014. Effects of high-speed rail and airline cooperation under hub airport capacity constraint. Transp. Res. B Methodol. 60, 33–49.

Jiang, C., Zhang, A., 2016. Airline network choice and market coverage under high-speed rail competition. Transp. Res. A Policy Pract. 92, 248–260.

Jiang, C., D'Alfonso, T., Wan, Y., 2017. Air-rail cooperation: partnership level, market structure and welfare implications. Transp. Res. B Methodol. 104, 461–482.

Li, X., Jiang, C., Wang, K., Ma, J., 2018. Determinants of partnership levels in air-rail cooperation. J. Air Transp. Manag. 71, 88–96.

Monios, J., 2016. Intermodal transport as a regional development strategy: the case of Italian freight villages. Growth Chang. 47 (3), 363–377.

Monios, J., Lambert, B., 2013. The Heartland Intermodal Corridor: public private partnerships and the transformation of institutional settings. J. Transp. Geogr. 27, 36–45.

Ng, A.K.Y., Jiang, C., Li, X., O'Connor, K., Lee, P.T.W., 2018. A conceptual overview on government initiatives and the transformation of transport and regional systems. J. Transp. Geogr. 71, 199–203.

Ng, A.K., Wong, K., Shou, E.C., Jiang, C., 2019. Geography and institutional change: insights from a container terminal operator. Marit. Econ. Logist. 21 (3), 334–352.

Ollivier, G., Sondhi, J., Zhou, N., 2014. High-speed railways in China: a look at construction costs. China Trans. Top. 9, 1–8.

Panayides, P.M., Parola, F., Lam, J.S.L., 2015. The effect of institutional factors on public–private partnership success in ports. Transp. Res. A Policy Pract. 71, 110–127.

Pfragner, P., 2011. Intermodal Hub Frankfurt Airport. Avion-TGV. Une alliance d'avenir, Paris.

Rothengatter, W., 2010. 18 Competition between airlines and high-speed rail. In: Critical Issues in Air Transport Economics and Business. Taylor & Francis, United Kingdom, pp. 319.

Saxenian, A., 2002. Transnational communities and the evolution of global production networks: the cases of Taiwan, China and India. Ind. Innov. 9 (3), 183–202.

Saxenian, A., 2006. The New Argonauts: Cambridge. Harvard University Press, MA.

The Economist, 2017. Railways: the lure of speed. The Economist (January 14), 27–28.

Ureña, J.M., Menerault, P., Garmendia, M., 2009. The high-speed rail challenge for big intermediate cities: a national, regional and local perspective. Cities 26 (5), 266–279.

Vickerman, R., 2017. Can high-speed rail have a transformative effect on the economy? Trans. Policy.

Wang, X., Huang, S., Zou, T., Yan, H., 2012. Effects of the high speed rail network on China's regional tourism development. Tour. Manag. Perspect. 1, 34–38.

World Bank, 2014a. Regional Economic Impact Analysis of High Speed Rail in China. Available at: http://www.worldbank.org/content/dam/Worldbank/document/EAP/China/high_speed-rail-%20in-china-en.pdf.

World Bank, 2014b. Cost of High Speed Rail in China One Third Lower than in Other Countries. Available at: http://www.worldbank.org/en/news/press-release/2014/07/10/cost-of-high-speed-rail-in-china-one-third-lower-than-in-other-countries.

Xia, W., Jiang, C., Wang, K., Zhang, A., 2019. Air-rail revenue sharing in a multi-airport system: effects on traffic and social welfare. Transp. Res. B Methodol. 121, 304–319.

Xiao, J., Yuan, Z., 2007. China will build ten major urban agglomerations to dominate the future development of the national economy (in Chinese) China Economic Times. Development Research Centre of the State Council, Beijing published online, 29.

Yin, M., Bertolini, L., Duan, J., 2015. The effects of the high-speed railway on urban development: international experience and potential implications for China. Prog. Plan. 98, 1–52.

Zheng, S., Kahn, M.E., 2013. China's bullet trains facilitate market integration and mitigate the cost of megacity growth. Proc. Natl. Acad. Sci. 110 (14), E1248–E1253.

Further reading

Chinese Ministry of Railways, 2016. Available at: http://www.china-railway.com.cn/.

Chapter 16

Connect or be connected strategy in the context of the Belt and Road Initiative: A Korean case

Paul Tae-Woo Lee[a], Jihong Chen[b]
[a]Ocean College, Zhejiang University, Zhoushan, China, [b]College of Transport and Communications, Shanghai Maritime University, Shanghai, China

1 Introduction

Six years have been passed since Chinese President Xi Jinping announced main idea of the Belt and Road Initiative (BRI) in 2013. Its contents and directions have been addressed in the "Vision and Actions on Jointly Building Silk Road Economic Belt and 21st-Century Maritime Silk Road" published on 28 March 2015 (National Development and Reform Commission: NDRC et al., 2015). The Initiative has been accelerated by the establishment of the Asian Infrastructure Investment Bank (AIIB) in 2016 when its members reached 57 countries, comprising of regional members and nonregional members as of December 2016. The document consists of Preface and eight chapters, i.e., "Background, Principles, Framework, Cooperation Priorities, Cooperation Mechanisms, China's Region in Pursuing Opening-up, China in Action, and Embracing a Brighter Future Together." Some key words related to connectivity issue can be drawn from it, among others, China's inland region, connectivity, economic corridor, transport corridor, city cluster, pilot free trade zone, marine economy development demonstration zone, marine economy pilot zone, core are, bay area, infrastructure, and financial mechanism (Lee et al., 2018a; Chhetri et al., 2018). Out of the above key words, the most important item for this article is economic and transport corridors. Figs. 1 and 2 show a summary of the corridors as follows:

- China-Pakistan Economic Corridor (CPEC): Kashgar (Xinjiang Uygur autonomous region)—Gwadar port (Pakistan)
- Bangladesh-China-India-Myanmar economic corridor: Kunming (Yunnan province, China)—Mandalay (Myanmar)—Dhaka (Bangladesh)—Kolkata (India)

Maritime Transport and Regional Sustainability. https://doi.org/10.1016/B978-0-12-819134-7.00017-4

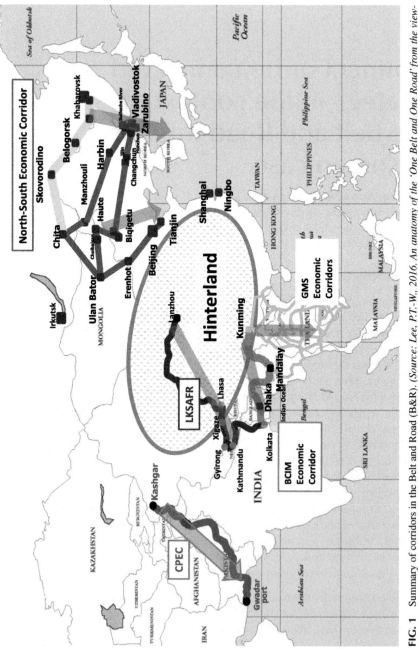

FIG. 1 Summary of corridors in the Belt and Road (B&R). *(Source: Lee, P.T.-W. 2016. An anatomy of the 'One Belt and One Road' from the viewpoint of structural changes in maritime transport. In: 2016 Annual Conference of International Association of Maritime Economists. Kuhne Logistics University, Hamburg Germany, August 23–26.)*

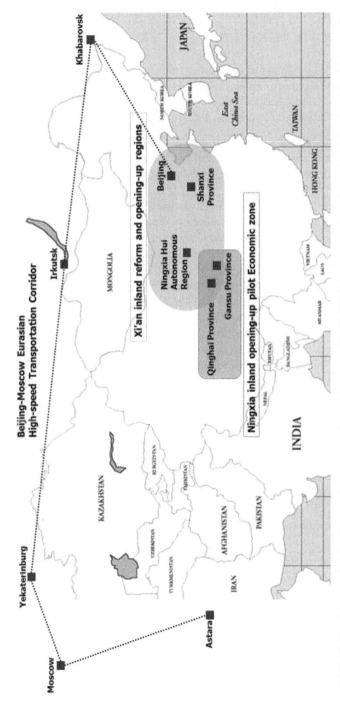

FIG. 2 Beijing-Moscow Eurasian high-speed transport corridor. (*Source: Lee, P.T.-W. 2016. An anatomy of the 'One Belt and One Road' from the viewpoint of structural changes in maritime transport. In: 2016 Annual Conference of International Association of Maritime Economists. Kuhne Logistics University, Hamburg Germany, August 23–26.*)

- Four subcorridors in Greater Mekong Sub-region Economic Cooperation (GMSEC)
- China, Mongolia, and Russia Economic (CMR) Corridor: Heilongjiang Silk Road Belt
- Beijing-Moscow Eurasian high-speed transport corridor: Beijing (China)—Khabarovsk (Russia)—Irkutsk (Russia)—Yekaterinburg (Russia)—Moscow (Russia)—Astara (Azerbaijan)
- Lanzhou-Kathmandu South Asia Freight Rail (LKSAFR).

The expected impacts of the corridors are multidimensional such as change of energy supply chain, Chinese global port chain development, dry ports development in inland China, interport competition, and infrastructure development (Lee et al., 2018a, b, c; Wei et al., 2018; Chen et al., 2019). Out of the above six corridors, CPEC is expected to play a key role in generating port regionalization and network reconfiguration along the ports in the Indian Ocean because Hambantota Port in Sri Lanka has been leased to China for 99 years in tandem with the development of Colombo port in Sri Lanka and Gwadar Port in Pakistan in the context of the BRI (Lee et al., 2018a; Ruan et al., 2019). 18 billion US dollars investment for building CPEC includes Karot hydropower project, which is the first project of the Silk Road Fund.

The CMR economic corridor so-called Heilongjiang Silk Road Belt is interrelated to Korea's "New Northward Policy" and Russia's "New East Asian Policy" because they have common aims to develop connectivity among China, South Korea, North Korea, Russia, and Mongolia. In addition, the economic corridor is linked to the East Sea Economic Rim (ESER) where the above countries and Japan except Mongolia are connected by ports and shipping routes. Therefore, the economic corridor contributes to developing trade transit transport corridor in northeast Asian region (Lim et al., 2017; Lee et al., 2018c).

All the above are interwoven in the BRI context in the northeast Asian region. Chinese government has announced the Polar Silk Road (PSR) in January 2018,[a] which is similar to Arctic shipping routes comprising of Northeast Passage, Northwest Passage, and Central Passage. As a result of global warming, the PSR is likely to become an important transport route for international trade. Therefore, there are three ways available for China/Korea to connect them

---

a. The document has the following contents:
  I. The Arctic Situation and Recent Changes
  II. China and the Arctic
  III. China's Policy Goals and Basic Principles on the Arctic
  IV. China's Policies and Positions on Participating in Arctic Affairs
  - Deepening the exploration and understanding of the Arctic
  - Protecting the eco-environment of the Arctic and addressing climate change
  - Utilizing Arctic Resources in a Lawful and Rational Manner
  - Participating Actively in Arctic governance and international cooperation
  - Promoting peace and stability in the Arctic Conclusion. (Source: The State Council Information Office of the People's Republic of China, 2018, *China's Arctic Policy*, Beijing.)

to Europe, i.e., existing south-west bound maritime routes, railway by Trans-China Railway (TCR) and Trans-Siberian Railway (TSR), and the PSR. China's interest in developing the PSR arises from her desire to establish a comprehensive Belt and Road as the "three Silk Roads," referring to the joint promotion of land (Belt), sea (Road), and the Arctic. The China, Mongolia, and Russia Economic (CMR) Corridor: Heilongjiang Silk Road Belt in tandem with the PSR enables logistics providers and carriers in the countries of the ESER to connect themselves to Europe by railway, i.e., TCR and TSR and by sea.

Having considered the above economic/transport corridors in the BRI, we define "connect strategy" when a country along the Belt and Road joins memorandum of understanding with China and/or become a member of AIIB. Or the country participates actively in constructing infrastructure in its country in collaboration with China and neighboring countries to develop corridors and connectivity along the B&R in the region. On the contrary, "be connected strategy" means that a country follows the above activities in a passive way. Therefore, it is worthwhile to investigate the impacts of the BRI from the above two strategy. This chapter aims to discuss some key points to make connectivity between Korea and Europe efficient in the context of the BRI and to draw its implications from the Korean perspective.

2   How to react Korean ports to Chinese overseas port developments[b]

2.1   China's investment in overseas port development along the Maritime Silk Road

China's ports are actively responding to the BRI and are stepping up efforts seeking development and cooperation opportunities with ports along the B&R. China has become the No. 1 seaport country, having 14 of the top 20 ports in the world in terms of cargo throughput and container throughput. The B&R construction has become one of the key elements of infrastructures attracting global investment. Data from the Ministry of Commerce show that Chinese enterprises invested a total of $14.53 billion in countries along the B&R in 2016. As of the end of 2016, Chinese enterprises set up 56 inchoate cooperation zones in countries along the B&R, with the accumulated investment totaling $18.55 billion. China has invested in ports and terminals in 23 countries and regions along the B&R, with the annual amount exceeding $10 billion. Chinese investors used to focus more on maritime enterprises and terminal operation. With their investment in leasing and constructing foreign ports answering the call of the BRI, Chinese port enterprises have made more solid strides in "going global" along

b. This section has been taken from Chen, J., Fei, Y., Lee, P.T.-W., Tao, X., 2019. Overseas port investment policy for China's central and local governments in the Belt and Road Initiative. J. Contemp. China 28(116), 196–215. The authors express their thanks to the publisher for granting the copyright permission for this book chapter.

the Maritime Silk Road (MSR). Such port investment is expected to generate favorable benefits and returns by improving port connectivity and establishing supply chain network for the nation's economy. For example, Shanghai Port has taken a lead in "going global" by connecting China to the world and developing its port alliance in foreign ports. For instance, the port in 2015 won the bid for running terminals at the Haifa New Port in Israel for 25 years starting from 2021. At the call of the BRI, many coastal ports in China are trying to tap into the global market and seek to develop investment opportunities for expanding port connectivity along the MSR. However, national preferential policies alone are not enough to promote the "going global strategy" because enterprises need to possess sound conditions and orderly prioritized investment policies. Not all ports are as powerful as Shanghai Port; smaller ports are in different locations with their different capacities and local urban economic strengths. For this reason, major coastal ports in China have to formulate reasonably prioritized development plans based on 11 variables (see Appendix A), including their port characteristics and economic environment to promote the "going global strategy" steadily.

To cope with this, the "going global strategy" of Chinese coastal ports has been proposed. The ports cover Shanghai, Tianjin, Ningbo-Zhoushan, Guangzhou, Shenzhen, Zhanjiang, Shantou, Qingdao, Yantai, Dalian, Fuzhou, Xiamen, Quanzhou, and Haikou as key gateways providing support to the "going global strategy" along the MSR, and the proposal regards these port cities as nodes to cobuild a smooth, safe, and efficient transportation passage. The 14 ports try to actively promote the "going global strategy," and respond to China's BRI (see Fig. 3). However, the conditions and strengths of their cities are diverse. Considering these factors, the ports are required to develop their "going global strategy" in a rational and orderly manner based on their own development status. There is also another matter of which port will "go global" first. This is related to governance issue because several stakeholders, including central, provincial, and local governments, engage in overseas investment and financing in the context of the BRI. In addition, some dry ports should be connected to seaports in the country (Wei et al., 2018). Therefore, a study on priority of China's overseas port investment is critical to avoid waste of national resources and provide a sound guideline to mitigate conflicts between the central government and the provincial government and among the provincial governments.

A mechanism that Chinese local governments and port authorities cooperate with foreign terminal operators with their capital investment has gradually become a practice in container terminal operation in China. This is a motivation for provincial and local governments to accelerate the "going global strategy." The BRI has triggered global port network developments along the MSR with a long-term port lease such as Darwin Port in Australia and Hambantota Port in Sri Lanka and purchasing shares of Piraeus Port in Greece (see Appendix B). These developments will contribute to connecting Chinese coastal ports into global port

FIG. 3 Key Ports under the "Going Global Strategy" along the Maritime Silk Road.

logistic network and strengthening the functions of ports in the supply value chain. In this regard, it is worth noting a newly proposed "New Maritime Silk Road" because the MSR does not cover current maritime network, missing sub-Saharan region, South America, and the Pacific Islands (Lee et al., 2018a).

Literature on BRI showed the following aspects of change in transport and logistics landscape:

- development of inland-inter regional rail and highway corridors and city clustering in China;
- connectivity of economic corridors to the Indian Ocean and East Sea in association with dry ports in China;
- development of sea-river combined transport and sea-railway/highway
- development of free trade economic zones along inland road corridors and in ports;
- structural changes in maritime cargo flow and shipping network in association with corridor developments and intermodal network; and
- alignment of participating countries' transport and logistics development along the B&R.

The above aspects provide an insight into how to implement the "going global strategy," reflecting the current supply chain and future plan of Chinese coastal ports as well as the comprehensive economic and geographical characteristics of the port cities. Enumerating key factors influencing the container port system in China, such as the "Go West" strategy, BRI, introduction of modern corporate governance principles, and free trade zones, Notteboom and Yang (2017) argued that the above factors have accelerated "port integration and co-operation and tried to attract foreign investments to Chinese ports with an internationalization of Chinese port-related companies through investments in foreign ports."

About 65% of overseas terminals established by Chinese port enterprises are located along the MSR. Guided by the "going global strategy," Chinese overseas port investment projects are not only conducive to building international shipping pivots along the B&R to boost China's foreign trade, but also help strengthen international cooperation and division of labor to extend ports' industry chains. Chinese port constructors are the earliest overseas investors. For example, China Communications Construction Company Limited built the Gwadar Deep-water Port in Pakistan in 2002 to fuel port construction along the MSR. Except China Merchants Port Holdings Company Limited and COSCO Shipping Ports Limited, domestic terminal operators in China, such as Shanghai International Port Group, Port of Qingdao Group and Guangxi Beibu Gulf International Port Group, have just begun to put the "going global strategy" onto their agenda. Their "going global strategy" focuses on exporting terminal operation technologies and management experience and acquiring terminal operation rights. At present, the strategy of Chinese port enterprises is primarily carried out in three ways: merger and acquisition (M&A), joint venture, and leasing.

Chinese port enterprises have accumulated some experience in overseas port investment and construction. Two global terminal operators as the first-tier group in China, i.e., China Merchants Port Holdings Company Limited and COSCO Shipping Ports Limited, have succeeded in improving their capabilities of merger and acquisition and operation and management of port terminals. Shanghai International Port Group as one of the second-tier port enterprises is a quasiterminal operator whose development potential cannot be underestimated. The third-tier enterprises in China mainly rely on domestic homeports to serve overseas upstream and downstream terminal companies. For example, Port of Yantai Group serves its homeport and cooperates with several large domestic and foreign port groups to jointly invest in building the Port of Boke in Guinea, enjoying full management over the port's terminal business. Chinese port enterprises' investment in overseas ports has been demonstrating potential growth (see Appendix B).

2.2 Ports' "going global strategy" advocated by the central and local governments

Thanks to the BRI, Chinese ports continue to evolve toward "going global" and internationalization, and reach out to the countries along the B&R. As far as shipping and port sector is concerned in the context of the BRI, the Chinese government has focused on port infrastructure, land-water combined transport and port cooperation, increase seagoing routes and liner frequency, and strengthen cooperation in maritime logistics information in accordance with the vision and plan of the BRI. This will not only activate and expand international trade and cooperation between China and the countries along the route but also encourage major coastal ports in China to step up overseas investment and construction. Consequently, China has strived to become new international shipping centers and transit hubs. Port plays a pivot and pioneering role in the joint effort of building the B&R. President Xi Jinping has described ports as "important pivots" and "important hubs" on many international occasions to highlight their importance in the BRI. As a carrier of global trade flow, ports have increasingly become a kernel of regional economic development, and remain a major driver of global city growth. To become new international shipping centers, Chinese ports must forge ahead in an orderly and steady manner, instead of rushing out.

The Chinese central government and local governments have distinct roles in terms of governance models and methods referring to China's port management system reform. Since the central government initiated the port system reform in 1998, local governments have played an increasingly important role in operation and management of ports, resulting in intense interport competition among neighboring regions. But the current situation lacks effective control over the excessive port competition, which causes waste of resources at the national and provincial level. Both the central government and local governments back ports' "going global strategy," but they take different governance models and methods. The central government looks at national interests and introduces macro-governance policies on investment in and cooperation with ports along the B&R, while local governments are responsible for governing the international cooperation of local port enterprises.

As can be seen in Appendix B and discussed above, Chinese central and local governments have made efforts to expand overseas port investment along the MSR. In particular, COSCO, China Merchant Groups and even local governments are major players to acquire major ports in along the MSR. It means that they are emerging world terminal operators and establish vertical operation in shipping lines and port services. On the contrary, Korea's shipping companies and port authorities are little players in overseas port investment. Korea is becoming a weak player in networking port operations compared to world terminal operators such as Port Singapore Authority, Hutchison, and DP World. In other words, if Korea's stakeholders in shipping and port industry intend to improve marketability, they will be considering choice of connect strategy in collaboration with Chinese partners along the MSR because the "going global strategy" of Chinese ports can be better aligned with Korea's ports in a sustainable way.

The next section deals with connecting Korea to Europe by land, namely railway connection because it is another transport mode for traffic cargoes in the context of the BRI.

3   Key factors for efficiently connecting Korea to Europe in the context of the BRI

As discussed in the previous section, there are three possible routes to connect Korea to Europe. Fig. 4 shows existing shipping route, railways by TCR and TSR, and Arctic shipping routes (so-called the PSR). The current shipping routes have been well established for the Korean logistics providers and carriers. The discussions about the Arctic shipping routes show that they are too early to commercially implement the service for the stake holders, although its feasibility and economic benefits are positive (Fu et al., 2018; Lee and Song, 2014; Lindstad et al., 2016; Liu and Kronbak, 2010; Meng et al., 2017; Tseng and Cullinane, 2018; Vavrus et al., 2012; Verny and Grigentin, 2009; Zhang et al., 2016; Zhu et al., 2018). A series of denuclearization talks among North Korea, South Korea, and United States have been progressing, despite its uncertainty is envisaged. Having considered this circumstance, it is worthwhile for this section to investigate the impacts of the BRI on connectivity between Korea and Europe, focusing on railway mode.

The discussion on railway mode to connect between Korean Peninsula and Europe requires a prerequisite assumption that the railways are connected between North Korea and South Korea and then connected to existing TCR and/ or TSR. The first point to consider arises out of capacity of the railway and demand for the service. This is also interrelated to modal competition between the railway service and shipping service in terms of freight and service quality. This aspect can be inferred from a Chinese cost comparison case as shown in Table 1.

Table 1 shows that China has five main railway service routes for Europe by origin city to use intermodal service. Their total freight costs range from 1500 to 2626US$ per 40 equivalent unit container (FEU). The highest freight 2626US$/ FEU by intermodal service is still higher than the lowest railway freight with

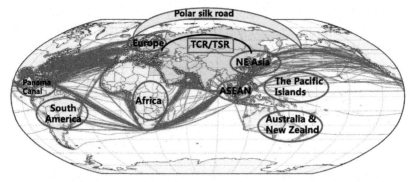

FIG. 4   Global network connecting Northeast Asia to the rest of the World. *(Source: Modified by the authors based on Lee, P.T.-W., Hu, Z.H., Lee, S.J., 2016. Research trend in the Belt and Road Initiatives. In: 2016 OBOR Conference. RMIT University, Melbourne, Australia, December 1–2. Note: The width in pink color indicates the amount of ship traffic flows. Source: Author modified the picture based on Hu, Z.-H. (zhhu@shmtu.edu.cn).)*

TABLE 1 Freight time and cost comparison between shipping routes and China Railway Express.

	Yu-Xin-Europe	Rong-Europe	Zheng-Europe	Han-Europe	Su-Man-Europe
Original city	Chongqing	Chengdu	Zhengzhou	Wuhan	Suzhou
Domestic freight time (days)	12	5	4	5	0
Shipping time (days)	25				
Total freight time (days)	37	30	29	30	25
Frequency	At least 7 per week				
Inland freight costs (USD/FEU)	645	1,126	443	258	–
Shipping cost (USD/FEU)	1500USD/FEU (A.P. Moller-Maersk A/S (MSK)'s freight)				
Total costs (USD/FEU)	2145	2626	1943	1758	1500

Source: Jiang, Y., 2018. Evolution of Hinterland patterns between China Railway Express and Seaborne Container Shipping under the B&R initiative. In: 2019 IAME Conference, Mombasa, Kenya, 11–14 Sept. 2018.

subsidy by central and/or local governments in China, while the former's total transportation time is longer than the latter's. Apparently, from the viewpoint of overall aspects of freight cost and service period between China and Europe, Chinese railway service is more competitive than shipping service, thanks to subsidy from central and local governments. However, we envisage the following questions regarding viability of Chinese railway service:

- How much subsidy can make Chinese railway service viable?
- How long will the subsidy by central and local governments last for Chinese railway service?
- How can China solve cargo imbalance between China and Europe?

The above Chinese case may raise some similar questions to Korean stakeholders, leaving aside technical issues related to railway operation. For example, is it possible for a Korean rail operator to provide competitive fright rate without subsidy compared to shipping service? Unlike Chinese railway service, the cargoes originated from Korea will take longer service time to Europe. It means that Korean railway service has less advantage in transportation days compared to Chinese railway. In addition, Chinese main five railways have weekly service with each block train having more than 42 wagons. It is questionable whether a Korean rail operator can arrange such block train with weekly service with enough cargoes. If he cannot, how can he make block train in collaboration with Chinese TCR operators? This article does not deal with TSR but TCR because it is concerned with the BRI. Moreover, if subsidy is necessary for Korean shippers to make the railway service feasible and competitive, who and how will provide such subsidy for the railway service? On top of that, subsidy for railway service will cause Busan Port to face a severe competition with railway operators to capture cargoes to be generated from the same hinterland because shipping networks and connectivity is interrelated to port competition (Lam et al., 2018).

As can be seen in Fig. 1, if all corridors are working well in the future, the size of the hinterland in the green circle would be encroaching against Shanghai and Ningbo Ports (Lee et al., 2018a, b, c). As a result, it would cause more severe interport competition between Shanghai and Busan Port to capture trans-shipment cargoes with price competition (Anderson et al., 2008; Ishii et al., 2013; Dong et al., 2018).

The second factor we need to consider efficient connectivity by railway between Korea and Europe is to find a distribution center in the European region. Fig. 5 shows a development strategy to connect China to Europe by establishing distribution center so-called Great Stone industrial park in 95 km^2 in Minsk in Belarus,[c] which was invested by China Merchants Group and is adjacent to Lithuania and Germany and Poland.

The Great Stone in Minsk, Belarus aims to provide distribution service for cargoes coming from China by railway as well as accommodate block trains. The characteristics and advantages of the location are as follows:

- Block train service available between China and Belarus (e.g., Changsha-Great Stone Park).
- Lithuanian rail system is the same as Russia, China, and other CIS countries so that the rail way operator does not need to change boggy system of the wagons.
- Border crossing time between Lithuania and Great Stone industrial park in Minsk, Belarus takes 30 min for one block train; it is so-called Shuttle train Viking.
- Lithuania offers free economic zone in Klaipeda Port and Kaunas with tax incentive policies to attract cargoes from China. Once the cargoes complete customs clearance in the country, they do no need any more customs clearance within EU market because the country is a member of EU.

c. This is based on the first author's field trip to Lithuania and interviews on 7–11 Oct. 2018: meetings with Vice Minister of Transportation, Lithuania; Director of Klaipeda port and CEO of Free Trade Zone (FTZ); CEO of Kaunas FTZ; Interviews with China Merchants Groups from Minsk, Belarus; Secretary-General of China International Freight Forwarders Association.

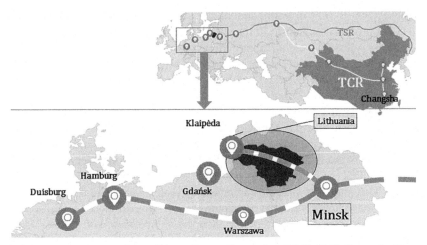

FIG. 5   A strategy connecting to China to Europe. *Source: Lee, P.T.-W., 2018. Connecting Korea to Europe in the context of the Belt and Road Initiative. KMI Int. J. Marit. Affairs Fish. 10(2), 43-54.*

- Klaipeda Port expansion by CMG id under negotiation with MOU.
- Port infrastructure by central government and superstructure by terminal operators
- Major shipping lines are calling in Klaipeda, e.g., MSC and Yang Ming. COSCO is planning subject to China Merchant Groups' investment decision.
- Minsk is expected to capture cargoes to and from the Scandinavia Peninsula through Klaipeda Port.

Despite the above advantages, China's block train cannot be commercialized for Lithuania because Russia charges 3–4 times higher than normal usage rate of the Russian railway section for the direct block trains bounding for the Lithuanian territory from China.

Having recognized the above merits and strategic location of Minsk, Lithuania has set up global logistics hub strategy in the Baltic Region aiming to not only connect China by railway and seaport to Lithuania through Great Stone industrial park in association with free economic zones in Klaipeda and Kaunas but also capture cargoes to and from West Europe and Scandinavian countries. Therefore, the two countries are trying to together solve the higher usage rate issue of the Russian railway section because both can get benefits from the direct block train services. This is an exemplary of challenges why governments are able to intervene in removing obstacles to implement the BRI (Lee et al., 2018b).

The above Lithuania-China case also gives stakeholders in Northeast Asia a couple of insights. First, Lithuania has adopted "connect strategy" to make her logistics hub for block train cargoes coming from Asia. Second, the rail operator in collaboration with logistics providers needs to consider establishing a strategic distribution center in Europe. Third, they have two options to choose TSR or TCR or both combination to achieve efficient railway service to connect Northeast Asia to Europe via China or Russia. It will be a matter of not only professional technical rail operation but also global logistics strategy in a comprehensive way. Fourth, Klaipeda and dry ports in Lithuania could be alternatively the last railway service stations from Northeast Asia to Europe instead of Duisburg. It is partly because the whole service from Northeast Asia to the country has the same railway system and partly because Lithuania is an EU member. Last but not least, cargo imbalance between Northeast Asia and Europe should be considered; it is one of the problems current Chinese rail operators are facing now. Klaipeda Port has singed the Memorandum of Understanding on constructing container terminals with China Merchants Group. This may help mitigate the imbalance issue in association with Chinese shipping companies which try to capture cargoes to and from the Baltic and Scandinavian regions and triggering interport competition with Gdansk Port. The above insights give Korea some useful lessons to activate railway service between Korean Peninsula and Europe.

4 Concluding remarks

The BRI has still a short history of six years, which may not be enough to observe its formidable outcome. However, it can be said that its impact on maritime transportation, trade, global logistics pattern, and railway service development between China and Europe are potentially large. In addition, the "Belt and Road Initiative" has been carved in General Program of the Constitution of the Communist Party of China at the 19th National Congress of the Communist Party of China[d] on 24 October 2017. It implies that China has paved a solid way for the BRI to implement it for coming years. In this circumstance, Korean stakeholders need to consider two options: either "Connect Strategy" or "Being Connected Strategy." The former is proactive, while the latter passive. Lithuania and Sri Lanka[e] follow the former. In November 2018, North Korea and South Korea have agreed to carry out joint investigation of feasibility to connect railways within the Korean Peninsula. As we have seen the China and Lithuania case, we have noticed that there are several key issues to make the rail connectivity between Korea and Europe efficient and feasible. As far as rail connectivity service between Korean Peninsula and Europe is concerned, we need to investigate railway service with or without subsidy for competitive edge

---

d. Quoted from the Constitution: "The Party shall constantly work to develop good neighborly relations between China and its surrounding countries and work to strengthen unity and cooperation between China and other developing countries. It shall follow the principle of achieving shared growth through discussion and collaboration, and pursue the Belt and Road Initiative."

e. On Sri Lanka case, see Ruan et al. (forthcoming), *Maritime Policy & Management*.

against shipping service, establishment of railway cargo distribution center in Europe, and collaboration with TCR operators for operating direct block train service, leaving aside the technical issue of rail operation. The railway connection and development between North Korea and South Korea requires social overhead capital as government infrastructure; therefore, the role of the central government is prerequisite and essential as China and Korea did for container port developments (Ng et al., 2018; Lee and Lam, 2017; Lee and Flynn, 2011) unless the UN sanction on it is mandate.

Acknowledgments

The authors would like to express their thanks to the publishers of *KMI International Journal of Maritime Affairs and Fisheries* and *Journal of Contemporary China* for granting them their copyright permission for this book chapter, respectively.

A   Variables of development priority of ports[f]

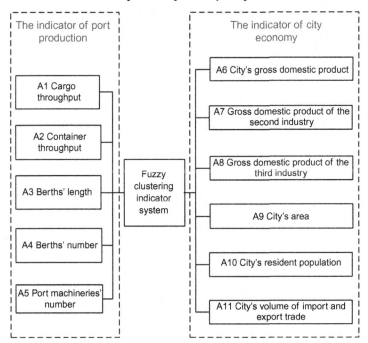

Source: Chen, J., Fei, Y., Lee, P.T.-W., Tao, X., 2019. Overseas port investment policy for China's central and local governments in the Belt and Road Initiative. J. Contemp. China 28(116), 196–215.

f. Appendices A and B are taken from Chen et al. (2019).

A1 and A2, cargo throughput and container throughput of the port, indicate the current scale of the port, given the fact the throughput is the most direct manifestation of a port's development scale and fundamental conditions. Of the two, the cargo throughput uses 10,000 tons as the unit, being an important quantitative indicator reflecting the port's scale and service level, and indicating the local construction and development status. The container throughput uses 10,000 TEUs as the unit, being an important indicator of the port scale.

A3, berth length at a port, refers to the actual length of berths for docking ships and shows the production infrastructure levels. It can also indicate the size of vessels allowed to dock at the port. Larger ships generally choose to call at ports with sufficient cargo sources, wide shipping routes, and complete port facilities. A4, the number of berths at a port, refers to the number of berths that are able to dock ships and complete cargo loading and unloading operations. The more berths a port owns, the more frequent the port is called at by ships. This article can say the number of berths impacts the port scale to a large extent.

Sufficient machineries (A5) can meet the requirements for high-intensity port production operations. It is an indicator of the port's productivity. The number of machineries impacts the labor intensity and cargo throughput capacity of a port and reflects the fundamental conditions of a port.

The annual gross product indicator of A6 of a city reflects the total value of all products produced by the city within a period of time. It is used to evaluate the overall economic development status of a city. The annual gross product of the secondary industries, A7, refers to the result of industry-related activities by all units or factories in a region, measured by the actual product prices. The annual gross product of the tertiary industries, A8, refers to the result of service-related activities by all units in a region, measured by the actual product prices. They are all indicators of a city's economy. Sound economic conditions are conducive to promoting the maritime shipping demands of a port.

A9, the urban area, directly impacts the division of land use. It is an indicator of urban land scale. Sufficient land areas are the foundation for port development. The larger the area allocated to ports, the more space for port plans to meet the city's development demands.

A10, urban population, refers to the population that has close relationships with urban activities. It is an indicator reflecting a city's economic potentials. Urban population constitutes the major part of a city. The larger the urban population, the larger the consumption potential of the city and, consequently, the larger the demands for both goods and shipping services. This sequence will boost the city's economic development and the demands for maritime shipping at the port.

A11, the import and export trade volume of a city, is used to observe the overall scale of a city in its foreign trade. It signals the existing economic and trade structures. The larger the import and export trade volume, the more foreign trade-oriented the city's economy, and the more favorable for the city to implement the "going global" development policies.

B Overseas port investment and operation of Chinese enterprises

Port enterprises	Year	Terminal projects	Investment
COSCO Shipping Ports Limited	2001	Port of Long Beach in the United States	Stake of 51%
	2003	Pasir Panjang Terminal in Brazil	Stake of 49%
	2004	Port of Antwerp in Belgium	Stake of 25%
	2005	Port of Naples in Italy	Stake of 46.25%
	2006	Port of Rotterdam in the Netherlands	Joint construction
	2007	Container terminal of Port Said (East) in Egypt	Stake of 20% in the Suez Canal Container Terminal
	2008	SSA Terminals in the United States	Stake of 33.33%
	2009	Terminal expansion of Piraeus Port in Greece	35-year franchise of Container Terminal 2# and 3#
	2012	Taiwan Kao Ming Container Terminal in China	Stake of 30%
	2015	Port of Busan in South Korea	Stake of 20%
	2015	Kumport Terminal in Turkey	Stake of 26%
	2016	Reefer Terminal S.P.A in Italy	Stake of 40%
	2016	Abu Dhabi Khalifa Port in the U.A.E.	Stake of 90%
	2016	Euromax Terminal in the Netherlands	Stake of 47.5%
	2016	Piraeus Port in Greece	Stake of 67%
	2016	Cosco-PSA Terminal in Singapore	Joint venture
	2017	Zeebrugge Terminal in Belgium	Stake of 100%
	2017	Noatum Port in Spain	Stake of 51%

Continued

Port enterprises	Year	Terminal projects	Investment
China Merchants Port Holdings Company Limited	2008	VICP in Vietnam	Stake of 49%
	2010	Port of Hambantota in Sri Lanka	Stake of 64.98% together with China Harbor Engineering Corporation (CHEC)
	2010	Vung Tau Container Terminal in Vietnam	Stake of 49%
	2010	TICT in Nigeria	Stake of 28.5%
	2011	CICT in Sri Lanka	Stake of 85%
	2012	Togo Container Terminal	Stake of 50%
	2013	Djibouti Container Terminal	Stake of 23.5%
	2013	Terminal Link	Stake of 49%
	2013	Comprehensive development project of Bagamoyo Port in Tanzania	Investment of $10 billion
	2014	Zarubino Port in Russia	Joint construction
	2014	Newcastle Port in Australia	Stake of 50%
	2015	Kumport Terminal in Turkey	Stake of 26%
	2015	Kyaukpyu Port in Myanmar	Build-operate-transfer (BOT) project
	2017	TCP in Brazil	Stake of 90%
	2017	Hambantota Port in Sri Lanka	Stake of 85%

China Communications Construction Company Limited	2002	Gwadar Deep-water Port in Pakistan	Providing 75% of port construction funds and obtaining operating rights in 2013
	2013	New Container Berth at Port Sudan	Providing assistance in construction
	2014	Mauritania Friendship Port Construction Project	—
	Commenced in 2014	Walvis Bay Container Terminal in Namibia	Providing assistance in construction
China Harbour Engineering Corporation (CHEC)	Commenced in 2013	Colombo South Harbor Container Terminal in Sri Lanka	—
	Commenced in 2014	International Bulk Cargo Terminal of Port Qasim in Pakistan	—
	Commenced in 2014	New harbor in Abaco Island in Bahamas	—
	Signed in 2014	Southern Port of Ashdod in Israel	—
	2015	Port of Ain Sukhna in Egypt	—
China Road & Bridge Corporation	Commenced in 2014	Mauritania Friendship Port Expansion Project	Providing assistance in construction
Shanghai International Port Group	2010	Zeebrugge Terminal in Belgium	Stake of 25%
	2015	Haifa Bayport in Israel	25-year franchise from 2021 onwards

Continued

Port enterprises	Year	Terminal projects	Investment
Qingdao Port Group	2011	Maday Island Terminal, Kyaukpyu Port in Myanmar	signed China-Myanmar Oil Pipeline and Oil Terminal Operation Strategic Framework Agreement with PetroChina
	2016	Vado Port in Italy	Acquisition
Guangxi Beibu Gulf International Port Group	2013	MCKIP in Malaysia	Joint construction
	2013	Kuantan Port in Malaysia	Joint construction
	2015	Kuantan Port in Malaysia	Stake of 40%
	2017	Muara Port in Brunei	Joint construction
Port of Yantai Group	2015	Port of Boke in Guinea	Stake of 10%
Dalian Port Group	2016	DMP and Djibouti Free Trade Zone	Joint construction
Shenzhen Yantian Port Group and Rizhao Port Group	2016	Melaka Gateway Port in Malaysia	Joint construction
Hebei Port Group	2016	Jambi Industrial Park Port in Indonesia	—

Source: Annual Report of COSCO Shipping Ports Limited, Annual Report of China Merchants Port Holdings Company Limited, "Belt and Road" Initiative and Interactive Development of China's Shipping Industry by Zhen Hong, Shanghai Pujiang Education Press, 2016.

References

Anderson, C.M., Park, Y.-A., Chang, Y.-T., Yang, C.-H., Lee, T.-W., Luo, M., 2008. A game-theoretic analysis of competition among container port hubs: the case of Busan and Shanghai. Marit. Policy Manag. 35 (1), 5–26.

Chen, J., Fei, Y., Lee, P.T.-W., Tao, X., 2019. Overseas port investment policy for China's central and local governments in the Belt and Road Initiative. J. Contemp. China 28 (116), 196–215.

Chhetri, P., Nkhoma, M., Peszynski, K., Chhetri, A., Lee, P.T.-W., 2018. Global logistics city concept: a cluster-led strategy under the belt and road initiative. Marit. Policy Manag. 45 (3), 319–335.

Dong, G., Zheng, S., Lee, P.T.-W., 2018. The effects of regional port integration: the case of Ningbo-Zhoushan Port. Transp. Res. E 120, 1–15.

Fu, S., Yan, X., Zhang, D., Zhang, M., 2018. Risk influencing factors analysis of Arctic maritime transportation systems: a Chinese perspective. Marit. Policy Manag. 45 (4), 439–455.

Ishii, M., Lee, P.T.-W., Tezuka, K., Chang, Y.-T., 2013. A game theoretical analysis of port competition. Transp. Res. E: Logist. Transp. Rev. 49 (1), 92–106.

Lam, J.S.L., Cullinane, K., Lee, P.T.-W., 2018. The 21st-century maritime silk road: challenges and opportunities for transport management and practice. Transp. Rev. 38 (4), 413–415.

Lee, P.T.-W., Flynn, M., 2011. Charting a new paradigm of container port development policy: the Asian Doctrine. Transp. Rev. 31 (6), 791–806.

Lee, P.T.-W., Lam, J.S.L., 2017. A review of port devolution and governance models with compound eyes approach. Transp. Rev. 37 (4), 507–520.

Lee, P.T.-W., Hu, Z.-H., Lee, S.-J., Choi, K.-S., Shin, S.-H., 2018a. Research trends and agenda on the Belt and Road (B&R) initiative with a focus on maritime transport. Marit. Policy Manag. 45 (3), 282–300.

Lee, P.T.-W., Feng, X., Lee, S.-W., 2018b. Challenges and chances of the Belt and Road Initiative at the maritime policy and management level. Marit. Policy Manag. 45 (3), 279–281.

Lee, P.T.-W., Lee, S.-W., Hu, Z.-H., Choi, K.-S., Choi, N.Y.H., Shin, S.-H., 2018c. Promoting Korean International Trade in the East Sea Economic Rim in the context of the Belt and Road Initiative. J. Korea Trade 22 (3), 212–227.

Lee, S.W., Song, J.M., 2014. Economic possibilities of shipping though Northern Sea route. Asian J. Ship. Logist. 30 (3), 415–430.

Lim, S.-W., Suthiwartnarueput, K., Abareshi, A., Lee, P.T.-W., Duval, Y., 2017. Key factors in developing transit trade corridors in Northeast Asia. J. Korea Trade 21 (3), 191–207.

Lindstad, H., Bright, R.M., Strømman, A.H., 2016. Economic savings linked to future arctic shipping trade are at odds with climate change mitigation. Transp. Policy 45, 24–30.

Liu, M., Kronbak, J., 2010. The potential economic viability of using the Northern Sea Route (NSR) as an alternative route between Asia and Europe. J. Transp. Geogr. 18 (3), 434–444.

Meng, Q., Zhang, Y., Xu, M., 2017. Viability of transarctic shipping routes: a literature review from the navigational and commercial perspectives. Marit. Policy Manag. 44 (1), 16–41.

National Development and Reform commission (NDRC), Ministry of Foreign Affairs and Ministry of Commerce of the People's Republic of China with State Council, 2015. Vision and Actions on Jointly Building Silk Road Economic Belt and 21st Century Maritime Silk Road. The National Development and Reform Commission (NDRC), Beijing.

Ng, A.K.Y., Jiang, C., Li, X., O'Connor, K., Lee, P.T.-W., 2018. A conceptual overview on government initiatives and the transformation of transport and regional systems. J. Transp. Geogr. 71, 199–203.

Notteboom, T., Yang, Z., 2017. Port governance in China since 2004: institutional layering and the growing impact of broader policies. Res. Transp. Bus. Manag. 22, 184–200.

Ruan, X., Bandara, Y.M., Lee, J.-Y., Lee, P.T.-W., Chhetri, P., 2019. Impacts of the Belt and Road Initiative in the Indian subcontinent under future port development scenarios. Marit. Policy Manage. https://doi.org/10.1080/03088839.2019.1594425.

State Council Information Office of the People's Republic of China, 2018. China's Arctic Policy. State Council Information Office, Beijing.

Tseng, P.H., Cullinane, K., 2018. Key criteria influencing the choice of Arctic shipping: a fuzzy analytic hierarchy process model. Marit. Policy Manag. 45 (4), 422–438.

Vavrus, S.J., Holland, M.M., Jahn, A., Bailey, D.A., Blazey, B.A., 2012. Container shipping on the Northern Sea Route. Int. J. Prod. Econ. 122 (1), 107–117.

Verny, J., Grigentin, C., 2009. Container shipping on the Northern Sea Route. Int. J. Prod. Econ. 122 (1), 107–117.

Wei, H., Sheng, Z., Lee, P.T.-W., 2018. The role of dry port in hub-and-spoke network under Belt and Road Initiative. Marit. Policy Manag. 45 (3), 370–387.

Zhang, Y., Meng, Q., Ng, S.H., 2016. Shipping efficiency comparison between Northern Sea Route and the conventional Asia-Europe shipping route via Suez Canal. J. Transp. Geogr. 57, 241–249.

Zhu, S., Fu, X., Ng, A.K.Y., Luo, M., Ge, Y.-E., 2018. The environmental costs and economic implications of container shipping on the Northern Sea Route. Marit. Policy Manag. 45 (4), 456–477.

Further reading

Jiang, Y., 2018. Evolution of Hinterland patterns between China Railway Express and Seaborne Container Shipping under the B&R initiative. In: 2019 IAME Conference, Mombasa, Kenya, 11–14 Sept. 2018.

Lee, P.T.-W., 2016. An anatomy of the 'One Belt and One Road' from the viewpoint of structural changes in maritime transport. In: 2016 Annual Conference of International Association of Maritime Economists. Kuhne Logistics University, Hamburg Germany, August 23–26.

Lee, P.T.-W., Hu, Z.H., Lee, S.J., 2016. Research trend in the Belt and Road Initiatives. In: 2016 OBOR Conference. RMIT University, Melbourne, Australia, December 1–2.

Lee, P.T.-W., 2018. Connecting Korea to Europe in the context of the Belt and Road Initiative. KMI Int. J. Marit. Affairs Fish. 10 (2), 43–54.

Port of Klaipeda, 2018. Klaipeda Port Expansion: Business Opportunities. Klaipeda, Lithuania, Port of Klaipeda.

Chapter 17

A multiobjective programming model for comparing existing and potential corridors between the Indian Ocean and China

Ying-En Ge, Lidan Du, Zhongyu Wang, Yong Zhou
College of Transport and Communications, Shanghai Maritime University, Shanghai, China

1 Introduction

1.1 Background

With the rapid development of China in the past four decades, today's China has a greater interest in establishing a stronger connectivity with the rest of the world than in the past, which is an important aim of the Belt and Road Initiative (BRI). This initiative is a great plan involving political, economic and regional development, and environmental factors around the world. It is of utmost importance to ensure the smoothness and safety of the connectivity for China's and regional economic development in a sustainable and efficient manner. The Malacca Strait, as a connectivity link between the Indian Ocean and the South China Sea, is vital to the Chinese economy since about half of Chinese cargo transport goes through it and enters the Indian Ocean. As an important country of global trade, China needs ensure the safety and smooth flow of cargo transport between the Indian Ocean region and China.

At present, the vast majority of maritime cargo transport between China and the Indian Ocean must pass through the Malacca Strait, often called the "Asian throat," and the performance of the Strait has a huge impact on China's marine transport lines and China's economy. It is because of this that China has to identify alternative transport corridors to move their cargoes smoothly and safely between the Indian Ocean region and China. Therefore, the implementation and boost of the BRI is targeted at establishing transport lines or logistics corridors between the Indian Ocean region and China.

The BRI's "Vision and Action" mentions the China-Pakistan Economic Corridor (CPEC), and China-Myanmar-Bangladesh-India (CMBI) Economic

Maritime Transport and Regional Sustainability. https://doi.org/10.1016/B978-0-12-819134-7.00018-6
289

Corridor, which are under construction or discussion. These two corridors will offer new opportunities or alternatives to China's maritime transport lines going through the Malacca Strait. The idea of constructing the Kra Canal in Southern Thailand has been discussed widely in Southeast Asia and in the rest of the world (Lau and Lee, 2016; Heng and Yip, 2018; Gao and Lu, 2019). If it is to be built, it would provide a direct alternative to the Malacca Strait.

In the literature, there is a great deal of work on evaluating these corridors or their alternatives from the political, economic, environmental, and other perspectives. Here is a brief overview of the literature on this topic.

1.2 Literature review

Having been stimulated by a real-life application, Caramia and Guerriero (2009) propose a multiobjective model to investigate a long-haul freight transport problem, where the two objectives, respectively associated with travel time and transport cost, are to be minimized together with the maximization of transportation mean sharing index. Vehicle capacity, time windows, and transportation jobs have to obey additional constraints related to mandatory and forbidden nodes. A heuristic algorithm is applied to solve their problem. Yang et al. (2011) present an intermodal network model to examine the competitiveness of 36 alternative routes for freight transport from China to and beyond the Indian Ocean, and a goal programming approach is proposed to handle the formulated model with multiple and conflicting objectives, such as minimizing transport cost, transit time, and variability of transit time simultaneously. Zhou (2012) proposes a logistics network method for a two-level logistics network with fixed topology, considering the determination of the capacity of each transport line to seek the optimal logistics network utilization, constrained by traffic distribution plan and the logistics node and transport line capacity. Shaikh et al. (2016) evaluate the current timetable, cost, energy consumption, and greenhouse gas emissions of existing and proposed oil supply routes from the Middle East and Africa to the border of China. They estimate the Capital Expenditure (CAPEX), Operating Expense (OPEX), and average cost per barrel of the pipeline along the CPEC based on a weighted model. Chen (2018) proposes four transport corridors for the trade between China and Europe, including the Maritime Silk Road, the Central European Banley, the Gwadar Port Iron-Sea Transport, the Ice Silk Road, and systematically compares their development trends, current situation, and potentials, and certain advantages and limitations of each of these corridors are identified and discussed.

The Indian Ocean region is an important region for the implementation of BRI to strengthen the connectivity between Africa/Europe and the Far East. In this context, the China-Pakistan Economic Corridor, the China-Myanmar-Bangladesh-Indian Corridor, and the Kra Canal proposal have been discussed widely, as transport or logistics corridors running parallel to the Malacca Strait. The investigations in the existing literature generally focus on a corridor only with analysis of the impacts on the local economy and environment if it runs. It is time to move on for us to carry out a comparative study of these alternative transport corridors between the Indian Ocean region and the Far East after we have reached a reasonable understanding of these corridors. This comparison will benefit further discussions on planning, constructing, and operating, as well as academically investigating these corridors.

When we plan these three alternatives running parallel to the Malacca Strait, it is necessary to consider the split of goods among them as well as the investment plus the rate of return.

1.3 Intellectual merits of this research with scenario settings

This chapter is to propose a multiobjective optimization model while different types of cargoes are transported along the corridors of interest between the Indian Ocean region and China. For the convenience of exposition, the cargoes are assumed to move separately from Saudi Arabia and South Africa to China or from China to Saudi Arabia and South Africa through the aforementioned four corridors, as illustrated in Fig. 1, in which the bottom route of the right side of the map is the existing one going through the Malacca Strait and the other three are potential ones. On the map in Fig. 1, routes 1–4 are also known as routes A, B, C, and D, respectively.

The proposed model has four objectives, respectively associated with transport cost, energy consumption, greenhouse emissions, and safety. In the model to be formulated, all the four objectives appear in one single objective function

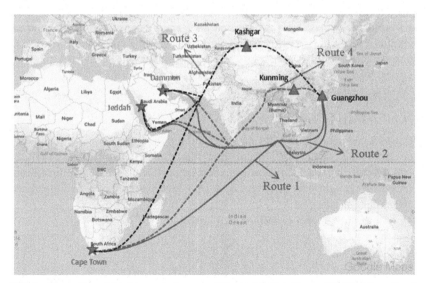

FIG. 1 Illustration of the four routes between China and Indian Ocean. (*Modified from Google Maps.*)

in the form of a weighted sum of them. Though, we still use "multiobjective programming model" in the title of the chapter. The solution to the model offers a satisfactory cargo (volume) allocation over the four corridors. The formulated model can be used as a tool of decision-making support since the results from it may provide meaningful suggestions for policy makers.

Table 1 lists the cargoes the four corridors of interest may transport. The key reason to choose Saudi Arabia as an end of the transport corridors is that a large amount of crude oil is transported to the Far East from this region and the main reason we choose South Africa as an end of the corridors is that a lot of iron ores are transported to the Far East from this place. The products from China to Saudi Arabia or South Africa are mainly electronic ones. It is noteworthy that all these cargoes may be transported via all these corridors.

1.4 Structure of this chapter

The rest of this chapter is organized as follows. Section 2 introduces a multiobjective programming model to capture the distribution of cargoes over the four transport corridors under discussion. Section 3 applies the formulated model to make a comparison of these corridors. Section 4 investigates the rate of return on investment in the three potential corridors. Section 5 gives a brief discussion on the regional impacts of transport from the analysis carried out in this chapter. Section 6 concludes the chapter.

2 Methodology

The resulting cargo allocation model consists of two components: objective system and constraints. The objective system consists of four objectives or indicators for each corridor, i.e., transport cost, energy consumption, greenhouse emissions, and safety. In this work, we use VLCCs of 200 million barrels to transport oil and full 40-foot containers for other cargoes.

2.1 Objective system

2.1.1 Transport cost

Transport cost mainly includes the cost incurred in the process of moving cargoes from their origin to their destination, regardless of other indirect activities, such as marketing, packaging, information support, and general administration. We consider freight rate, inventory cost, shipment-handling cost, and insurance cost in the transport cost.

First, the freight rate is defined as follows:

$$F_i = \sum_{n=1}^{3} d_{in} \cdot V \cdot x_i \tag{1}$$

where F_i is the total freight rate from origin to destination along route i by transport mode n (n = 1, 2, 3, respectively corresponding to marine, rail and

TABLE 1 Description of the cargoes transported.

Cargoes	Origins	Destinations	Corridors	Transport modes
Crude oil	Damman	Guangzhou	Malacca Strait	Marine
		Guangzhou	Kra Canal	Marine
		Kashgar	CPEC[a]	Marine cum pipeline
		Kunming	CMBI[b]	Marine cum pipeline
Iron ores	Cape Town	Guangzhou	Malacca Strait	Marine
		Guangzhou	Kra Canal	Marine
		Kashgar	CPEC[a]	Marine cum rail
		Kunming	CMBI[b]	Marine cum rail
Electronic products	Guangzhou	Jeddah	Malacca Strait	Marine
			Kra Canal	Marine
			CPEC[a]	Marine cum rail
			CMBI[b]	Marine cum rail
	Guangzhou	Damman	Malacca Strait	Marine
			Kra Canal	Marine
			CPEC[a]	Marine cum rail
			CMBI[b]	Marine cum rail
	Guangzhou	Cape Town	Malacca Strait	Marine
			Kra Canal	Marine
			CPEC[a]	Marine cum rail
			CMBI[b]	Marine cum rail

[a]*China to Pakistan Economic Corridor.*
[b]*China-Myanmar-Bangladesh-India Economic Corridor.*

pipeline), d_{in} is the unit freight rate to transport a full 40-foot container or same volume of cargo by route i, V is the total cargo volume, and x_i is the ratio of cargo volume for route i to the total cargo volume.

The three different modes of transport are considered because not all the four alternative corridors use the same transport modes.

Second, according to Min (1990), the inventory cost occurs in three places: the consignor, in-transit, and the consignee. The inventory costs to the consignor and the consignee are generally considered as parts of the cost of manufacture and sale, respectively. The cost of in-transit inventory is considered a crucial part of the cost in the whole transport. For simplicity, we only need treat the in-transit inventory cost as the inventory cost (IIC_i), which is related to the total freight value on route i and transport time and written mathematically in the following expression:

$$IIC_i = FV_i \cdot \sum_{n=1}^{3} T_{in} \cdot IR_n \tag{2}$$

where T_{in} denotes the total time for cargoes to be transported on route i by transport mode n, IR_n is the inventory-holding cost rate (in percentage) in mode n, and FV_i represents the total value of freight moving on route i and is defined below:

$$FV_i = fv \cdot V \cdot x_i \tag{3}$$

where fv denotes the production value of a cargo in a full 40-foot container or a full barrel.

The total time for cargoes to be transported from origin to destination on route i by transport mode n is given by the ratio of distance to speed, i.e.:

$$T_{in} = \frac{D_{in}}{v_n} \tag{4}$$

where D_{in} and v_n denote the distance and velocity on route i by transport n, respectively.

Third, the insurance cost of transport is similar to the inventory cost and related to the freight value, transport time, and the insurance cost rate β_n (%):

$$IS_i = FV_i \cdot \sum_{n=1}^{3} T_{in} \cdot \beta_n \tag{5}$$

where β_n varies according to mode n, which is classified into marine, rail and pipeline transport.

The fourth type of the transport cost is shipment-handling cost, which mainly occurs in the process of loading and unloading and is related to the total freight value and the cargo volume:

$$H_i = FV_i \cdot \sqrt[3]{V} \cdot x_i \cdot \eta \tag{6}$$

where η represents the specified factor of the shipment-handling cost.

To sum up, the total cost (C_i) to the transport of a cargo can be written as follows:

$$C_i = F_i + IIC_i + H_i + IS_i \tag{7}$$

In particular, for oil transport, F_i is just defined as marine cost, and varies as a route differs, depending on the distance between two ports of route i (D_i). As the proposed corridors currently remain conceptual, we are short of data related to them. The marine cost per barrel is estimated by means of the known marine cost (F_A) and the distance of route 1 (denoted as D_A). The freight rate for the other routes (i) is calculated as in Eq. (8) and it is assumed that the pipeline cost is \$4.00/barrel of ESPO pipeline, irrespective of the pipeline length.

$$F_i = \frac{D_i}{D_A} \times F_A \tag{8}$$

2.1.2 Energy consumption and GHG emissions

The energy consumption (E_i) and GHG emissions (G_i) are estimated respectively by means of the following expressions, which are proposed on the basis of the work in Shaikh (2016):

$$E_i = V \cdot x_i \cdot m \cdot \sum_{n=1}^{3} \phi_{En} \cdot D_{in} \tag{9}$$

$$G_i = V \cdot x_i \cdot m \cdot \sum_{n=1}^{3} \phi_{Gn} \cdot D_{in} \tag{10}$$

where m means the weight of a full 40-foot container or a full barrel oil, φ_{En} and φ_{Gn} represent respectively the standard values per unit energy consumed and per GHG emitted in transporting 1-ton cargo for the distance of 1 km by mode n.

It is noted that the standard values may differ from one transport mode to the other.

2.1.3 Safety

In this chapter, the risk value is adopted to evaluate safety, which implies that the higher risk value represents the lower safety level of a route. To calculate the risk of a route, the following kinds of risk events that may occur to transport have been assumed:

(1) Natural disasters, such as lightning, tsunami, earthquake, flood.
(2) Traffic accidents, mainly happening to sea transportation, such as stranding, collision, explosion, capsize.
(3) General extraneous risks, such as stealing by pirates, cargo clash and rust in the process of transportation, natural damage of pipe.
(4) Special extraneous risks, such as war, man-made damage of hardware facilities, and closure of corridors due to political disputes among the neighboring countries.

The resulting risk value can be calculated in the following equation:

$$R_i = \sum_{n=1}^{3} \left(D_{in} \cdot \sum_{r=1}^{4} L_{irn} \cdot P_{irn} \right) \tag{11}$$

where L_{irn} means the loss when risk r happens to transport mode n on route i (the percentage of the total freight value on route i), and P_{irn} is the probability of risk r happening to transport mode n on route i.

2.2 Resulting model

A mathematical programming model can be formulated for our problem and written in the following form:

$$\min Z = w_1 C_i + w_2 E_i + w_3 G_i + w_4 R_i \qquad (12)$$

for all $i = 1, 2, 3, 4$ ··Subject to:

$$\begin{cases} \sum_{i=1}^{4} x_i = 1, \\ x_i \geq 0, \qquad i = 1, 2, 3, 4 \end{cases} \qquad (13)$$

where the objective function in Eq. (12) minimizes the weighted sum of total transport cost, energy consumption, greenhouse gas emissions, and risks, with the degree of priority of each type of costs indicated by the weights w_j ($j = 1, 2, 3, 4$), and C_i, E_i, G_i and R_i given in Eqs. (7), (9)–(11), respectively.

In addition, the sum of w_j for $j = 1, 2, 3$, and 4 is equal to 1.

3 Model application and analysis

The previously formulated mathematical programming model (12) and (13) is a nonlinear programming problem. Since the objective function of the model contains four components with different units and scales, we normalize the four components by dividing each of them with their respective potential maximum values. Considering that the four components exhibit significantly different distributions, such a normalization procedure retains their respective original distributions and transforms the value of each of the four components into a range between 0 and 1. Our model is solved by means of Cplex and Yalmip.

3.1 Scenario settings

To proceed, we are making the following assumptions:

(1) The original weight vector for the objective function in Eq. (12) is set to be:

$$\left(w_1, w_2, w_3, w_4\right) = \left(0.4, 0.25, 0.25, 0.1\right)$$

(2) The unit freight value (fv) of a full 40-foot container or a full barrel of crude oil is assumed to be US\$800.00.

(3) The inventory cost rate and insurance cost rate both vary from one transport mode to the other and are set respectively as follows:

$$IR_n = \begin{bmatrix} 0.025 & 0.015 & 0.01 \end{bmatrix}$$
$$\beta_n = \begin{bmatrix} 0.015 & 0.02 & 0.01 \end{bmatrix}$$

(4) The shipment-handling cost rate is set to be 2.5% on both routes 1 and 2, and 3.75% for the others.

(5) The probability of risk r by transport mode n is assumed as follows:

$$P_{rn} = \begin{bmatrix} 0.3 & 0.4 & 0.6 \\ 0.4 & 0.3 & 0.05 \\ 0.2 & 0.2 & 0.15 \\ 0.1 & 0.1 & 0.2 \end{bmatrix}$$

$$P_{arn} = P_{brn}, \quad a \neq b$$

3.2 Scenario analysis

The cargo volume allocation given by the previously formulated model may vary as the total volume varies. In this section, we will carry out a series of numerical experiments to test the sensitivity and robustness of the cargo volume allocation to the changes in various parameters. For the sake of exposition, we take cargo transport from South Africa to China as a case with sensitivity analysis of the cargo volume allocation to the weights and safety cost in the objective function (12).

3.2.1 Sensitivity to the weight w_j

An increase in the weight of a cost component will inevitably lead to a reduction in other weights since their sum is equal to one. The objective function of the previously formulated model (12)-(13) can be rewritten in the following form:

$$\min Z = \sum_{i=1}^{I} \sum_{j=1}^{J} w_j f_{ij} \tag{14}$$

where $f_{ij} = \sum_r K_{ij}^r$, r denotes one of the four alternative routes, K indicates a cost component; K represents C, E, G and R when $j = 1, 2, 3, 4$. If w_{j_0} increases by θ then a new set of weights is given by $w'_{j_0} = w_{j_0} + \theta$ and

$$w'_j = w_j - \theta/3 \quad \text{at} \quad j \neq j_0 \tag{15}$$

Figs. 2–5 show the results of a series of experiments regarding the sensitivity of the cargo volume allocation over the four routes to the weight w_j. It can be seen from Fig. 2 that, as the decrease of w_1 makes route 3 undertake more and more cargo volume and even all cargo when w_1 falls down to 0.1, which implies that the economic factor is an inferior strength of route 3 (China to Pakistan Economic Corridor). In other words, route 3 may be the most costly one among the four ones in terms of the transport cost.

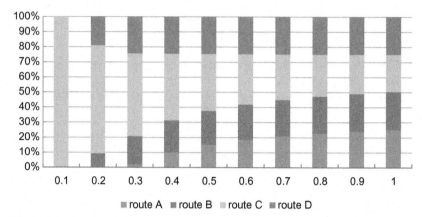

FIG. 2 Volume ratio variation as the weight associated with the transport cost varies.

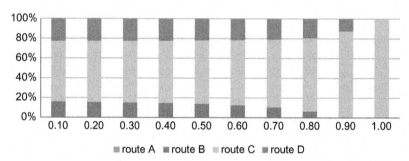

FIG. 3 Volume ratio variation as the weight associated with the energy consumption varies.

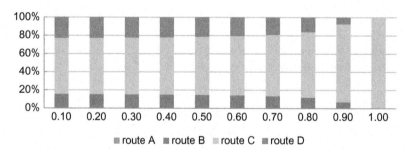

FIG. 4 Volume ratio variation as the weight associated with the GHG emissions changes.

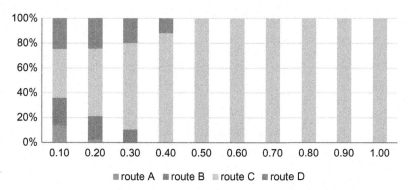

FIG. 5 Volume ratio variation as the weight associated with the safety changes.

As shown in Figs. 3 and 4, energy consumption and GHG emissions have a similar effect on the four routes. While $w_2 = 1$ or $w_3 = 1$, the CPEC is an absolutely preferred alternative. As w_2 or w_3 increases from 0, the volume allocated over the CPEC increases gradually. The volume onto the CPEC, on the other hand, is still quite high (more than 50%) even though w_j (j=2, 3) approaches zero. This implies that the CPEC is a good choice in the sense of sustainability (energy consumption, GHG emissions). As for safety, the CPEC is in a relative dominant position. Before w_4 approaches 0.5, route 3 has taken up all cargo volume, which implies that it dominates the other three alternatives in terms of safety in the given set of scenario settings. However, to have more convincing opinions of this, we need more real-life data to make our set scenario more like the real-life one.

3.2.2 Sensitivity to the safety cost

The probability of occurrence of each risk is assumed, based on the data from the existing corridors of the type. It is noteworthy that the variation in the probability may lead to different results. The maintenance cost of railways and pipelines in response to various risks is not too small to be counted. Hence, the sensitivity analysis in this subsection mainly considers two factors: the probability of occurrence of risks and the maintenance cost of pipelines.

3.2.2.1 The probability of risks to occur

The CPEC and the CMIB both consist of marine and pipeline transport. Since the pipeline is more vulnerable and subject to more unpredictable challenges, we are carrying out the sensitivity analysis of the cargo volume allocation to the probability of each risk to occur while oil transport is carried out by pipeline. Figs. 6 and 7 show the results from the variation in the probability of risks to occur to the pipeline transport.

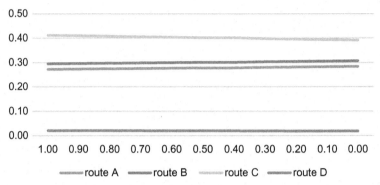

FIG. 6 Volume ratio variation as risk 1 varies.

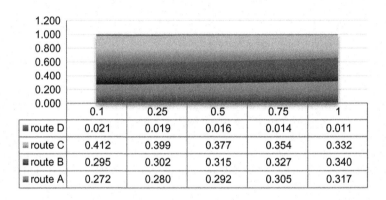

	0.1	0.25	0.5	0.75	1
route D	0.021	0.019	0.016	0.014	0.011
route C	0.412	0.399	0.377	0.354	0.332
route B	0.295	0.302	0.315	0.327	0.340
route A	0.272	0.280	0.292	0.305	0.317

FIG. 7 Volume ratio variation as the pipeline maintenance cost varies.

It is readily seen that the probability of occurence of risks has a little effect on the volume allocation, so the variation trend is stable and negligible.

3.2.2.2 The maintenance cost of pipeline

Due to the unavailability of exact pipeline maintenance cost, a certain proportion of the freight value is proposed to be the maintenance cost, and denoted by the value of lost freight. Keep the maintenance cost of marine L_{ir1} as constant and make the pipeline maintenance cost set as follows:

$$L_{ir3} = 0.1FV_i, \ L_{ir3} = 0.25FV_i, \ L_{ir3} = 0.5FV_i, \ L_{ir3} = 0.75FV_i, \ L_{ir3} = FV_i$$

We may then solve the model under these settings and compare the results, which are displayed in Fig. 7.

It can be seen that the increase in the percentage of lost freight value (i.e., maintenance cost of pipelines) has a great effect on volume allocation, and that the volume allocated onto the CPEC has just decreased slightly, eliminating inaccuracy due to data inevitability.

4 Analysis of return on investment

Except the Malacca Strait, the other three transport corridors have not been built or not completed yet. The necessity of building these alternatives has always been a controversial issue. It has not yet been decided in favor of the Kra Canal, after having been discussed for decades. It is undoubted that many factors have affected the process, but getting sufficient funds to complete this mega-project is a problem that cannot be avoided at all. In addition, the CPEC and the CMBIEC are the two important corridors proposed in the context of the BRI, which certainly needs a huge amount of financial investment along the Belt and/or Road. Therefore, it is imperial to investigate the rate of return on investment in these corridors.

It is assumed that the CPEC will be completed in 2030, so we choose 2030 as a time point for investment to start to return. For the sake of exposition, this analysis only considers three typical cargoes being transported between China and Saudi Arabia or South Africa, which are respectively crude oil from Dammam to China, iron ores from Cape Town to China, and plastic products from Guangzhou to Cape Town.

4.1 Predicting volume of imports and exports

From 2000 to 2017, as shown in Figs. 8–10, the volume of imports and exports of China's foreign trade has witnessed a rising trend, which could be approximately considered a linear rise. Therefore, this chapter assumes to make

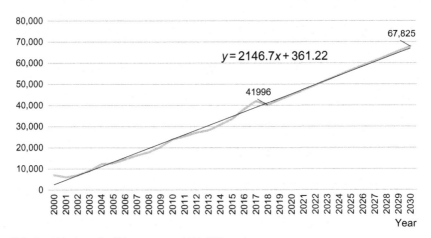

FIG. 8 China's crude oil imports up to 2030 (10k tons).

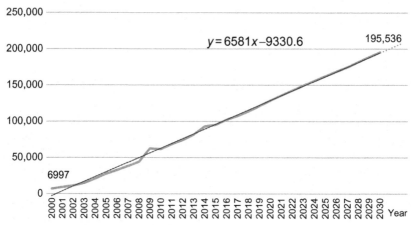

FIG. 9 China's iron ore imports up to 2030 (10k tons).

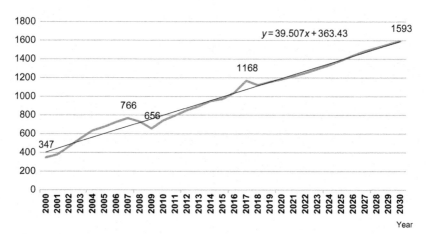

FIG. 10 China's plastic product exports up to 2030 (10k tons).

a linear prediction for the volume of imports and exports of China in 2030 (see Figs. 8–10 again).

It is assumed that the percentage of crude oil imports from Saudi Arabia to the three cities of China will be up to 12%, and that 8% of the total iron ores will be transported from South Africa to China. Plastic product exports from Guangzhou to Cape Town account for 80% of the total.

4.2 Optimal volume allocation

According to the sensitivity analysis in the previous section, it is known that the impact of risks on the volume allocation results is negligible. In addition, the inventory cost rates, insurance cost rates, and shipment-handling cost rates

TABLE 2 The optimal cargo volume allocation.

Cargo	Crude oil	Iron ores	Plastic products
Volume ratio	0.03:0.17:0.62:0.18	0.21:0.41:0.28:0.10	0:1:0:0

are all very small, whose variations play only a minor role in the calculation of the transport cost. Therefore, keep the values of those parameters in transport cost and safety the same as in the previous assumptions. In addition, the different weights of the four components in the objective function (12) will lead to different results. We have seen that the weight of transport costs plays the most sensitive role in the cargo volume allocation, and that the effects of rest of the cost components are not very large. Therefore, in investment analysis, the weight of transport cost is assumed to be 0.4, and the weights for the rest of the costs are all set to 0.2.

Moreover, the freight value of a full 40-foot container is different when different cargoes are loaded. Combining the information found, we assume that a full barrel of crude oil as is worth 50 US dollars, and that a full 40-foot container of plastic products is worth 500 US dollars. In addition, it is known that the price per ton of iron ores is about $76, so the fv of a full 40-foot container filled with iron ores can be set as $1950. Considering the reality of imports and exports, the total cargo volume of a full year is simply divided into 12 for 12 months. Therefore, the volume in the formulated model is the value equal to the annual total forecast volume divided by 12.

Based on the aforementioned assumptions, the results of three types of cargo volume allocation are listed in Table 2.

4.3 Analysis of return on investment

In order to display more intuitively the superiority of optimal allocation from the previously formulated model, we make a comparison of these corridors in terms of the aforementioned performance indices (i.e. four cost components in the objective function) between pre and postoptimization. The values of the indices are given in Tables 3–5.

It is suggested that almost all the four cost components have been improved more or less, and especially for crude oil, the values of the three out of the four have been improved by 50% or so. Although for the other two cargoes, the improvements of the four objectives are not so large, there have still been significant advantages in most cases because of the very large basis value.

In the transport cost evaluation, this chapter adds transport time cost so the transport cost not only includes economic cost but also time cost, which leads to the larger value of transport cost. It is known that the completed construction of

TABLE 3 A comparison of crude oil before and after optimization.

Route	Ratio	Transport cost (×10⁷ US$)	Energy consumption (×10¹¹ KJ)	GHG emissions (×10¹⁰ g)	Safety (×10¹¹ US$)
Route 1	0.03	2.5	4.6	3.6	2.7
Route 2	0.17	68.0	23.3	18.3	14.0
Route 3	0.62	1360.0	33.7	27.2	11.0
Route 4	0.18	208.0	20.3	16.0	11.0
After (total)	1	1638.5	81.9	65.1	38.7
Before	–	2270.0	153.0	12.0	91.5
Optimization value	–	631.5	71.1	54.9	52.8
Optimization rate	–	38.5%	46.5%	45.8%	57.7%

TABLE 4 A comparison of iron ores before and after optimization.

Route	Ratio	Transport cost (×10⁸ US$)	Energy consumption (×10¹¹ KJ)	GHG emissions (×10¹⁰ g)	Safety (×10¹¹ US$)
Route 1	0.21	1.6	5.2	4.1	6.7
Route 2	0.41	4.0	9.6	7.5	12.3
Route 3	0.28	5.5	5.4	3.7	6.6
Route 4	0.10	3.1	2.3	1.7	3.0
After (total)	1	14.2	22.5	17.0	28.6
Before	—	16.3	22.2	17.3	32.4
Optimization value	—	2.1	−0.3	0.3	3.8
Optimization rate	—	14.8%	−1.3%	1.8%	13.3%

TABLE 5 A comparison of plastic products before and after optimization.

Route	Ratio	Transport cost (×10⁷ US$)	Energy consumption (×10¹² KJ)	GHG emissions (×10¹¹ g)	Safety (×10¹¹ US$)
Route 1	0	0	0	0	0
Route 2	1	9.8	3.1	2.4	1.2
Route 3	0	0	0	0	0
Route 4	0	0	0	0	0
After (total)	1	9.8	3.1	2.4	1.2
Before	–	10.4	3.3	2.6	1.3
Optimization value	–	0.6	0.2	0.2	0.1
Optimization rate	–	6.1%	6.5%	8.3%	8.3%

the CPEC is expected to invest $46 billion, and that $28 billion will be invested in constructing the Kra Canal. It is defined that the rate of return on investment in the three corridors is the total optimization value of transport cost, so the return rate is approximately 50%. This chapter only considers three types of cargoes, and for more cargoes there must be more cost advantages. Therefore, under the premise of comprehensively considering economic cost and time cost, the rate of return on investment in the three corridors is a little higher, to some extent. The transport time is an important aspect in the process of foreign trade that we need consider. Based on our inevstigation in this chapter, it may be feasible and rational to invest in the three corridors so that cargo imports and exports can be finished more effectively and economically.

5   Transport and regional impacts

The rapid development of China's economy in the past four decades itself is a good evidence of positive impacts of transportation development on economic growth. Having made a long-lasting great success by building roads before making more wealth, China took the initiative in 2013 and proposed the BRI, which is another practice of this idea. That is, it is hoped that the BRI will promote more transportation infrastructure to be built or improved so that this world or the regions the Belt or the Road goes through may be connected in a better way, which promotes economic development due to lower transport cost and better, inspiring, efficiency.

If all three new corridors that are discussed in this paper are built, South Asia will certainly be more connected, so that the logistics cost will be reduced significantly and resources or products can be transported in a more efficient and effective manner. Then, a business operator may optimize his or her business in a wider area.

6   Concluding remarks

This chapter proposes a mathematical programming model to allocate a set of given cargoes over the four transport corridors or routes of interest between China and Saudi Arabia or South Africa in the Indian Ocean region, which handles the four objectives: minimized transport cost, minimized energy consumption, minimized GHG emissions, and maximized safety by means of a weighted sum of them. Then a scenario analysis is carried out to test and verify the newly proposely model, and the sensitivity analysis of the cargo volume allocation over the four alternative routes to the weights and safety factor in the formulated model is carried out to identify the degrees of influence of each type of objectives.

A series of experiments show that the resulting cargo volume allocation from the formulated model may be greatly affected by the weights of the four objectives, especially by the weight of transport cost, whose variation can make a big difference to the resulting cargo allocation among the corridors. Moreover, as the parameter values in the safety definition vary, the allocation results have been almost unaffected. Subsequently, the rate of return on investment in the three potential corridors is analyzed. This work may not provide a good or final set of suggestions for policy makers, but certainly will promote the implementation of Belt and Road Initiative and encourage more convining analysis of potential ideas or projects related to this initiative.

Acknowledgments

The authors feel grateful for the financial support of the Shanghai Municipal Science and Technology Commission via Grant no.: 17040501800. The anonymous referees' comments on the early version of this chapter are also greatly appreciated.

References

Caramia, M., Guerriero, F., 2009. A heuristic approach to long-haul freight transportation with multiple objective functions. Omega 37 (3), 600–614.

Cen, X., 2018. A comparative study of various Sino-European trade transportation channels under the Belt and Road strategy. Logist. Eng. Manag. 8, 7–11.

Gao, T.H., Lu, J., 2019. The impacts of strait and canal blockages on the transportation costs of the Chinese fleet in the shipping network. Marit. Policy Manag. 46 (6), 669–686. https://doi.org/1 0.1080/03088839.2019.1594423.

Heng, Z., Yip, T.L., 2018. Impacts of Kra Canal and its toll structures on tanker traffic. Marit. Policy Manag. 45 (1), 125–139. https://doi.org/10.1080/03088839.2017.1407043.

Lau, C.Y., Lee, J.W.C., 2016. The Kra Isthmus Canal: a new strategic solution for China's energy consumption scenario? Environ. Manag. 57 (1), 1–20. https://doi.org/10.1007/s00267-015-0591-0.

Min, H., 1990. International intermodal choices via chance-constrained goal programming. Transportation Research Part A: Policy and Practice 25 (6), 351–362.

Shaikh, F., Ji, Q., Fan, Y., 2016. Prospects of Pakistan–China energy and economic corridor. Renew. Sustain. Energy Rev. 59, 253–263.

Yang, X., Low, J.M.W., Tang, L.C., 2011. Analysis of intermodal freight from China to Indian Ocean: a goal programming approach. J. Transp. Geogr. 19 (4), 515–527.

Zhou, X., 2012. Multi-Objective Optimization Method Based on Variable Weights for Flow Allocation in Logistics Network. Southwest Jiaotong University, Chengdu, China.

Further reading

Cao, W., Bluth, C., 2013. Challenges and countermeasures of China's energy security. Energy Policy 53, 381.

China Railway 12306. http://www.12306.cn/yjcx/hybj.jsp.

Hao, H., Geng, Y., Li, W., Guo, B., 2015. Energy consumption and GHG emissions from China's freight transport sector: scenarios through 2050. Energy Policy 85, 94–101.

Jincheng Logistics Network, http://www.jctrans.com.

Khan, S.A., 2013. Geo-economic imperatives of Gwadar Sea Port and Kashgar economic zone for Pakistan and China. IPRI J. XIII (2), 87–100.

Leung, G.C.K., 2011. China's energy security: perception and reality. Energy Policy 39, 1330.

Lin, W., Chen, B., Xie, L., Pan, H., 2015. Estimating energy consumption of transport modes in China using DEA. Sustainability 7, 4225.

Martin, B., Stein, O.E., Sören, E., 2016. Assessment of the applicability of goal- and risk-based design on Arctic sea transport systems. Ocean Eng. 128, 183–198.

National Bureau of Statistics of China, 2002–2017, http://www.stats.gov.cn/english/.

Online Software, 2015. Marine Sea Route Estimation. http://ports.com/sea-route/.

Zhang, H.Y., Ji, Q., Fan, Y., 2013. An evaluation framework for oil import security based on the supply chain with a case study focused on China. Energy Econ. 38, 87–95.

Chapter 18

Mehar method for solving unbalanced generalized interval-valued trapezoidal fuzzy number transportation problems

Akansha Mishra[a], Amit Kumar[a], S.S. Appadoo[b]

[a]*School of Mathematics, Thapar Institute of Engineering & Technology, Patiala, India,*
[b]*Department of Supply Chain Management, University of Manitoba, Winnipeg, MB, Canada*

1 Introduction

Ebrahimnejad (2016) pointed out that several methods have been proposed for solving fuzzy/fully fuzzy transportation problems (transportation problems in which each parameter is represented by a triangular/trapezoidal fuzzy number) (Ebrahimnejad, 2014, 2015a,b,c, 2016; Oheigeartaigh, 1982; Chanas et al., 1984, 1993; Chanas and Kuchta, 1996; Jimenez and Verdegay, 1998, 1999; Liu and Kao, 2004; Dinagar and Palanivel, 2009; Pandian and Natarajan, 2010; Kumar and Kaur, 2010, 2011a,b, 2014; Gupta et al., 2012; Shanmugasundari and Ganesan, 2013; Sudhagar and Ganesan, 2012; Kaur and Kumar, 2012; Chiang, 2005; Gupta and Kumar, 2012). However, no method has been proposed to find the solution of generalized interval-valued fuzzy transportation problems (transportation problems in which each parameter is represented by a triangular/trapezoidal fuzzy number). To fill this gap, Ebrahimnejad (2016) proposed the following methods:

1. A method to transform an unbalanced generalized IVTrFTP into a balanced generalized IVTrFTP.
2. A linear programming method to find the solution of a balanced generalized IVTrFTP.

Ebrahimnejad (2016) considered an unbalanced generalized IVTrFNTP and applied the methods, proposed by him, to transform it into a balanced generalized IVTrFNTP as well as to find its solution.

Maritime Transport and Regional Sustainability. https://doi.org/10.1016/B978-0-12-819134-7.00019-8

311

In this paper, it is shown that on applying Ebrahimnejad's method for transforming an unbalanced generalized IVTrFNTP into a balanced generalized IVTrFNTP (Ebrahimnejad, 2016), the obtained dummy supply and/or dummy demand is not a generalized interval-valued trapezoidal fuzzy number (IVTrFN), and therefore this method is not valid. In addition, a new method (known as the Mehar method) is proposed to transform an unbalanced generalized IVTrFNTP into a balanced generalized IVTrFNTP. Furthermore, the validity of the proposed Mehar method is discussed.

This paper is organized as follows:

2   Ebrahimnejad's method for transforming an unbalanced generalized IVTrFNTP into a balanced generalized IVTrFNTP

Ebrahimnejad proposed the following method to transform an unbalanced generalized IVTrFNTP into a balanced generalized IVTrFNTP, i.e., to transform $\sum_{i=1}^{m} \tilde{\tilde{a}}_i \neq \sum_{j=1}^{n} \tilde{\tilde{b}}_j$ into $\sum_{i=1}^{m} \tilde{\tilde{a}}_i = \sum_{j=1}^{n} \tilde{\tilde{b}}_j$, where, m represents number of sources, n represents number of destinations,

$$\sum_{i=1}^{m} \tilde{\tilde{a}}_i = \left(\left(\sum_{i=1}^{m} a_{i1}^{L}, \sum_{i=1}^{m} a_{i2}^{L}, \sum_{i=1}^{m} a_{i3}^{L}, \sum_{i=1}^{m} a_{i4}^{L}; \omega^{L} \right), \left(\sum_{i=1}^{m} a_{i1}^{U}, \sum_{i=1}^{m} a_{i2}^{U}, \sum_{i=1}^{m} a_{i3}^{U}, \sum_{i=1}^{m} a_{i4}^{U}; \omega^{U} \right) \right)$$

represents total interval-valued fuzzy supply, and

$$\sum_{j=1}^{n} \tilde{\tilde{b}}_j = \left(\left(\sum_{j=1}^{n} b_{j1}^{L}, \sum_{j=1}^{n} b_{j2}^{L}, \sum_{j=1}^{n} b_{j3}^{L}, \sum_{j=1}^{n} b_{j4}^{L}; \omega^{L} \right), \left(\sum_{j=1}^{n} b_{j1}^{U}, \sum_{j=1}^{n} b_{j2}^{U}, \sum_{j=1}^{n} b_{j3}^{U}, \sum_{j=1}^{n} b_{j4}^{U}; \omega^{U} \right) \right)$$

represents total interval-valued fuzzy demand. The generalized IVTrFN $\langle (a_{i1}^{L}, a_{i2}^{L}, a_{i3}^{L}, a_{i4}^{L}; \omega^{L}), (a_{i1}^{U}, a_{i2}^{U}, a_{i3}^{U}, a_{i4}^{U}; \omega^{U}) \rangle$ represents the supply of the product at ith source S_i. The generalized IVTrFN $\langle (b_{j1}^{L}, b_{j2}^{L}, b_{j3}^{L}, b_{j4}^{L}; \omega^{L}), (b_{j1}^{U}, b_{j2}^{U}, b_{j3}^{U}, b_{j4}^{U}; \omega^{U}) \rangle$ represents the demand of the product at jth destination D_j.

Case 1: If $\sum_{i=1}^{m} a_{i1}^{L} \leq \sum_{j=1}^{n} b_{j1}^{L}$, $\sum_{i=1}^{m} a_{i2}^{L} \leq \sum_{j=1}^{n} b_{j2}^{L}$, $\sum_{i=1}^{m} a_{i3}^{L} \leq \sum_{j=1}^{n} b_{j3}^{L}$, $\sum_{i=1}^{m} a_{i4}^{L} \leq \sum_{j=1}^{n} b_{j4}^{L}$,

$\sum_{i=1}^{m} a_{i1}^{U} \leq \sum_{j=1}^{n} b_{j1}^{U}$, $\sum_{i=1}^{m} a_{i2}^{U} \leq \sum_{j=1}^{n} b_{j2}^{U}$, $\sum_{i=1}^{m} a_{i3}^{U} \leq \sum_{j=1}^{n} b_{j3}^{U}$, $\sum_{i=1}^{m} a_{i4}^{U} \leq \sum_{j=1}^{n} b_{j4}^{U}$, then add a

dummy source S_{m+1} having dummy supply

$$\left\langle \left(\sum_{j=1}^{n} b_{j1}^{L} - \sum_{i=1}^{m} a_{i1}^{L}, \sum_{j=1}^{n} b_{j2}^{L} - \sum_{i=1}^{m} a_{i2}^{L}, \sum_{j=1}^{n} b_{j3}^{L} - \sum_{i=1}^{m} a_{i3}^{L}, \sum_{j=1}^{n} b_{j4}^{L} - \sum_{i=1}^{m} a_{i4}^{L}; \omega^{L} \right), \right.$$
$$\left. \left(\sum_{j=1}^{n} b_{j1}^{U} - \sum_{i=1}^{m} a_{i1}^{U}, \sum_{j=1}^{n} b_{j2}^{U} - \sum_{i=1}^{m} a_{i2}^{U}, \sum_{j=1}^{n} b_{j3}^{U} - \sum_{i=1}^{m} a_{i3}^{U}, \sum_{j=1}^{n} b_{j4}^{U} - \sum_{i=1}^{m} a_{i4}^{U}; \omega^{U} \right) \right\rangle$$

by considering the cost for supplying the unit quantity of the product from the dummy source S_{m+1} to all the destinations as a generalized IVTrFN $\tilde{\tilde{0}} = \big((0,0,0,0;1),(0,0,0,0;1)\big)$.

Case 2: If $\displaystyle\sum_{j=1}^{n} b_{j1}^{L} \leq \sum_{i=1}^{m} a_{i1}^{L}$, $\displaystyle\sum_{j=1}^{n} b_{j2}^{L} \leq \sum_{i=1}^{m} a_{i2}^{L}$, $\displaystyle\sum_{j=1}^{n} b_{j3}^{L} \leq \sum_{i=1}^{m} a_{i3}^{L}$, $\displaystyle\sum_{j=1}^{n} b_{j4}^{L} \leq \sum_{i=1}^{m} a_{i4}^{L}$,

$\displaystyle\sum_{j=1}^{n} b_{j1}^{U} \leq \sum_{i=1}^{m} a_{i1}^{U}$, $\displaystyle\sum_{j=1}^{n} b_{j2}^{U} \leq \sum_{i=1}^{m} a_{i2}^{U}$, $\displaystyle\sum_{j=1}^{n} b_{j3}^{U} \leq \sum_{i=1}^{m} a_{i3}^{U}$, $\displaystyle\sum_{j=1}^{n} b_{j4}^{U} \leq \sum_{i=1}^{m} a_{i4}^{U}$,

then add a dummy destination D_{n+1} having dummy demand

$$\left\langle \begin{array}{l} \left(\displaystyle\sum_{i=1}^{m} a_{i1}^{L} - \sum_{j=1}^{n} b_{j1}^{L}, \sum_{i=1}^{m} a_{i2}^{L} - \sum_{j=1}^{n} b_{j2}^{L}, \sum_{i=1}^{m} a_{i3}^{L} - \sum_{j=1}^{n} b_{j3}^{L}, \sum_{i=1}^{m} a_{i4}^{L} - \sum_{j=1}^{n} b_{j4}^{L}; \omega^{L}\right), \\ \left(\displaystyle\sum_{i=1}^{m} a_{i1}^{U} - \sum_{j=1}^{n} b_{j1}^{U}, \sum_{i=1}^{m} a_{i2}^{U} - \sum_{j=1}^{n} b_{j2}^{U}, \sum_{i=1}^{m} a_{i3}^{U} - \sum_{j=1}^{n} b_{j3}^{U}, \sum_{i=1}^{m} a_{i4}^{U} - \sum_{j=1}^{n} b_{j4}^{U}; \omega^{U}\right) \end{array} \right\rangle$$

by considering the cost for supplying the unit quantity of the product from the dummy destination D_{n+1} to all the sources as a generalized IVTrFN, $\tilde{\tilde{0}} = \big((0,0,0,0;1),(0,0,0,0;1)\big)$.

Case 3: If neither Case 1 nor Case 2 is satisfied, then carry out the following:

(i) Add a dummy source S_{m+1} having the dummy supply $\big\langle (A_{(m+1)1}^{L}, A_{(m+1)2}^{L},$

$A_{(m+1)3}^{L}, A_{(m+1)4}^{L}), (A_{(m+1)1}^{U}, A_{(m+1)2}^{U}, A_{(m+1)3}^{U}, A_{(m+1)4}^{U}) \big\rangle$ by considering the cost for supplying the unit quantity of the product from all the dummy sources S_{m+1} to all the dummy destinations as generalized IVTrFN $\tilde{\tilde{0}} = \big((0,0,0,0;1),(0,0,0,0;1)\big)$

$$A_{(m+1)1}^{L} = \left| \sum_{j=1}^{n} b_{j1}^{U} - \sum_{i=1}^{m} a_{i1}^{U} \right| + \max\left\{ 0, \sum_{j=1}^{n} b_{j1}^{L} - \sum_{i=1}^{m} a_{i1}^{L} \right\},$$

$$A_{(m+1)2}^{L} = A_{(m+1)1}^{L} + \max\left\{ 0, \left(\sum_{j=1}^{n} b_{j2}^{L} - \sum_{j=1}^{n} b_{j1}^{L} \right) - \left(\sum_{i=1}^{m} a_{i2}^{L} - \sum_{i=1}^{m} a_{i1}^{L} \right) \right\},$$

$$A_{(m+1)3}^{L} = A_{(m+1)2}^{L} + \max\left\{ 0, \left(\sum_{j=1}^{n} b_{j3}^{L} - \sum_{j=1}^{n} b_{j2}^{L} \right) - \left(\sum_{i=1}^{m} a_{i3}^{L} - \sum_{i=1}^{m} a_{i2}^{L} \right) \right\},$$

$$A_{(m+1)4}^{L} = A_{(m+1)3}^{L} + \max\left\{ 0, \left(\sum_{j=1}^{n} b_{j4}^{L} - \sum_{j=1}^{n} b_{j3}^{L} \right) - \left(\sum_{i=1}^{m} a_{i4}^{L} - \sum_{i=1}^{m} a_{i3}^{L} \right) \right\},$$

$$A_{(m+1)1}^{U} = \max\left\{ 0, \sum_{j=1}^{n} b_{j1}^{U} - \sum_{i=1}^{m} a_{i1}^{U} \right\},$$

$$A^U_{(m+1)2} = \sum_{j=1}^{n} b^U_{j1} - \sum_{i=1}^{m} a^U_{i1} + \max\left\{0, \sum_{j=1}^{n} b^U_{j1} - \sum_{i=1}^{m} a^U_{i1}\right\}$$
$$+ \max\left\{0, \left(\sum_{j=1}^{n} b^U_{j2} - \sum_{j=1}^{n} b^U_{j1}\right) - \left(\sum_{i=1}^{m} a^U_{i2} - \sum_{i=1}^{m} a^U_{i1}\right)\right\},$$

$$A^U_{(m+1)3} = A^U_{(m+1)2} + \max\left\{0, \left(\sum_{j=1}^{n} b^U_{j3} - \sum_{j=1}^{n} b^U_{j2}\right) - \left(\sum_{i=1}^{m} a^U_{i3} - \sum_{i=1}^{m} a^U_{i2}\right)\right\},$$

$$A^U_{(m+1)4} = A^U_{(m+1)3} + \max\left\{0, \left(\sum_{j=1}^{n} b^U_{j4} - \sum_{j=1}^{n} b^U_{j3}\right) - \left(\sum_{i=1}^{m} a^U_{i4} - \sum_{i=1}^{m} a^U_{i3}\right)\right\}.$$

(ii) Add a dummy destination D_{n+1} having the dummy demand $\left\langle (B^L_{(n+1)1}, B^L_{(n+1)2}, B^L_{(n+1)3}, B^L_{(n+1)4}), (B^U_{(n+1)1}, B^U_{(n+1)2}, B^U_{(n+1)3}, B^U_{(n+1)4}) \right\rangle$ by considering the cost for supplying the unit quantity of the product from all the sources to the dummy destination D_{n+1} as generalized IVTrFN $\tilde{\tilde{0}} = \big((0,0,0,0;1),(0,0,0,0;1)\big)$.

$$B^L_{(n+1)1} = \left|\sum_{j=1}^{n} b^U_{j1} - \sum_{i=1}^{m} a^U_{i1}\right| + \max\left\{0, \sum_{i=1}^{m} a^L_{i1} - \sum_{j=1}^{n} b^L_{j1}\right\},$$

$$B^L_{(n+1)2} = B^L_{(n+1)1} + \max\left\{0, \left(\sum_{i=1}^{m} a^L_{i2} - \sum_{i=1}^{m} a^L_{i1}\right) - \left(\sum_{j=1}^{n} b^L_{j2} - \sum_{j=1}^{n} b^L_{j1}\right)\right\},$$

$$B^L_{(n+1)3} = B^L_{(n+1)2} + \max\left\{0, \left(\sum_{i=1}^{m} a^L_{i3} - \sum_{i=1}^{m} a^L_{i2}\right) - \left(\sum_{j=1}^{n} b^L_{j3} - \sum_{j=1}^{n} b^L_{j2}\right)\right\},$$

$$B^L_{(n+1)4} = B^L_{(n+1)3} + \max\left\{0, \left(\sum_{i=1}^{m} a^L_{i4} - \sum_{i=1}^{m} a^L_{i3}\right) - \left(\sum_{j=1}^{n} b^L_{j4} - \sum_{j=1}^{n} b^L_{j3}\right)\right\},$$

$$B^U_{(n+1)1} = \max\left\{0, \sum_{i=1}^{m} a^U_{i1} - \sum_{j=1}^{n} b^U_{j1}\right\},$$

$$B^U_{(n+1)2} = \left|\sum_{i=1}^{m} a^U_{i1} - \sum_{j=1}^{n} b^U_{j1}\right| + \max\left\{0, \sum_{i=1}^{m} a^U_{i1} - \sum_{j=1}^{n} b^U_{j1}\right\}$$
$$+ \max\left\{0, \left(\sum_{i=1}^{m} a^U_{i2} - \sum_{i=1}^{m} a^U_{i1}\right) - \left(\sum_{j=1}^{n} b^U_{j2} - \sum_{j=1}^{n} b^U_{j1}\right)\right\},$$

$$B^U_{(n+1)3} = B^U_{(n+1)2} + \max\left\{0, \left(\sum_{i=1}^{m} a^U_{i3} - \sum_{i=1}^{m} a^U_{i2}\right) - \left(\sum_{j=1}^{n} b^U_{j3} - \sum_{j=1}^{n} b^U_{j2}\right)\right\},$$

$$B^U_{(n+1)4} = B^U_{(n+1)3} + \max\left\{0, \left(\sum_{i=1}^{m} a^U_{i4} - \sum_{i=1}^{m} a^U_{i3}\right) - \left(\sum_{j=1}^{n} b^U_{j4} - \sum_{j=1}^{n} b^U_{j3}\right)\right\}.$$

3 Flaws of Ebrahimnejad's method for transforming an unbalanced generalized IVTrFNTP into a balanced generalized IVTrFNTP

In this section, some numerical values are considered to show that Ebrahimnejad's method for transforming an unbalanced generalized IVTrFNTP into a balanced generalized IVTrFNTP (Ebrahimnejad, 2016) is not valid.

(1) Let
$$\left\langle\left(\sum_{i=1}^{m} a^L_{i1}, \sum_{i=1}^{m} a^L_{i2}, \sum_{i=1}^{m} a^L_{i3}, \sum_{i=1}^{m} a^L_{i4}; \omega^L\right), \left(\sum_{i=1}^{m} a^U_{i1}, \sum_{i=1}^{m} a^U_{i2}, \sum_{i=1}^{m} a^U_{i3}, \sum_{i=1}^{m} a^U_{i4}; \omega^U\right)\right\rangle \quad \text{and}$$
$$= \left\langle (1,2,7,11;1),(0,3,8,7;1)\right\rangle$$
$$\left\langle\left(\sum_{j=1}^{n} b^L_{j1}, \sum_{j=1}^{n} b^L_{j2}, \sum_{j=1}^{n} b^L_{j3}, \sum_{j=1}^{n} b^L_{j4}; \omega^L\right), \left(\sum_{j=1}^{n} b^U_{j1}, \sum_{j=1}^{n} b^U_{j2}, \sum_{j=1}^{n} b^U_{j3}, \sum_{j=1}^{n} b^U_{j4}; \omega^U\right)\right\rangle$$
$$= \left\langle (4,8,9,13;1),(1,6,10,18;1)\right\rangle.$$

Since $\quad \sum_{i=1}^{m} a^L_{i1} \leq \sum_{j=1}^{n} b^L_{j1}, \quad \sum_{i=1}^{m} a^L_{i2} \leq \sum_{j=1}^{n} b^L_{j2}, \quad \sum_{i=1}^{m} a^L_{i3} \leq \sum_{j=1}^{n} b^L_{j3}, \quad \sum_{i=1}^{m} a^L_{i4} \leq \sum_{j=1}^{n} b^L_{j4},$

$\sum_{i=1}^{m} a^U_{i1} \leq \sum_{j=1}^{n} b^U_{j1}, \quad \sum_{i=1}^{m} a^U_{i2} \leq \sum_{j=1}^{n} b^U_{j2}, \quad \sum_{i=1}^{m} a^U_{i3} \leq \sum_{j=1}^{n} b^U_{j3}, \quad \sum_{i=1}^{m} a^U_{i4} \leq \sum_{j=1}^{n} b^U_{j4},$ i.e., Case 1 of Ebrahimnejad's method (Ebrahimnejad, 2016), discussed in Section 2, is satisfied. So, according to (Ebrahimnejad's (2016) method, the dummy supply will be $\langle (4-1, 8-2, 9-7, 13-11; 1), (1-0, 6-3, 10-8, 18-7; 1)\rangle = \langle (3,6,2,2;1), (1,3,2,11;1)\rangle$.

However, it is not a generalized IVTrFN for the following reason: It can be easily verified from the graphical representation of generalized IVTrFN (Ebrahimnejad, 2016) as well as from the existing definition of a generalized IVTrFN (Ebrahimnejad, 2016) that in a generalized IVTrFN $\langle (a_1^L, a_2^L, a_3^L, a_4^L; \omega^L), (a_1^U, a_2^U, a_3^U, a_4^U; \omega^U)\rangle$ the condition $a_1^U \leq a_1^L \leq a_2^U \leq a_2^L \leq a_3^L \leq a_3^U \leq a_4^L \leq a_4^U$ should always be satisfied. While, for the obtained dummy supply $\langle (3,6,2,2;1), (1,3,2,11;1)\rangle$, this condition is not satisfied. Hence, Ebrahimnejad's method (Ebrahimnejad, 2016) to obtain the dummy supply is not correct.

(2) Let $\left\langle \left(\sum_{i=1}^{m} a_{i1}^{L}, \sum_{i=1}^{m} a_{i2}^{L}, \sum_{i=1}^{m} a_{i3}^{L}, \sum_{i=1}^{m} a_{i4}^{L}; \omega^{L} \right), \left(\sum_{i=1}^{m} a_{i1}^{U}, \sum_{i=1}^{m} a_{i2}^{U}, \sum_{i=1}^{m} a_{i3}^{U}, \sum_{i=1}^{m} a_{i4}^{U}; \omega^{U} \right) \right\rangle$ and

$$= \left\langle (4,8,9,13;1), (1,6,10,18;1) \right\rangle$$

$$\left\langle \left(\left(\sum_{j=1}^{n} b_{j1}^{L}, \sum_{j=1}^{n} b_{j2}^{L}, \sum_{j=1}^{n} b_{j3}^{L}, \sum_{j=1}^{n} b_{j4}^{L}; \omega^{L} \right), \left(\sum_{j=1}^{n} b_{j1}^{U}, \sum_{j=1}^{n} b_{j2}^{U}, 0 \sum_{j=1}^{n} b_{j3}^{U}, \sum_{j=1}^{n} b_{j4}^{U}; \omega^{U} \right) \right\rangle$$

$$= \left\langle (1,2,7,11;1), (0,3,8,7;1) \right\rangle.$$

Since $\sum_{j=1}^{n} b_{j1}^{L} \le \sum_{i=1}^{m} a_{i1}^{L}$, $\sum_{j=1}^{n} b_{j2}^{L} \le \sum_{i=1}^{m} a_{i2}^{L}$, $\sum_{j=1}^{n} b_{j3}^{L} \le \sum_{i=1}^{m} a_{i3}^{L}$, $\sum_{j=1}^{n} b_{j4}^{L} \le \sum_{i=1}^{m} a_{i4}^{L}$,

$\sum_{j=1}^{n} b_{j1}^{U} \le \sum_{i=1}^{m} a_{i1}^{U}$, $\sum_{j=1}^{n} b_{j2}^{U} \le \sum_{i=1}^{m} a_{i2}^{U}, \sum_{j=1}^{n} b_{j3}^{U} \le \sum_{i=1}^{m} a_{i3}^{U}$, $\sum_{j=1}^{n} b_{j4}^{U} \le \sum_{i=1}^{m} a_{i4}^{U}$,

i.e., Case 2 of Ebrahimnejad's method (Ebrahimnejad, 2016), discussed in Section 2, is satisfied. So, according to Ebrahimnejad's method (Ebrahimnejad, 2016), the dummy demand will be $\langle (4-1, 8-2, 9-7, 13-11; 1), (1-0, 6-3, 10-8, 18-7; 1) \rangle = \langle (3,6,2,2;1), (1,3,2,11;1) \rangle$.

However, it is not a generalized IVTrFN for the following reason. It can be easily verified from the graphical representation of generalized IVTrFN (Ebrahimnejad, 2016) as well as from the existing definition of a generalized IVTrFN (Ebrahimnejad, 2016) that in a generalized IVTrFN $\langle (a_1^{L}, a_2^{L}, a_3^{L}, a_4^{L}; \omega^{L}), (a_1^{U}, a_2^{U}, a_3^{U}, a_4^{U}; \omega^{U}) \rangle$ the condition $a_1^{U} \le a_1^{L} \le a_2^{U} \le a_2^{L} \le a_3^{L} \le a_3^{U} \le a_4^{L} \le a_4^{U}$ should always be satisfied. While, for the obtained dummy demand $\langle (3,6,2,2;1), (1,3,2,11;1) \rangle$, this condition is not satisfied. Hence, Ebrahimnejad's method (Ebrahimnejad, 2016) to obtain the dummy demand is not correct.

(3) Let $\left\langle \left(\sum_{i=1}^{m} a_{i1}^{L}, \sum_{i=1}^{m} a_{i2}^{L}, \sum_{i=1}^{m} a_{i3}^{L}, \sum_{i=1}^{m} a_{i4}^{L}; \omega^{L} \right), \left(\sum_{i=1}^{m} a_{i1}^{U}, \sum_{i=1}^{m} a_{i2}^{U}, \sum_{i=1}^{m} a_{i3}^{U}, \sum_{i=1}^{m} a_{i4}^{U}; \omega^{U} \right) \right\rangle$

$$= \left\langle \left(110, 150, 160, 180; \frac{2}{3} \right), (100, 140, 170, 190; 1) \right\rangle$$

and $\left\langle \left(\sum_{j=1}^{n} b_{j1}^{L}, \sum_{j=1}^{n} b_{j2}^{L}, \sum_{j=1}^{n} b_{j3}^{L}, \sum_{j=1}^{n} b_{j4}^{L}; \omega^{L} \right), \left(\sum_{j=1}^{n} b_{j1}^{U}, \sum_{j=1}^{n} b_{j2}^{U}, \sum_{j=1}^{n} b_{j3}^{U}, \sum_{j=1}^{n} b_{j4}^{U}; \omega^{U} \right) \right\rangle$

$$= \left\langle \left(90, 120, 140, 200; \frac{2}{3} \right), (75, 105, 155, 215; 1) \right\rangle.$$

Since $\sum_{j=1}^{n} b_{j1}^{L} \le \sum_{i=1}^{m} a_{i1}^{L}$, $\sum_{j=1}^{n} b_{j2}^{L} \le \sum_{i=1}^{m} a_{i2}^{L}$, $\sum_{j=1}^{n} b_{j3}^{L} \le \sum_{i=1}^{m} a_{i3}^{L}$, $\sum_{i=1}^{m} a_{i4}^{L} \le \sum_{j=1}^{n} b_{j4}^{L}$,

$\sum_{j=1}^{n} b_{j1}^{U} \le \sum_{i=1}^{m} a_{i1}^{U}$, $\sum_{j=1}^{n} b_{j2}^{U} \le \sum_{i=1}^{m} a_{i2}^{U}, \sum_{j=1}^{n} b_{j3}^{U} \le \sum_{i=1}^{m} a_{i3}^{U}$, $\sum_{i=1}^{m} a_{i4}^{U} \le \sum_{j=1}^{n} b_{j4}^{U}$, i.e., Case 3 of

Ebrahimnejad's method (Ebrahimnejad, 2016), discussed in Section 2, is satisfied. So, according to Ebrahimnejad's method (Ebrahimnejad, 2016), the dummy demand will be $\left\langle \left(45,55,55,55; \frac{2}{3} \right), \left(25,60,60,60; 1 \right) \right\rangle$.

However, the obtained dummy demand is not a generalized IVTrFN for the following reason. It can be easily verified from the graphical representation of generalized IVTrFN (Ebrahimnejad, 2016) as well as from the existing definition of a generalized IVTrFN (Ebrahimnejad, 2016) that in a generalized IVTrFN $\langle (a_1^L, a_2^L, a_3^L, a_4^L \varpi^L), (a_1^U, a_2^U, a_3^U, a_4^U; \omega^U) \rangle$ the condition $a_1^U \leq a_1^L \leq a_2^U \leq a_2^L \leq a_3^L \leq a_3^U \leq a_4^L \leq a_4^U$ should always be satisfied. While, for the obtained dummy demand $\left\langle \left(45,55,55,55; \frac{2}{3} \right), \left(25,60,60,60; 1 \right) \right\rangle$, this condition is not satisfied. Hence, Ebrahimnejad's method (Ebrahimnejad, 2016) to obtain the dummy demand is not correct.

4   Proposed Mehar method for transforming an unbalanced generalized IVTrFNTP into a balanced generalized IVTrFNTP

In this section, a new method (known as the Mehar method) is proposed to transform an unbalanced generalized IVTrFNTP into a balanced generalized IVTrFNTP.

Let us consider an unbalanced generalized IVTrFNTP having m sources and n destinations such that the availability of the product and the demand of the product at ith source (S_i) and jth destination (D_j) be represented by generalized IVTrFNTP $\tilde{\tilde{a}}_i = \left\langle \left(a_{i1}^L, a_{i2}^L, a_{i3}^L, a_{i4}^L; \omega^L \right), \left(a_{i1}^U, a_{i2}^U, a_{i3}^U, a_{i4}^U; \omega^U \right) \right\rangle$ and $\tilde{\tilde{b}}_j = \left\langle \left(b_{j1}^L, b_{j2}^L, b_{j3}^L, b_{j4}^L; \omega^L \right), \left(b_{j1}^U, b_{j2}^U, b_{j3}^U, b_{j4}^U; \omega^U \right) \right\rangle$, respectively. Then, this unbalanced generalized IVTrFNTP can be transformed into a balanced generalized IVTrFNTP as follows:

Case 1: If $\sum_{i=1}^{m} a_{i1}^U \leq \sum_{j=1}^{n} b_{j1}^U$, $\sum_{i=1}^{m} a_{i1}^L - \sum_{i=1}^{m} a_{i1}^U \leq \sum_{j=1}^{n} b_{j1}^L - \sum_{j=1}^{n} b_{j1}^U$,

$\sum_{i=1}^{m} a_{i2}^U - \sum_{i=1}^{m} a_{i1}^L \leq \sum_{j=1}^{n} b_{j2}^U - \sum_{j=1}^{n} b_{j1}^L$, $\sum_{i=1}^{m} a_{i2}^L - \sum_{i=1}^{m} a_{i2}^U \leq \sum_{j=1}^{n} b_{j2}^L - \sum_{j=1}^{n} b_{j2}^U$,

$\sum_{i=1}^{m} a_{i3}^L - \sum_{i=1}^{m} a_{i2}^L \leq \sum_{j=1}^{n} b_{j3}^L - \sum_{j=1}^{n} b_{j2}^L$, $\sum_{i=1}^{m} a_{i3}^U - \sum_{i=1}^{m} a_{i3}^L \leq \sum_{j=1}^{n} b_{j3}^U - \sum_{j=1}^{n} b_{j3}^L$,

$\sum_{i=1}^{m} a_{i4}^L - \sum_{i=1}^{m} a_{i3}^U \leq \sum_{j=1}^{n} b_{j4}^L - \sum_{j=1}^{n} b_{j3}^U$, $\sum_{i=1}^{m} a_{i4}^U - \sum_{i=1}^{m} a_{i4}^L \leq \sum_{j=1}^{n} b_{j4}^U - \sum_{j=1}^{n} b_{j4}^L$ then add a

dummy source S_{m+1} having dummy supply $\left\langle \left(A_{(m+1)1}^L, A_{(m+1)2}^L, A_{(m+1)3}^L, A_{(m+1)4}^L \right), \left(A_{(m+1)1}^U, A_{(m+1)2}^U, A_{(m+1)3}^U, A_{(m+1)4}^U \right) \right\rangle$ by considering the cost for supplying the unit quantity of the product from the dummy source S_{m+1} to all the destinations D_{n+1} as a generalized IVTrFN $\tilde{\tilde{0}} = \left\langle (0,0,0,0;1), (0,0,0,0;1) \right\rangle$.

$$A^U_{(m+1)1} = \sum_{j=1}^{n} b^U_{j1} - \sum_{i=1}^{m} a^U_{i1}, A^L_{(m+1)1} = A^U_{(m+1)1} + \left(\sum_{j=1}^{n} b^L_{j1} - \sum_{j=1}^{n} b^U_{j1} \right) - \left(\sum_{i=1}^{m} a^L_{i1} - \sum_{i=1}^{m} a^U_{i1} \right),$$

$$A^U_{(m+1)2} = A^L_{(m+1)1} + \left(\sum_{j=1}^{n} b^U_{j2} - \sum_{j=1}^{n} b^L_{j1} \right) - \left(\sum_{i=1}^{m} a^U_{i2} - \sum_{i=1}^{m} a^L_{i1} \right),$$

$$A^L_{(m+1)2} = A^U_{(m+1)2} + \left(\sum_{j=1}^{n} b^L_{j2} - \sum_{j=1}^{n} b^U_{j2} \right) - \left(\sum_{i=1}^{m} a^L_{i2} - \sum_{i=1}^{m} a^U_{i2} \right),$$

$$A^L_{(m+1)3} = A^L_{(m+1)2} + \left(\sum_{j=1}^{n} b^L_{j3} - \sum_{j=1}^{n} b^L_{j2} \right) - \left(\sum_{i=1}^{m} a^L_{i3} - \sum_{i=1}^{m} a^L_{i2} \right),$$

$$A^U_{(m+1)3} = A^L_{(m+1)3} + \left(\sum_{j=1}^{n} b^U_{j3} - \sum_{j=1}^{n} b^L_{j3} \right) - \left(\sum_{i=1}^{m} a^U_{i3} - \sum_{i=1}^{m} a^L_{i3} \right),$$

$$A^L_{(m+1)4} = A^U_{(m+1)3} + \left(\sum_{j=1}^{n} b^L_{j4} - \sum_{j=1}^{n} b^U_{j3} \right) - \left(\sum_{i=1}^{m} a^L_{i4} - \sum_{i=1}^{m} a^U_{i3} \right),$$

$$A^U_{(m+1)4} = A^L_{(m+1)4} + \left(\sum_{j=1}^{n} b^U_{j4} - \sum_{j=1}^{n} b^L_{j4} \right) - \left(\sum_{i=1}^{m} a^U_{i4} - \sum_{i=1}^{m} a^L_{i4} \right).$$

Case 2: If $\sum_{j=1}^{n} b^U_{j1} \le \sum_{i=1}^{m} a^U_{i1}$, $\sum_{j=1}^{n} b^U_{j1} - \sum_{j=1}^{n} b^U_{j1} \le \sum_{i=1}^{m} a^L_{i1} - \sum_{i=1}^{m} a^U_{i1}$,

$$\sum_{j=1}^{n} b^U_{j2} - \sum_{j=1}^{n} b^L_{j1} \le \sum_{i=1}^{m} a^U_{i2} - \sum_{i=1}^{m} a^L_{i1}, \ \sum_{j=1}^{n} b^L_{j2} - \sum_{j=1}^{n} b^U_{j2} \le \sum_{i=1}^{m} a^L_{i2} - \sum_{i=1}^{m} a^U_{i2},$$

$$\sum_{j=1}^{n} b^L_{j3} - \sum_{j=1}^{n} b^L_{j2} \le \sum_{i=1}^{m} a^L_{i3} - \sum_{i=1}^{m} a^L_{i2}, \ \sum_{j=1}^{n} b^U_{j3} - \sum_{j=1}^{n} b^L_{j3} \le \sum_{i=1}^{m} a^U_{i3} - \sum_{i=1}^{m} a^L_{i3},$$

$$\sum_{j=1}^{n} b^L_{j4} - \sum_{j=1}^{n} b^U_{j3} \le \sum_{i=1}^{m} a^L_{i4} - \sum_{i=1}^{m} a^U_{i3}, \ \sum_{j=1}^{n} b^U_{j4} - \sum_{j=1}^{n} b^L_{j4} \le \sum_{i=1}^{m} a^U_{i4} - \sum_{i=1}^{m} a^L_{i4} \ \text{then add a}$$

dummy destination D_{n+1} having dummy demand $\langle (B^L_{(n+1)1}, B^L_{(n+1)2}, B^L_{(n+1)3}, B^L_{(n+1)4}), (B^U_{(n+1)1}, B^U_{(n+1)2}, B^U_{(n+1)3}, B^U_{(n+1)4}) \rangle$ by considering the cost for supplying the unit quantity of the product from all the sources to the dummy destination D_{n+1} as a generalized IVTrFN $\tilde{\tilde{0}} = ((0,0,0,0; 1),(0,0,0,0; 1))$.

$$B^U_{(n+1)1} = \sum_{i=1}^{m} a^U_{i1} - \sum_{j=1}^{n} b^U_{j1}, B^L_{(n+1)1} = B^U_{(n+1)1} + \left(\sum_{i=1}^{m} a^L_{i1} - \sum_{i=1}^{m} a^U_{i1} \right) - \left(\sum_{j=1}^{n} b^L_{j1} - \sum_{j=1}^{n} b^U_{j1} \right),$$

$$B^U_{(n+1)2} = B^L_{(n+1)1} + \left(\sum_{i=1}^{m} a^U_{i2} - \sum_{i=1}^{m} a^L_{i1} \right) - \left(\sum_{j=1}^{n} b^U_{j2} - \sum_{j=1}^{n} b^L_{j1} \right),$$

$$B^L_{(n+1)2} = B^U_{(n+1)2} + \left(\sum_{i=1}^{m} a^L_{i2} - \sum_{i=1}^{m} a^U_{i2} \right) - \left(\sum_{j=1}^{n} b^L_{j2} - \sum_{j=1}^{n} b^U_{j2} \right),$$

$$B^L_{(n+1)3} = B^L_{(n+1)2} + \left(\sum_{i=1}^{m} a^L_{i3} - \sum_{i=1}^{m} a^L_{i2} \right) - \left(\sum_{j=1}^{n} b^L_{j3} - \sum_{j=1}^{n} b^L_{j2} \right),$$

$$B^U_{(n+1)3} = B^L_{(n+1)3} + \left(\sum_{i=1}^{m} a^U_{i3} - \sum_{i=1}^{m} a^L_{i3} \right) - \left(\sum_{j=1}^{n} b^U_{j3} - \sum_{j=1}^{n} b^L_{j3} \right),$$

$$B^L_{(n+1)4} = B^U_{(n+1)3} + \left(\sum_{i=1}^{m} a^L_{i4} - \sum_{i=1}^{m} a^U_{i3} \right) - \left(\sum_{j=1}^{n} b^L_{j4} - \sum_{j=1}^{n} b^U_{j3} \right),$$

$$B^U_{(n+1)4} = B^L_{(n+1)4} + \left(\sum_{i=1}^{m} a^U_{i4} - \sum_{i=1}^{m} a^L_{i4} \right) - \left(\sum_{j=1}^{n} b^U_{j4} - \sum_{j=1}^{n} b^L_{j4} \right).$$

Case 3: If neither Case 1 nor Case 2 is satisfied, then carry out the following:
(1) Add a dummy source S_{m+1} having dummy supply $\langle (A^L_{(m+1)1}, A^L_{(m+1)2}, A^L_{(m+1)3}, A^L_{(m+1)4}), (A^U_{(m+1)1}, A^U_{(m+1)2}, A^U_{(m+1)3}, A^U_{(m+1)4}) \rangle$ by considering the cost for supplying the unit quantity of the product from the dummy source S_{m+1} to all the destinations as a generalized IVTrFN $\tilde{\tilde{0}} = \left((0,0,0,0;1), (0,0,0,0;1) \right)$ ·

$$A^U_{(m+1)1} = \max\left\{ 0, \left(\sum_{j=1}^{n} b^U_{j1} - \sum_{i=1}^{m} a^U_{i1} \right) \right\},$$

$$A^L_{(m+1)1} = A^U_{(m+1)1} + \max\left\{ 0, \left(\left(\sum_{j=1}^{n} b^L_{j1} - \sum_{j=1}^{n} b^U_{j1} \right) - \left(\sum_{i=1}^{m} a^L_{i1} - \sum_{i=1}^{m} a^U_{i1} \right) \right) \right\},$$

$$A^U_{(m+1)2} = A^L_{(m+1)1} + \max\left\{ 0, \left(\left(\sum_{j=1}^{n} b^U_{j2} - \sum_{j=1}^{n} b^L_{j1} \right) - \left(\sum_{i=1}^{m} a^U_{i2} - \sum_{i=1}^{m} a^L_{i1} \right) \right) \right\},$$

$$A^L_{(m+1)2} = A^U_{(m+1)2} + \max\left\{ 0, \left(\left(\sum_{j=1}^{n} b^L_{j2} - \sum_{j=1}^{n} b^U_{j2} \right) - \left(\sum_{i=1}^{m} a^L_{i2} - \sum_{i=1}^{m} a^U_{i2} \right) \right) \right\},$$

$$A^L_{(m+1)3} = A^L_{(m+1)2} + \max\left\{ 0, \left(\left(\sum_{j=1}^{n} b^L_{j3} - \sum_{j=1}^{n} b^L_{j2} \right) - \left(\sum_{i=1}^{m} a^L_{i3} - \sum_{i=1}^{m} a^L_{i2} \right) \right) \right\},$$

$$A^U_{(m+1)3} = A^L_{(m+1)3} + \max\left\{ 0, \left(\left(\sum_{j=1}^{n} b^U_{j3} - \sum_{j=1}^{n} b^L_{j3} \right) - \left(\sum_{i=1}^{m} a^U_{i3} - \sum_{i=1}^{m} a^L_{i3} \right) \right) \right\},$$

$$A^L_{(m+1)4} = A^U_{(m+1)3} + \max\left\{ 0, \left(\left(\sum_{j=1}^{n} b^L_{j4} - \sum_{j=1}^{n} b^U_{j3} \right) - \left(\sum_{i=1}^{m} a^L_{i4} - \sum_{i=1}^{m} a^U_{i3} \right) \right) \right\},$$

$$A^U_{(m+1)4} = A^L_{(m+1)4} + \max\left\{ 0, \left(\left(\sum_{j=1}^{n} b^U_{j4} - \sum_{j=1}^{n} b^L_{j4} \right) - \left(\sum_{i=1}^{m} a^U_{i4} - \sum_{i=1}^{m} a^L_{i4} \right) \right) \right\}$$

(2) Add a dummy destination D_{n+1} having the dummy demand $\left\langle (B^L_{(m+1)1}, B^L_{(m+1)2}, B^L_{(m+1)3}, B^L_{(m+1)4}), (B^U_{(m+1)1}, B^U_{(m+1)2}, B^U_{(m+1)3}, B^U_{(m+1)4}) \right\rangle$ by considering the cost for supplying the unit quantity of the product from all the sources to the dummy destination D_{n+1} as a generalized IVTrFN $\tilde{\tilde{0}} = \left((0,0,0,0;1), (0,0,0,0;1) \right)$.

$$B^U_{(n+1)1} = \max\left\{ 0, \left(\sum_{i=1}^{m} a^U_{i1} - \sum_{j=1}^{n} b^U_{j1} \right) \right\},$$

$$B^L_{(n+1)1} = B^U_{(n+1)1} + \max\left\{ 0, \left(\left(\sum_{i=1}^{m} a^L_{i1} - \sum_{i=1}^{m} a^U_{i1} \right) - \left(\sum_{j=1}^{n} b^L_{j1} - \sum_{j=1}^{n} b^U_{j1} \right) \right) \right\},$$

$$B^U_{(n+1)2} = B^L_{(n+1)1} + \max\left\{ 0, \left(\left(\sum_{i=1}^{m} a^U_{i2} - \sum_{i=1}^{m} a^L_{i1} \right) - \left(\sum_{j=1}^{n} b^U_{j2} - \sum_{j=1}^{n} b^L_{j1} \right) \right) \right\},$$

$$B^L_{(n+1)2} = B^U_{(n+1)2} + \max\left\{ 0, \left(\left(\sum_{i=1}^{m} a^L_{i2} - \sum_{i=1}^{m} a^U_{i2} \right) - \left(\sum_{j=1}^{n} b^L_{j2} - \sum_{j=1}^{n} b^U_{j2} \right) \right) \right\},$$

$$B^L_{(n+1)3} = B^L_{(n+1)2} + \max\left\{ 0, \left(\left(\sum_{i=1}^{m} a^L_{i3} - \sum_{i=1}^{m} a^L_{i2} \right) - \left(\sum_{j=1}^{n} b^L_{j3} - \sum_{j=1}^{n} b^L_{j2} \right) \right) \right\},$$

$$B^U_{(n+1)3} = B^L_{(n+1)3} + \max\left\{ 0, \left(\left(\sum_{i=1}^{m} a^U_{i3} - \sum_{i=1}^{m} a^L_{i3} \right) - \left(\sum_{j=1}^{n} b^U_{j3} - \sum_{j=1}^{n} b^L_{j3} \right) \right) \right\},$$

$$B^L_{(n+1)4} = B^U_{(n+1)3} + \max\left\{ 0, \left(\left(\sum_{i=1}^{m} a^L_{i4} - \sum_{i=1}^{m} a^U_{i3} \right) - \left(\sum_{j=1}^{n} b^L_{j4} - \sum_{j=1}^{n} b^U_{j3} \right) \right) \right\},$$

$$B^U_{(n+1)4} = B^L_{(n+1)4} + \max\left\{ 0, \left(\left(\sum_{i=1}^{m} a^U_{i4} - \sum_{i=1}^{m} a^L_{i4} \right) - \left(\sum_{j=1}^{n} b^U_{j4} - \sum_{j=1}^{n} b^L_{j4} \right) \right) \right\}.$$

5 Validity of the proposed Mehar method

To prove the validity of the proposed Mehar method, it is sufficient to prove that:

1. The obtained dummy supply will be a generalized IVTrFN.
2. The obtained dummy demand will be a generalized IVTrFN.
3. In Case 1, $\sum_{i=1}^{m} \tilde{\tilde{a}}_i + \tilde{\tilde{A}}_{m+1} = \sum_{j=1}^{n} \tilde{\tilde{b}}_j$.

4. In Case 2, $\sum_{i=1}^{m} \tilde{\tilde{a}}_i = \sum_{j=1}^{n} \tilde{\tilde{b}}_j + \tilde{\tilde{B}}_{n+1}$.

5. In Case 3, $\sum_{i=1}^{m} \tilde{\tilde{a}}_i + \tilde{\tilde{A}}_{m+1} = \sum_{j=1}^{n} \tilde{\tilde{b}}_j + \tilde{\tilde{B}}_{n+1}$.

Therefore, the same is proved in this section.

5.1 The obtained dummy supply will be a generalized IVTrFN

It is obvious from Case 1 of the proposed Mehar method that for the obtained dummy supply
$\left\langle (A_{(m+1)1}^{L}, A_{(m+1)2}^{L}, A_{(m+1)3}^{L}, A_{(m+1)4}^{L}), (A_{(m+1)1}^{U}, A_{(m+1)2}^{U}, A_{(m+1)3}^{U}, A_{(m+1)4}^{U}) \right\rangle$, the
conditions $A_{(m+1)1}^{U} \leq A_{(m+1)1}^{L} \leq A_{(m+1)2}^{U} \leq A_{(m+1)2}^{L} \leq A_{(m+1)3}^{L} \leq A_{(m+1)3}^{U} \leq A_{(m+1)4}^{L}$
$\leq A_{(m+1)4}^{U}$, will be satisfied. Therefore, the dummy supply $\left\langle (A_{(m+1)1}^{L}, A_{(m+1)2}^{L}, \right.$
$A_{(m+1)3}^{L}, A_{(m+1)4}^{L}), (A_{(m+1)1}^{U}, A_{(m+1)2}^{U}, A_{(m+1)3}^{U}, A_{(m+1)4}^{U}) \rangle$, obtained from Case 1 of
the proposed Mehar method, will always be a generalized IVTrFN.

5.2 The obtained dummy demand will be a generalized IVTrFN

It is obvious from Case 2 of the proposed Mehar method that for the obtained dummy demand $\left\langle (B_{(n+1)1}^{L}, B_{(n+1)2}^{L}, B_{(n+1)3}^{L}, B_{(n+1)4}^{L}), (B_{(n+1)1}^{U}, B_{(n+1)2}^{U}, B_{(n+1)3}^{U}, \right.$
$B_{(n+1)4}^{U}) \rangle$, the conditions $B_{(n+1)1}^{U} \leq B_{(n+1)1}^{L} \leq B_{(n+1)2}^{U} \leq B_{(n+1)2}^{L} \leq B_{(n+1)3}^{U} \leq B_{(n+1)3}^{L}$
$\leq B_{(n+1)4}^{U} \leq B_{(n+1)4}^{L}$, will be satisfied. Therefore, the dummy demand $\left\langle (B_{(n+1)1}^{L}, \right.$
$B_{(n+1)2}^{L}, B_{(n+1)3}^{L}, B_{(n+1)4}^{L}), (B_{(n+1)1}^{U}, B_{(n+1)2}^{U}, B_{(n+1)3}^{U}, B_{(n+1)4}^{U}) \rangle$, obtained from the
Case 2 of the proposed Mehar method, will always be a generalized IVTrFN.

5.3 The obtained dummy supply and dummy demand will be a generalized IVTrFNs

It is obvious from Case 3 of the proposed Mehar method that for the obtained dummy supply $\left\langle (A_{(m+1)1}^{L}, A_{(m+1)2}^{L}, A_{(m+1)3}^{L}, A_{(m+1)4}^{L}), (A_{(m+1)1}^{U}, A_{(m+1)2}^{U}, A_{(m+1)3}^{U}, \right.$
$A_{(m+1)4}^{U}) \rangle$, the conditions $A_{(m+1)1}^{U} \leq A_{(m+1)1}^{L} \leq A_{(m+1)2}^{U} \leq A_{(m+1)2}^{L} \leq A_{(m+1)3}^{L} \leq$
$A_{(m+1)3}^{U} \leq A_{(m+1)4}^{L} \leq A_{(m+1)4}^{U}$, will be satisfied. In addition, for the obtained dummy
demand $\left\langle (B_{(n+1)1}^{L}, B_{(n+1)2}^{L}, B_{(n+1)3}^{L}, B_{(n+1)4}^{L}), (B_{(n+1)1}^{U}, B_{(n+1)2}^{U}, B_{(n+1)3}^{U}, B_{(n+1)4}^{U}) \right\rangle$,
the conditions $B_{(n+1)1}^{U} \leq B_{(n+1)1}^{L} \leq B_{(n+1)2}^{U} \leq B_{(n+1)2}^{L} \leq B_{(n+1)3}^{L} \leq B_{(n+1)3}^{U} \leq B_{(n+1)4}^{L} \leq$
$B_{(n+1)4}^{U}$, will be satisfied. Therefore, the dummy supply $\left\langle (A_{(m+1)1}^{L}, A_{(m+1)2}^{L}, A_{(m+1)3}^{L}, \right.$
$A_{(m+1)4}^{L}), (A_{(m+1)1}^{U}, A_{(m+1)2}^{U}, A_{(m+1)3}^{U}, A_{(m+1)4}^{U}) \rangle$ and the dummy demand $\left\langle (B_{(n+1)1}^{L}, \right.$
$B_{(n+1)2}^{L}, B_{(n+1)3}^{L}, B_{(n+1)4}^{L}), (B_{(n+1)1}^{U}, B_{(n+1)2}^{U}, B_{(n+1)3}^{U}, B_{(n+1)4}^{U}) \rangle$, obtained from Case
3 of the proposed Mehar method, will always be a generalized IVTrFNs.

5.4 Validity of the condition $\sum\limits_{i=1}^{m}\tilde{a}_i + \tilde{A}_{m+1} = \sum\limits_{j=1}^{n}\tilde{b}_j$

$$\sum_{i=1}^{m}\tilde{a}_i + \tilde{A}_{m+1} = \sum_{j=1}^{n}\tilde{b}_j$$

$$\Rightarrow \left\langle \left(\sum_{i=1}^{m}a_{i1}^{L}, \sum_{i=1}^{m}a_{i2}^{L}, \sum_{i=1}^{m}a_{i3}^{L}, \sum_{i=1}^{m}a_{i4}^{L}; \omega^{L} \right), \left(\sum_{i=1}^{m}a_{i1}^{U}, \sum_{i=1}^{m}a_{i2}^{U}, \sum_{i=1}^{m}a_{i3}^{U}, \sum_{i=1}^{m}a_{i4}^{U}; \omega^{U} \right) \right\rangle$$

$$+ \left\langle \left(A_{(m+1)1}^{L}, A_{(m+1)2}^{L}, A_{(m+1)3}^{L}, A_{(m+1)4}^{L} \right), \left(A_{(m+1)1}^{U}, A_{(m+1)2}^{U}, A_{(m+1)3}^{U}, A_{(m+1)4}^{U} \right) \right\rangle$$

$$= \left\langle \left(\sum_{j=1}^{n}b_{j1}^{L}, \sum_{j=1}^{n}b_{j2}^{L}, \sum_{j=1}^{n}b_{j3}^{L}, \sum_{j=1}^{n}b_{j4}^{L}; \omega^{L} \right), \left(\sum_{j=1}^{n}b_{j1}^{U}, \sum_{j=1}^{n}b_{j2}^{U}, \sum_{j=1}^{n}b_{j3}^{U}, \sum_{j=1}^{n}b_{j4}^{U}; \omega^{U} \right) \right\rangle.$$

$$\Rightarrow \left\langle \begin{array}{l} \left(\sum\limits_{i=1}^{m}a_{i1}^{L} + A_{(m+1)1}^{L}, \sum\limits_{i=1}^{m}a_{i2}^{L} + A_{(m+1)2}^{L}, \sum\limits_{i=1}^{m}a_{i3}^{L} + A_{(m+1)3}^{L}, \sum\limits_{i=1}^{m}a_{i4}^{L} + A_{(m+1)4}^{L}; \omega^{L} \right), \\ \left(\sum\limits_{i=1}^{m}a_{i1}^{U} + A_{(m+1)1}^{U}, \sum\limits_{i=1}^{m}a_{i2}^{U} + A_{(m+1)2}^{U}, \sum\limits_{i=1}^{m}a_{i3}^{U} + A_{(m+1)3}^{U}, \sum\limits_{i=1}^{m}a_{i4}^{U} + A_{(m+1)}^{U}; \omega^{U} \right) \end{array} \right\rangle$$

$$= \left\langle \left(\sum_{j=1}^{n}b_{j1}^{L}, \sum_{j=1}^{n}b_{j2}^{L}, \sum_{j=1}^{n}b_{j3}^{L}, \sum_{j=1}^{n}b_{j4}^{L}; \omega^{L} \right), \left(\sum_{j=1}^{n}b_{j1}^{U}, \sum_{j=1}^{n}b_{j2}^{U}, \sum_{j=1}^{n}b_{j3}^{U}, \sum_{j=1}^{n}b_{j4}^{U}; \omega^{U} \right) \right\rangle.$$

$$\Rightarrow \sum_{i=1}^{m}a_{i1}^{L} + A_{(m+1)1}^{L} = \sum_{j=1}^{n}b_{j1}^{L}, \tag{1}$$

$$\sum_{i=1}^{m}a_{i2}^{L} + A_{(m+1)2}^{L} = \sum_{j=1}^{n}b_{j2}^{L}, \tag{2}$$

$$\sum_{i=1}^{m}a_{i3}^{L} + A_{(m+1)3}^{L} = \sum_{j=1}^{n}b_{j3}^{L}, \tag{3}$$

$$\sum_{i=1}^{m}a_{i4}^{L} + A_{(m+1)4}^{L} = \sum_{j=1}^{n}b_{j4}^{L}, \tag{4}$$

$$\sum_{i=1}^{m}a_{i1}^{U} + A_{(m+1)1}^{U} = \sum_{j=1}^{n}b_{j1}^{U}, \tag{5}$$

$$\sum_{i=1}^{m}a_{i2}^{U} + A_{(m+1)2}^{U} = \sum_{j=1}^{n}b_{j2}^{U}, \tag{6}$$

$$\sum_{i=1}^{m}a_{i3}^{U} + A_{(m+1)3}^{U} = \sum_{j=1}^{n}b_{j3}^{U}, \tag{7}$$

$$\sum_{i=1}^{m} a_{i4}^{U} + A_{(m+1)4}^{U} = \sum_{j=1}^{n} b_{j4}^{U}. \tag{8}$$

It is obvious that in order to prove $\sum_{i=1}^{m} \tilde{a}_i + \tilde{A}_{m+1} = \sum_{j=1}^{n} \tilde{b}_j$, there is a need to prove that Eqs. (1)–(8) are satisfied. Here, only the validity of Eq. (1) is proved. The validity of the remaining equations can be proved in the same manner.

$$\sum_{i=1}^{m} a_{i1}^{L} + A_{(m+1)1}^{L} = \sum_{i=1}^{m} a_{i1}^{L} + A_{(m+1)1}^{U} + \left(\sum_{j=1}^{n} b_{j1}^{L} - \sum_{j=1}^{n} b_{j1}^{U} \right) - \left(\sum_{i=1}^{m} a_{i1}^{L} - \sum_{i=1}^{m} a_{i1}^{U} \right)$$

$$= \sum_{i=1}^{m} a_{i1}^{L} + \sum_{j=1}^{n} b_{j1}^{U} - \sum_{i=1}^{m} a_{i1}^{U} + \left(\sum_{j=1}^{n} b_{j1}^{L} - \sum_{j=1}^{n} b_{j1}^{U} \right) - \left(\sum_{i=1}^{m} a_{i1}^{L} - \sum_{i=1}^{m} a_{i1}^{U} \right)$$

$$= \sum_{j=1}^{n} b_{j1}^{L}.$$

5.5 Validity of the condition $\sum_{i=1}^{m} \tilde{a}_i = \sum_{j=1}^{n} \tilde{b}_j + \tilde{B}_{n+1}$

$$\sum_{i=1}^{m} \tilde{a}_i = \sum_{j=1}^{n} \tilde{b}_j + \tilde{B}_{n+1}$$

$$\Rightarrow \left\langle \left(\sum_{i=1}^{m} a_{i1}^{L}, \sum_{i=1}^{m} a_{i2}^{L}, \sum_{i=1}^{m} a_{i3}^{L}, \sum_{i=1}^{m} a_{i4}^{L}; \omega^{L} \right), \left(\sum_{i=1}^{m} a_{i1}^{U}, \sum_{i=1}^{m} a_{i2}^{U}, \sum_{i=1}^{m} a_{i3}^{U}, \sum_{i=1}^{m} a_{i4}^{U}; \omega^{U} \right) \right\rangle$$

$$= \left\langle \left(\sum_{j=1}^{n} b_{j1}^{L}, \sum_{j=1}^{n} b_{j2}^{L}, \sum_{j=1}^{n} b_{j3}^{L}, \sum_{j=1}^{n} b_{j4}^{L}; \omega^{L} \right), \left(\sum_{j=1}^{n} b_{j1}^{U}, \sum_{j=1}^{n} b_{j2}^{U}, \sum_{j=1}^{n} b_{j3}^{U}, \sum_{j=1}^{n} b_{j4}^{U}; \omega^{U} \right) \right\rangle$$

$$+ \left\langle \left(B_{(n+1)1}^{L}, B_{(n+1)2}^{L}, B_{(n+1)3}^{L}, B_{(n+1)4}^{L} \right), \left(B_{(n+1)1}^{U}, B_{(n+1)2}^{U}, B_{(n+1)3}^{U}, B_{(n+1)4}^{U} \right) \right\rangle.$$

$$\Rightarrow \left\langle \left(\sum_{i=1}^{m} a_{i1}^{L}, \sum_{i=1}^{m} a_{i2}^{L}, \sum_{i=1}^{m} a_{i3}^{L}, \sum_{i=1}^{m} a_{i4}^{L}; \omega^{L} \right), \left(\sum_{i=1}^{m} a_{i1}^{U}, \sum_{i=1}^{m} a_{i2}^{U}, \sum_{i=1}^{m} a_{i3}^{U}, \sum_{i=1}^{m} a_{i4}^{U}; \omega^{U} \right) \right\rangle$$

$$= \left\langle \left(\sum_{j=1}^{n} b_{j1}^{L} + B_{(n+1)1}^{L}, \sum_{j=1}^{n} b_{j2}^{L} + B_{(n+1)2}^{L}, \sum_{j=1}^{n} b_{j3}^{L} + B_{(n+1)3}^{L}, \sum_{j=1}^{n} b_{j4}^{L} + B_{(n+1)4}^{L}; \omega^{L} \right), \atop \left(\sum_{j=1}^{n} b_{j1}^{U} + B_{(n+1)1}^{U}, \sum_{j=1}^{n} b_{j2}^{U} + B_{(n+1)2}^{U}, \sum_{j=1}^{n} b_{j3}^{U} + B_{(n+1)3}^{U}, \sum_{j=1}^{n} b_{j4}^{U} + B_{(n+1)4}^{U}; \omega^{U} \right) \right\rangle.$$

$$\Rightarrow \sum_{i=1}^{m} a_{i1}^{L} = \sum_{j=1}^{n} b_{j1}^{L} + B_{(n+1)1}^{L}, \tag{9}$$

$$\sum_{i=1}^{m} a_{i2}^{L} = \sum_{j=1}^{n} b_{j2}^{L} + B_{(n+1)2}^{L}, \tag{10}$$

$$\sum_{i=1}^{m} a_{i3}^{L} = \sum_{j=1}^{n} b_{j3}^{L} + B_{(n+1)3}^{L}, \tag{11}$$

$$\sum_{i=1}^{m} a_{i4}^{L} = \sum_{j=1}^{n} b_{j4}^{L} + B_{(n+1)4}^{L}, \tag{12}$$

$$\sum_{i=1}^{m} a_{i1}^{U} = \sum_{j=1}^{n} b_{j1}^{U} + B_{(n+1)1}^{U}, \tag{13}$$

$$\sum_{i=1}^{m} a_{i2}^{U} = \sum_{j=1}^{n} b_{j2}^{U} + B_{(n+1)2}^{U}, \tag{14}$$

$$\sum_{i=1}^{m} a_{i3}^{U} = \sum_{j=1}^{n} b_{j3}^{U} + B_{(n+1)3}^{U}, \tag{15}$$

$$\sum_{i=1}^{m} a_{i4}^{U} = \sum_{j=1}^{n} b_{j4}^{U} + B_{(n+1)4}^{U}. \tag{16}$$

It is obvious that in order to prove $\sum_{i=1}^{m} \tilde{a}_i = \sum_{j=1}^{n} \tilde{b}_j + \tilde{B}_{n+1}$, there is a need to prove that Eqs. (9)–(16) are satisfied. Here, only the validity of Eq. (9) is proved. The validity of the remaining equations can be proved in the same manner.

$$\sum_{j=1}^{n} b_{j1}^{L} + B_{(n+1)1}^{L} = \sum_{j=1}^{n} b_{j1}^{L} + B_{(n+1)1}^{U} + \left(\sum_{i=1}^{m} a_{i1}^{L} - \sum_{i=1}^{m} a_{i1}^{U} \right) - \left(\sum_{j=1}^{n} b_{j1}^{L} - \sum_{j=1}^{n} b_{j1}^{U} \right)$$

$$= \sum_{j=1}^{n} b_{j1}^{L} + \sum_{i=1}^{m} a_{i1}^{U} - \sum_{j=1}^{n} b_{j1}^{U} + \left(\sum_{i=1}^{m} a_{i1}^{L} - \sum_{i=1}^{m} a_{i1}^{U} \right) - \left(\sum_{j=1}^{n} b_{j1}^{L} - \sum_{j=1}^{n} b_{j1}^{U} \right) = \sum_{i=1}^{m} a_{i1}^{L}.$$

5.6 Validity of the condition $\sum_{i=1}^{m} \tilde{a}_i + \tilde{A}_{m+1} = \sum_{j=1}^{n} \tilde{b}_j + \tilde{B}_{n+1}$

$$\sum_{i=1}^{m} \tilde{a}_i + \tilde{A}_{m+1} = \sum_{j=1}^{n} \tilde{b}_j + \tilde{B}_{n+1}.$$

$$\Rightarrow \left\langle \left(\left(\sum_{i=1}^{m} a_{i1}^{L}, \sum_{i=1}^{m} a_{i2}^{L}, \sum_{i=1}^{m} a_{i3}^{L}, \sum_{i=1}^{m} a_{i4}^{L}; \omega^{L} \right), \left(\sum_{i=1}^{m} a_{i1}^{U}, \sum_{i=1}^{m} a_{i2}^{U}, \sum_{i=1}^{m} a_{i3}^{U}, \sum_{i=1}^{m} a_{i4}^{U}; \omega^{U} \right) \right\rangle \right.$$

$$+ \left\langle \left(\left(A_{(m+1)1}^{L}, A_{(m+1)2}^{L}, A_{(m+1)3}^{L}, A_{(m+1)4}^{L} \right), \left(A_{(m+1)1}^{U}, A_{(m+1)2}^{U}, A_{(m+1)3}^{U}, A_{(m+1)4}^{U} \right) \right\rangle \right.$$

$$= \left\langle \left(\left(\sum_{j=1}^{n} b_{j1}^{L}, \sum_{j=1}^{n} b_{j2}^{L}, \sum_{j=1}^{n} b_{j3}^{L}, \sum_{j=1}^{n} b_{j4}^{L}; \omega^{L} \right), \left(\sum_{j=1}^{n} b_{j1}^{U}, \sum_{j=1}^{n} b_{j2}^{U}, \sum_{j=1}^{n} b_{j3}^{U}, \sum_{j=1}^{n} b_{j4}^{U}; \omega^{U} \right) \right\rangle \right.$$

$$+ \left\langle \left(B_{(n+1)1}^{L}, B_{(n+1)2}^{L}, B_{(n+1)3}^{L}, B_{(n+1)4}^{L} \right), \left(B_{(n+1)1}^{U}, B_{(n+1)2}^{U}, B_{(n+1)3}^{U}, B_{(n+1)4}^{U} \right) \right\rangle.$$

$$\Rightarrow \left\langle \begin{pmatrix} \sum_{i=1}^{m} a_{i1}^{L} + A_{(m+1)1}^{L}, \sum_{i=1}^{m} a_{i2}^{L} + A_{(m+1)2}^{L}, \sum_{i=1}^{m} a_{i3}^{L} + A_{(m+1)3}^{L}, \sum_{i=1}^{m} a_{i4}^{L} + A_{(m+1)4}^{L}; \omega^{L} \end{pmatrix}, \\ \begin{pmatrix} \sum_{i=1}^{m} a_{i1}^{U} + A_{(m+1)1}^{U}, \sum_{i=1}^{m} a_{i2}^{U} + A_{(m+1)2}^{U}, \sum_{i=1}^{m} a_{i3}^{U} + A_{(m+1)3}^{U}, \sum_{i=1}^{m} a_{i4}^{U} + A_{(m+1)}^{U}; \omega^{U} \end{pmatrix} \right\rangle$$

$$= \left\langle \begin{pmatrix} \sum_{j=1}^{n} b_{j1}^{L} + B_{(n+1)1}^{L}, \sum_{j=1}^{n} b_{j2}^{L} + B_{(n+1)2}^{L}, \sum_{j=1}^{n} b_{j3}^{L} + B_{(n+1)3}^{L}, \sum_{j=1}^{n} b_{j4}^{L} + B_{(n+1)4}^{L}; \omega^{L} \end{pmatrix}, \\ \begin{pmatrix} \sum_{j=1}^{n} b_{j1}^{U} + B_{(n+1)1}^{U}, \sum_{j=1}^{n} b_{j2}^{U} + B_{(n+1)2}^{U}, \sum_{j=1}^{n} b_{j3}^{U} + B_{(n+1)3}^{U}, \sum_{j=1}^{n} b_{j4}^{U} + B_{(n+1)4}^{U}; \omega^{U} \end{pmatrix} \right\rangle .$$

$$\Rightarrow \sum_{i=1}^{m} a_{i1}^{L} + A_{(m+1)1}^{L} = \sum_{j=1}^{n} b_{j1}^{L} + B_{(n+1)1}^{L}, \tag{17}$$

$$\sum_{i=1}^{m} a_{i2}^{L} + A_{(m+1)2}^{L} = \sum_{j=1}^{n} b_{j2}^{L} + B_{(n+1)2}^{L}, \tag{18}$$

$$\sum_{i=1}^{m} a_{i3}^{L} + A_{(m+1)3}^{L} = \sum_{j=1}^{n} b_{j3}^{L} + B_{(n+1)3}^{L}, \tag{19}$$

$$\sum_{i=1}^{m} a_{i4}^{L} + A_{(m+1)4}^{L} = \sum_{j=1}^{n} b_{j4}^{L} + B_{(n+1)4}^{L}, \tag{20}$$

$$\sum_{i=1}^{m} a_{i1}^{U} + A_{(m+1)1}^{U} = \sum_{j=1}^{n} b_{j1}^{U} + B_{(n+1)1}^{U}, \tag{21}$$

$$\sum_{i=1}^{m} a_{i2}^{U} + A_{(m+1)2}^{U} = \sum_{j=1}^{n} b_{j2}^{U} + B_{(n+1)2}^{U}, \tag{22}$$

$$\sum_{i=1}^{m} a_{i3}^{U} + A_{(m+1)3}^{U} = \sum_{j=1}^{n} b_{j3}^{U} + B_{(n+1)3}^{U}, \tag{23}$$

$$\sum_{i=1}^{m} a_{i4}^{U} + A_{(m+1)4}^{U} = \sum_{j=1}^{n} b_{j4}^{U} + B_{(n+1)4}^{U}. \tag{24}$$

It is obvious that in order to prove $\sum_{i=1}^{m} \tilde{a}_i + \tilde{A}_{m+1} = \sum_{j=1}^{n} \tilde{b}_j + \tilde{B}_{n+1}$, there is a need to prove that Eqs. (17)–(24) are satisfied. Here, only the validity of Eq. (17) is proved. The validity of the remaining equations can be proved in the same manner.

Putting the values of $A_{(m+1)1}^{L}$ and $B_{(n+1)1}^{L}$ in Eq. (17), it will be transformed into Eq. (25).

$$\sum_{i=1}^{m} a_{i1}^{L} + \max\left\{0, \left(\sum_{j=1}^{n} b_{j1}^{U} - \sum_{i=1}^{m} a_{i1}^{U}\right)\right\} + \max\left\{0, \left(\left(\sum_{j=1}^{n} b_{j1}^{L} - \sum_{j=1}^{n} b_{j1}^{U}\right) - \left(\sum_{i=1}^{m} a_{i1}^{L} - \sum_{i=1}^{m} a_{i1}^{U}\right)\right)\right\}$$
$$= \sum_{j=1}^{n} b_{j1}^{L} + \max\left\{0, \left(\sum_{i=1}^{m} a_{i1}^{U} - \sum_{j=1}^{n} b_{j1}^{U}\right)\right\} + \max\left\{0, \left(\left(\sum_{i=1}^{m} a_{i1}^{L} - \sum_{i=1}^{m} a_{i1}^{U}\right) - \left(\sum_{j=1}^{n} b_{j1}^{L} - \sum_{j=1}^{n} b_{j1}^{U}\right)\right)\right\} \tag{25}$$

There may be the following four cases:

Case 1: $\left(\sum_{j=1}^{n} b_{j1}^{L} - \sum_{j=1}^{n} b_{j1}^{U}\right) \geq \left(\sum_{i=1}^{m} a_{i1}^{L} - \sum_{i=1}^{m} a_{i1}^{U}\right)$ and $\sum_{j=1}^{n} b_{j1}^{U} \geq \sum_{i=1}^{m} a_{i1}^{U}$.

Case 2: $\left(\sum_{j=1}^{n} b_{j1}^{L} - \sum_{j=1}^{n} b_{j1}^{U}\right) \leq \left(\sum_{i=1}^{m} a_{i1}^{L} - \sum_{i=1}^{m} a_{i1}^{U}\right)$ and $\sum_{j=1}^{n} b_{j1}^{U} \leq \sum_{i=1}^{m} a_{i1}^{U}$.

Case 3: $\left(\sum_{j=1}^{n} b_{j1}^{L} - \sum_{j=1}^{n} b_{j1}^{U}\right) \leq \left(\sum_{i=1}^{m} a_{i1}^{L} - \sum_{i=1}^{m} a_{i1}^{U}\right)$ and $\sum_{j=1}^{n} b_{j1}^{U} \geq \sum_{i=1}^{m} a_{i1}^{U}$.

Case 4: $\left(\sum_{j=1}^{n} b_{j1}^{L} - \sum_{j=1}^{n} b_{j1}^{U}\right) \geq \left(\sum_{i=1}^{m} a_{i1}^{L} - \sum_{i=1}^{m} a_{i1}^{U}\right)$ and $\sum_{j=1}^{n} b_{j1}^{U} \leq \sum_{i=1}^{m} a_{i1}^{U}$.

If $\left(\sum_{j=1}^{n} b_{j1}^{L} - \sum_{j=1}^{n} b_{j1}^{U}\right) \geq \left(\sum_{i=1}^{m} a_{i1}^{L} - \sum_{i=1}^{m} a_{i1}^{U}\right)$ and $\sum_{j=1}^{n} b_{j1}^{U} \geq \sum_{i=1}^{m} a_{i1}^{U}$ then

$\max\left\{0, \left(\left(\sum_{i=1}^{m} a_{i1}^{L} - \sum_{i=1}^{m} a_{i1}^{U}\right) - \left(\sum_{j=1}^{n} b_{j1}^{L} - \sum_{j=1}^{n} b_{j1}^{U}\right)\right)\right\} = 0$, $\quad \max\left\{0, \left(\sum_{i=1}^{m} a_{i1}^{U} - \sum_{j=1}^{n} b_{j1}^{U}\right)\right\} = 0$,

$\max\left\{0, \left(\left(\sum_{j=1}^{n} b_{j1}^{L} - \sum_{j=1}^{n} b_{j1}^{U}\right) - \left(\sum_{i=1}^{m} a_{i1}^{L} - \sum_{i=1}^{m} a_{i1}^{U}\right)\right)\right\}$, $\quad \max\left\{0, \left(\sum_{j=1}^{n} b_{j1}^{U} - \sum_{i=1}^{m} a_{i1}^{U}\right)\right\} = \sum_{j=1}^{n} b_{j1}^{U} - \sum_{i=1}^{m} a_{i1}^{U}$

$= \left(\sum_{j=1}^{n} b_{j1}^{L} - \sum_{j=1}^{n} b_{j1}^{U}\right) - \left(\sum_{i=1}^{m} a_{i1}^{L} - \sum_{i=1}^{m} a_{i1}^{U}\right)$

and therefore Eq. (25) will be satisfied.

If $\left(\sum_{j=1}^{n} b_{j1}^{L} - \sum_{j=1}^{n} b_{j1}^{U}\right) \leq \left(\sum_{i=1}^{m} a_{i1}^{L} - \sum_{i=1}^{m} a_{i1}^{U}\right)$ and $\sum_{j=1}^{n} b_{j1}^{U} \leq \sum_{i=1}^{m} a_{i1}^{U}$ then

$\max\left\{0, \left(\left(\sum_{j=1}^{n} b_{j1}^{L} - \sum_{j=1}^{n} b_{j1}^{U}\right) - \left(\sum_{i=1}^{m} a_{i1}^{L} - \sum_{i=1}^{m} a_{i1}^{U}\right)\right)\right\} = 0$, $\quad \max\left\{0, \left(\sum_{j=1}^{n} b_{j1}^{U} - \sum_{i=1}^{m} a_{i1}^{U}\right)\right\} = 0$,

$\max\left\{0, \left(\left(\sum_{i=1}^{m} a_{i1}^{L} - \sum_{i=1}^{m} a_{i1}^{U}\right) - \left(\sum_{j=1}^{n} b_{j1}^{L} - \sum_{j=1}^{n} b_{j1}^{U}\right)\right)\right\} = \left(\sum_{i=1}^{m} a_{i1}^{L} - \sum_{i=1}^{m} a_{i1}^{U}\right) - \left(\sum_{j=1}^{n} b_{j1}^{L} - \sum_{j=1}^{n} b_{j1}^{U}\right)$,

$\max\left\{0, \left(\sum_{i=1}^{m} a_{i1}^{U} - \sum_{j=1}^{n} b_{j1}^{U}\right)\right\} = \sum_{i=1}^{m} a_{i1}^{U} - \sum_{j=1}^{n} b_{j1}^{U}$ and therefore Eq. (25) will be satisfied.

If $\left(\sum_{j=1}^{n} b_{j1}^{L} - \sum_{j=1}^{n} b_{j1}^{U}\right) \leq \left(\sum_{i=1}^{m} a_{i1}^{L} - \sum_{i=1}^{m} a_{i1}^{U}\right)$ and $\sum_{j=1}^{n} b_{j1}^{U} \geq \sum_{i=1}^{m} a_{i1}^{U}$ then

$\max\left\{0, \left(\left(\sum_{j=1}^{n} b_{j1}^{L} - \sum_{j=1}^{n} b_{j1}^{U}\right) - \left(\sum_{i=1}^{m} a_{i1}^{L} - \sum_{i=1}^{m} a_{i1}^{U}\right)\right)\right\} = 0$, $\quad \max\left\{0, \left(\sum_{j=1}^{n} b_{j1}^{U} - \sum_{i=1}^{m} a_{i1}^{U}\right)\right\} = 0$,

$$\max\left\{0,\left(\left(\sum_{i=1}^{m}a_{i1}^{L}-\sum_{i=1}^{m}a_{i1}^{U}\right)-\left(\sum_{j=1}^{n}b_{j1}^{L}-\sum_{j=1}^{n}b_{j1}^{U}\right)\right)\right\}=\left(\sum_{i=1}^{m}a_{i1}^{L}-\sum_{i=1}^{m}a_{i1}^{U}\right)-\left(\sum_{j=1}^{n}b_{j1}^{L}-\sum_{j=1}^{n}b_{j1}^{U}\right),$$

$$\max\left\{0,\left(\sum_{i=1}^{m}a_{i1}^{U}-\sum_{j=1}^{n}b_{j1}^{U}\right)\right\}=\sum_{i=1}^{m}a_{i1}^{U}-\sum_{j=1}^{n}b_{j1}^{U}$$ and therefore Eq. (25) will be

satisfied.

If $\left(\sum_{j=1}^{n}b_{j1}^{L}-\sum_{j=1}^{n}b_{j1}^{U}\right)\geq\left(\sum_{i=1}^{m}a_{i1}^{L}-\sum_{i=1}^{m}a_{i1}^{U}\right)$ and $\sum_{j=1}^{n}b_{j1}^{U}\leq\sum_{i=1}^{m}a_{i1}^{U}$ then the values of

$$\max\left\{0,\left(\left(\sum_{j=1}^{n}b_{j1}^{L}-\sum_{j=1}^{n}b_{j1}^{U}\right)-\left(\sum_{i=1}^{m}a_{i1}^{L}-\sum_{i=1}^{m}a_{i1}^{U}\right)\right)\right\}=0, \qquad \max\left\{0,\left(\sum_{j=1}^{n}b_{j1}^{U}-\sum_{i=1}^{m}a_{i1}^{U}\right)\right\}=0,$$

$$\max\left\{0,\left(\left(\sum_{i=1}^{m}a_{i1}^{L}-\sum_{i=1}^{m}a_{i1}^{U}\right)-\left(\sum_{j=1}^{n}b_{j1}^{L}-\sum_{j=1}^{n}b_{j1}^{U}\right)\right)\right\}=\left(\sum_{i=1}^{m}a_{i1}^{L}-\sum_{i=1}^{m}a_{i1}^{U}\right)-\left(\sum_{j=1}^{n}b_{j1}^{L}-\sum_{j=1}^{n}b_{j1}^{U}\right),$$

$$\max\left\{0,\left(\sum_{i=1}^{m}a_{i1}^{U}-\sum_{j=1}^{n}b_{j1}^{U}\right)\right\}=\sum_{i=1}^{m}a_{i1}^{U}-\sum_{j=1}^{n}b_{j1}^{U}$$ and therefore Eq. (25) will be

satisfied.

6   Invalidity of the existing result

Ebrahimnejad (2016) considered a generalized IVTrFNTP having two sources, S_1, S_2, and three destinations, D_1, D_2, D_3, such that:

(i) The generalized IVTrF supplies at sources S_1 and S_2 are

$$\left\langle\left(70,90,90,100;\frac{2}{3}\right),(65,85,95,105;1)\right\rangle, \qquad \left\langle\left(40,60,70,80;\frac{2}{3}\right),(35,55,75,85;1)\right\rangle,$$

respectively.

(ii) The generalized IVTrF demands at destinations D_1, D_2 and D_3 are

$$\left\langle\left(30,40,50,70;\frac{2}{3}\right),(25,35,55,75;1)\right\rangle, \quad \left\langle\left(20,30,40,50;\frac{2}{3}\right),(15,25,45,55;1)\right\rangle$$

and $\left\langle\left(40,50,50,80;\frac{2}{3}\right),(35,45,55,85;1)\right\rangle$, respectively.

Ebrahimnejad (2016) claimed that

$$\left\langle\left(70,90,90,100;\frac{2}{3}\right),(65,85,95,105;1)\right\rangle+\left\langle\left(40,60,70,80;\frac{2}{3}\right),(35,55,75,85;1)\right\rangle$$ is not equal to

$$=\left\langle\left(110,150,160,180;\frac{2}{3}\right),(35,55,75,85;1)\right\rangle$$
$$\left\langle\left(30,40,50,70;\frac{2}{3}\right),(25,35,55,75,1)\right\rangle+\left\langle\left(20,30,40,50,\frac{2}{3}\right),(15,25,45,55;1)\right\rangle$$

$$+\left\langle\left(40,50,50,80;\frac{2}{3}\right),(35,45,55,85;1)\right\rangle=\left\langle\left(190,120,140,200;\frac{2}{3}\right),(75,105,155,215;1)\right\rangle,$$

i.e., the considered generalized IVTrFNTP is an unbalanced generalized IVTrFNTP. Therefore, there is a need to add a dummy source S_3 having a dummy generalized IVTrF supply $\left\langle\left(25,25,35,75;\frac{2}{3}\right),(0,25,45,85;1)\right\rangle$ and a dummy destination D_4 having a dummy generalized IVTrF demand

$\left\langle\left(45,55,55,55;\frac{2}{3}\right),(25,60,60,60;1)\right\rangle$ in order to solve the considered generalized IVTrFNTP.

However, the dummy generalized IVTrF demand $\left\langle\left(45,55,55,55;\frac{2}{3}\right),(25,60,60,60;1)\right\rangle$, obtained by Ebrahimnejad (2016),

is not a generalized IVTrFN for the following reason. It can be easily verified from the graphical representation of generalized IVTrFN (Ebrahimnejad, 2016) as well as from the existing definition of a generalized IVTrFN (Ebrahimnejad, 2016) that in a generalized IVTrFN $\langle(b_1^L, b_2^L, b_3^L, b_4^L; \omega^L), (b_1^U, b_2^U, b_3^U, b_4^U; \omega^U)\rangle$, the condition $b_1^U \leq b_1^L \leq b_2^L \leq b_2^L \leq b_3^L \leq b_3^U \leq b_4^L \leq b_4^U$ should always be

satisfied. While it can be easily verified that if the generalized IVTrF demand

$\left\langle\left(45,55,55,55;\frac{2}{3}\right),(25,60,60,60;1)\right\rangle$ is compared with a generalized IVTrFN

$\langle(b_1^L, b_2^L, b_3^L, b_4^L; \omega^L), (b_1^U, b_2^U, b_3^U, b_4^U; \omega^U)\rangle$, then $b_1^L = 45$, $b_2^L = 55$, $b_3^L = 55$, $b_4^L = 55$, $b_1^U = 25$, $b_2^U = 60$, $b_3^U = 60$, $b_4^U = 60$.

It is obvious that $b_2^U \geq b_2^L$, i.e., the necessary condition $b_2^U \leq b_2^L$ is not satisfied. Therefore, the obtained dummy demand is not a generalized IVTrFN. Hence, the result of this problem, obtained by Ebrahimnejad (2016), is not correct.

7 Exact dummy supply and dummy demand for the existing generalized IVTrFNTP

Ebrahimnejad (2016) solved the generalized IVTrFNTP, to illustrate the proposed method. Since, in the considered generalized IVTrFNTP, total supply $\left\langle\left(110,150,160,180;\frac{2}{3}\right),(100,140,170,190;1)\right\rangle$ is not equal to the total demand $\left\langle\left(90,120,140,200;\frac{2}{3}\right),(75,105,155,215;1)\right\rangle$, Ebrahimnejad applied

the proposed method to transform this problem into a balanced generalized IVTrFNTP and claimed that there is a need to add a dummy source having a dummy generalized interval-valued trapezoidal fuzzy (IVTrF) supply $\left\langle\left(25,25,35,75;\frac{2}{3}\right),(0,25,45,85;1)\right\rangle$ as well as a dummy destination having a dummy generalized IVTrF demand $\left\langle\left(45,55,55,55;\frac{2}{3}\right),(25,60,60,60;1)\right\rangle$.

However, as discussed in Section 3, the result of this problem, obtained by Ebrahimnejad (2016), is not correct. In this section, the proposed Mehar method is used to find the correct dummy supply and dummy demand.

Using the proposed Mehar method, the exact dummy supply and dummy demand can be obtained as follows:

$$\sum_{i=1}^{2}\tilde{a}_i = \left\langle \left(\sum_{i=1}^{2}a_{i1}^L, \sum_{i=1}^{2}a_{i2}^L, \sum_{i=1}^{2}a_{i3}^L, \sum_{i=1}^{2}a_{i4}^L; \omega^L \right), \left(\sum_{i=1}^{2}a_{i1}^U, \sum_{i=1}^{2}a_{i2}^U, \sum_{i=1}^{2}a_{i3}^U, \sum_{i=1}^{2}a_{i4}^U; \omega^U \right) \right\rangle$$

$$= \left\langle \left(110,150,160,180; \frac{2}{3} \right), (100,140,170,190; 1) \right\rangle$$

and

$$\sum_{j=1}^{3}\tilde{b}_j = \left\langle \left(\sum_{j=1}^{3}b_{j1}^L, \sum_{j=1}^{3}b_{j2}^L, \sum_{j=1}^{3}b_{j3}^L, \sum_{j=1}^{3}b_{j4}^L; \omega^L \right), \left(\sum_{j=1}^{3}b_{j1}^U, \sum_{j=1}^{3}b_{j2}^U, \sum_{j=1}^{3}b_{j3}^U, \sum_{j=1}^{3}b_{j4}^U; \omega^U \right) \right\rangle$$

$$= \left\langle \left(90,120,140,200; \frac{2}{3} \right), (75,105,155,215; 1) \right\rangle.$$

Case 3 of the proposed Mehar method is satisfied. Therefore, according to Case 3 of the proposed Mehar method,

$$A_{33}^U = A_{33}^L + \max\left\{ 0, \left(\left(\sum_{j=1}^{3}b_{j3}^U - \sum_{j=1}^{3}b_{j3}^L \right) - \left(\sum_{i=1}^{2}a_{i3}^U - \sum_{i=1}^{2}a_{i3}^L \right) \right) \right\}$$
$$= 20 + \max\left\{ 0,15-10 \right\} = 25,$$

$$A_{34}^L = A_{33}^U + \max\left\{ 0, \left(\left(\sum_{j=1}^{3}b_{j4}^L - \sum_{j=1}^{3}b_{j3}^U \right) - \left(\sum_{i=1}^{2}a_{i4}^L - \sum_{i=1}^{2}a_{i3}^U \right) \right) \right\}$$
$$= 25 + \max\left\{ 0,45-10 \right\} = 60,$$

$$A_{34}^U = A_{34}^L + \max\left\{ 0, \left(\left(\sum_{j=1}^{3}b_{j4}^U - \sum_{j=1}^{3}b_{j4}^L \right) - \left(\sum_{i=1}^{2}a_{i4}^U - \sum_{i=1}^{2}a_{i4}^L \right) \right) \right\}$$
$$= 60 + \max\left\{ 0,15-10 \right\} = 65,$$

$$B_{41}^U = \max\left\{ 0, \left(\sum_{i=1}^{2}a_{i1}^U - \sum_{j=1}^{3}b_{j1}^U \right) \right\}$$
$$= \max\left\{ 0,100-75 \right\} = 25,$$

$$B_{41}^{L} = B_{41}^{U} + \max\left\{0, \left(\left(\sum_{i=1}^{2} a_{i1}^{L} - \sum_{i=1}^{2} a_{i1}^{U}\right) - \left(\sum_{j=1}^{3} b_{j1}^{L} - \sum_{j=1}^{3} b_{j1}^{U}\right)\right)\right\},$$

$$= 25 + \max\{0,10-15\} = 25,$$

$$B_{42}^{U} = B_{41}^{L} + \max\left\{0, \left(\left(\sum_{i=1}^{2} a_{i2}^{U} - \sum_{i=1}^{2} a_{i1}^{L}\right) - \left(\sum_{j=1}^{3} b_{j2}^{U} - \sum_{j=1}^{3} b_{j1}^{L}\right)\right)\right\}$$

$$= 25 + \max\{0,30-15\} = 40,$$

$$B_{42}^{L} = B_{42}^{U} + \max\left\{0, \left(\left(\sum_{i=1}^{2} a_{i2}^{L} - \sum_{i=1}^{2} a_{i2}^{U}\right) - \left(\sum_{j=1}^{3} b_{j2}^{L} - \sum_{j=1}^{3} b_{j2}^{U}\right)\right)\right\}$$

$$= 40 + \max\{0,10-15\} = 40,$$

$$B_{43}^{L} = B_{42}^{L} + \max\left\{0, \left(\left(\sum_{i=1}^{2} a_{i3}^{L} - \sum_{i=1}^{2} a_{i2}^{L}\right) - \left(\sum_{j=1}^{3} b_{j3}^{L} - \sum_{j=1}^{3} b_{j2}^{L}\right)\right)\right\}$$

$$= 40 + \max\{0,10-20\} = 40,$$

$$B_{43}^{U} = B_{43}^{L} + \max\left\{0, \left(\left(\sum_{i=1}^{2} a_{i3}^{U} - \sum_{i=1}^{2} a_{i3}^{L}\right) - \left(\sum_{j=1}^{3} b_{j3}^{U} - \sum_{j=1}^{3} b_{j3}^{L}\right)\right)\right\}$$

$$= 40 + \max\{0,10-15\} = 40,$$

$$B_{44}^{L} = B_{43}^{U} + \max\left\{0, \left(\left(\sum_{i=1}^{2} a_{i4}^{L} - \sum_{i=1}^{2} a_{i3}^{U}\right) - \left(\sum_{j=1}^{3} b_{j4}^{L} - \sum_{j=1}^{3} b_{j3}^{U}\right)\right)\right\}$$

$$= 40 + \max\{0,10-45\} = 40,$$

$$B_{44}^{U} = B_{44}^{L} + \max\left\{0, \left(\left(\sum_{i=1}^{2} a_{i4}^{U} - \sum_{i=1}^{2} a_{i4}^{L}\right) - \left(\sum_{j=1}^{3} b_{j4}^{U} - \sum_{j=1}^{3} b_{j4}^{L}\right)\right)\right\}$$

$$= 40 + \max\{0,10-15\} = 40.$$

Therefore, the dummy fuzzy supply

$$\left\langle \left(A_{(m+1)1}^{L}, A_{(m+1)2}^{L}, A_{(m+1)3}^{L}, A_{(m+1)4}^{L}; \omega^{L}\right), \left(A_{(m+1)1}^{U}, A_{(m+1)2}^{U}, A_{(m+1)3}^{U}, A_{(m+1)4}^{U}; \omega^{U}\right)\right\rangle$$

$$= \left\langle \left(5,10,20,60; \frac{2}{3}\right), (0,5,25,65; 1)\right\rangle \text{ and the dummy fuzzy demand}$$

$$\left\langle \left(B^L_{(n+1)1}, B^L_{(n+1)2}, B^L_{(n+1)3}, BF^L_{(n+1)4}; \omega^L \right), \left(B^U_{(n+1)1}, B^U_{(n+1)2}, B^U_{(n+1)3}, B^U_{(n+1)4}; \omega^U \right) \right\rangle$$

$$= \left\langle \left(25, 40, 40, 40; \frac{2}{3} \right), \left(25, 40, 40, 40; 1 \right) \right\rangle.$$

8   Exact results of the existing generalized IVTrFNTP

To find the solution of the generalized IVTrFNTP, presented in Ebrahimnejad (2016) has solved a crisp linear programming problem having 56 equality constraints. Out of these 56 equality constraints, 8 constraints correspond to dummy supply and 8 constraints correspond to dummy demand.

However, as discussed in Section 3, the dummy supply and dummy demand, obtained by Ebrahimnejad (2016), is not correct. Therefore, the result of this problem, obtained by Ebrahimnejad (2016), is not correct.

To find the exact result of this problem, the equality constraints C_1 of the existing crisp linear programming problem have been replaced by the constraints C_2.

$$\left. \begin{array}{l} x^L_{31,1} + x^L_{32,1} + x^L_{33,1} + x^L_{34,1} = 25, \\ x^L_{31,2} + x^L_{32,2} + x^L_{33,2} + x^L_{34,2} = 25, \\ x^L_{31,3} + x^L_{32,3} + x^L_{33,3} + x^L_{34,3} = 35, \\ x^L_{31,4} + x^L_{32,4} + x^L_{33,4} + x^L_{34,4} = 75, \\ x^U_{31,1} + x^U_{32,1} + x^U_{33,1} + x^U_{34,1} = 0, \\ x^U_{31,2} + x^U_{32,2} + x^U_{33,2} + x^U_{34,2} = 25, \\ x^U_{31,3} + x^U_{32,3} + x^U_{33,3} + x^U_{34,3} = 45, \\ x^U_{31,4} + x^U_{32,4} + x^U_{33,4} + x^U_{34,4} = 85, \\ x^L_{14,1} + x^L_{24,1} + x^L_{34,1} = 45, \\ x^L_{14,2} + x^L_{24,2} + x^L_{34,2} = 55, \\ x^L_{14,3} + x^L_{24,3} + x^L_{34,3} = 55, \\ x^L_{14,4} + x^L_{24,4} + x^L_{34,4} = 55, \\ x^U_{14,1} + x^U_{24,1} + x^U_{34,1} = 25, \\ x^U_{14,2} + x^U_{24,2} + x^U_{34,2} = 60, \\ x^U_{14,3} + x^U_{24,3} + x^U_{34,3} = 60, \\ x^U_{14,4} + x^U_{24,4} + x^U_{34,4} = 60. \end{array} \right\} (C_1)$$

$$
\left.
\begin{aligned}
x^L_{31,1} + x^L_{32,1} + x^L_{33,1} + x^L_{34,1} &= 5, \\
x^L_{31,2} + x^L_{32,2} + x^L_{33,2} + x^L_{34,2} &= 10, \\
x^L_{31,3} + x^L_{32,3} + x^L_{33,3} + x^L_{34,3} &= 20, \\
x^L_{31,4} + x^L_{32,4} + x^L_{33,4} + x^L_{34,4} &= 60, \\
x^U_{31,1} + x^U_{32,1} + x^U_{33,1} + x^U_{34,1} &= 0, \\
x^U_{31,2} + x^U_{32,2} + x^U_{33,2} + x^U_{34,2} &= 5, \\
x^U_{31,3} + x^U_{32,3} + x^U_{33,3} + x^U_{34,3} &= 25, \\
x^U_{31,4} + x^U_{32,4} + x^U_{33,4} + x^U_{34,4} &= 65, \\
x^L_{14,1} + x^L_{24,1} + x^L_{34,1} &= 25, \\
x^L_{14,2} + x^L_{24,2} + x^L_{34,2} &= 40, \\
x^L_{14,3} + x^L_{24,3} + x^L_{34,3} &= 40, \\
x^L_{14,4} + x^L_{24,4} + x^L_{34,4} &= 40, \\
x^U_{14,1} + x^U_{24,1} + x^U_{34,1} &= 25, \\
x^U_{14,2} + x^U_{24,2} + x^U_{34,2} &= 40, \\
x^U_{14,3} + x^U_{24,3} + x^U_{34,3} &= 40, \\
x^U_{14,4} + x^U_{24,4} + x^U_{34,4} &= 40.
\end{aligned}
\right\} (C_2)
$$

On solving the existing crisp linear programming problem (Ebrahimnejad, 2016) with this modification, the obtained exact optimal solution and the optimal cost of the existing generalized IVTrFNTP, presented by Table 1, is

$$
\tilde{x}_{11} = \left\langle \left(30, 40, 40, 50; \frac{2}{3}\right), (25, 35, 45, 55; 1) \right\rangle,
$$

$$
\tilde{x}_{12} = \left\langle \left(15, 20, 20, 20; \frac{2}{3}\right), (15, 20, 20, 20; 1) \right\rangle,
$$

$$
\tilde{x}_{13} = \left\langle \left(0, 0, 0, 0; \frac{2}{3}\right), (0, 0, 0, 0; 1) \right\rangle,
$$

$$
\tilde{x}_{14} = \left\langle \left(25, 30, 30, 30; \frac{2}{3}\right), (25, 30, 30, 30; 1) \right\rangle
$$

$$
\tilde{x}_{21} = \left\langle \left(0, 0, 10, 10; \frac{2}{3}\right), (0, 0, 10, 10; 1) \right\rangle, \tilde{x}_{22} = \left\langle \left(0, 0, 0, 0; \frac{2}{3}\right), (0, 0, 0, 0; 1) \right\rangle,
$$

$$
\tilde{x}_{23} = \left\langle \left(40, 50, 50, 50; \frac{2}{3}\right), (35, 45, 55, 65; 1) \right\rangle, \tilde{x}_{24} = \left\langle \left(0, 10, 10, 10; \frac{2}{3}\right), (0, 10, 10, 10; 1) \right\rangle,
$$

$$
\tilde{x}_{31} = \left\langle \left(0, 0, 0, 10; \frac{2}{3}\right), (0, 0, 0, 10; 1) \right\rangle, \tilde{x}_{32} = \left\langle \left(5, 10, 20, 30; \frac{2}{3}\right), (0, 5, 25, 35; 1) \right\rangle,
$$

$$\tilde{\tilde{x}}_{33} = \left\langle \left(0,0,0,20; \frac{2}{3}\right), (0,0,0,20; 1) \right\rangle, \ \tilde{\tilde{x}}_{34} = \left\langle \left(0,0,0,0; \frac{2}{3}\right), (0,0,0,0; 1) \right\rangle.$$

9 Conclusion

In this chapter it has been shown that the existing method for transforming an unbalanced generalized IVTrFNTP into a balanced generalized IVTrFNTP is not valid. In addition, a new method (known as the Mehar method) was proposed for the same purpose and it was proved that this method is valid. Furthermore, the exact result of the existing unbalanced generalized IVTrFNTP was obtained.

References

Chanas, S., Kuchta, D., 1996. A concept of the optimal solution of the transportation problem with fuzzy cost coefficients. Fuzzy Sets Syst. 82 (2), 299–305.

Chanas, S., Kolodziejczyk, W., Machaj, A., 1984. A fuzzy approach to the transportation problem. Fuzzy Sets Syst. 13 (3), 211–221.

Chanas, S., Delgado, M., Verdegay, J.L., Vila, M.A., 1993. Interval and fuzzy extensions of classical transportation problems. Transp. Plan. Technol. 17 (2), 203–218.

Chiang, J., 2005. The optimal solution of the transportation problem with fuzzy demand and fuzzy product. J. Inf. Sci. Eng. 21, 439–451.

Dinagar, D.S., Palanivel, K., 2009. The transportation problem in fuzzy environment. Int. J. Algorithms Comput. Math. 2 (3), 65–71.

Ebrahimnejad, A., 2014. A simplified new approach for solving fuzzy transportation problems with generalized trapezoidal fuzzy numbers. Appl. Soft Comput. 19, 171–176.

Ebrahimnejad, A., 2015a. A duality approach for solving bounded linear programming problems with fuzzy variables based on ranking functions and its application in bounded transportation problems. Int. J. Syst. Sci. 46 (11), 2048–2060.

Ebrahimnejad, A., 2015b. Note on a fuzzy approach to transport optimization problem. Optim. Eng. https://doi.org/10.1007/s11081-015-9277-y.

Ebrahimnejad, A., 2015c. An improved approach for solving transportation problem with triangular fuzzy numbers. J. Intell. Fuzzy Syst. 29 (2), 963–974.

Ebrahimnejad, A., 2016. Fuzzy linear programming approach for solving transportation problems with interval-valued trapezoidal fuzzy numbers. Sadhana 41 (3), 299–316.

Gupta, A., Kumar, A., 2012. A new method for solving linear multi-objective transportation problems with fuzzy parameters. Appl. Math. Model. 36, 1421–1430.

Gupta, A., Kumar, A., Kaur, A., 2012. Mehar's method to find exact fuzzy optimal solution of unbalanced fully fuzzy multiobjective transportation problems. Optim. Lett. 6, 1737–1751.

Jimenez, F., Verdegay, J.L., 1998. Uncertain solid transportation problem. Fuzzy Sets Syst. 100 (13), 45–57.

Jimenez, F., Verdegay, J.L., 1999. Solving fuzzy solid transportation problems by an evolutionary algorithm based parametric approach. Eur. J. Oper. Res. 117 (3), 485–510.

Kaur, A., Kumar, A., 2012. A new approach for solving fuzzy transportation problems using generalized trapezoidal fuzzy numbers. Appl. Soft Comput. 12 (3), 1201–1213.

Kumar, A., Kaur, A., 2010. Application of linear programming for solving fuzzy transportation problems. J. Appl. Math. Inform. 29 (3–4), 831–846.

Kumar, A., Kaur, A., 2011a. A new method for solving fuzzy transportation problems using ranking function. Appl. Math. Model. 35 (12), 5652–5661.

Kumar, A., Kaur, A., 2011b. Application of classical transportation methods to find the fuzzy optimal solution of fuzzy transportation problems. Fuzzy Inf. Eng. 3 (1), 81–99.

Kumar, A., Kaur, A., 2014. Optimal way of selecting cities and conveyances for supplying coal in uncertain environment. Sadhana. https://doi.org/10.1007/s12046-013-0207-4.

Liu, S.T., Kao, C., 2004. Solving fuzzy transportation problems based on extension principle. Eur. J. Oper. Res. 153 (3), 661–674.

Oheigeartaigh, M., 1982. A fuzzy transportation algorithm. Fuzzy Sets Syst. 8 (3), 235–243.

Pandian, P., Natarajan, G., 2010. A new algorithm for finding a fuzzy optimal solution for fuzzy transportation problems. Appl. Math. Sci. 4 (2), 79–90.

Shanmugasundari, M., Ganesan, K., 2013. A novel approach for the fuzzy optimal solution of fuzzy transportation problem. Int. J. Eng. Res. Appl. 3 (1), 1416–1421.

Sudhagar, S., Ganesan, K., 2012. A fuzzy approach to transport optimization problem. Optim. Eng. https://doi.org/10.1007/s11081-012-9202-6.

Further reading

Kumar, A., Kaur, A., 2012. Methods for solving unbalanced fuzzy transportation problems. Oper. Res. 12 (3), 287–316.

Index

Note: Page numbers followed by *f* indicate figures, *t* indicate tables, and *np* indicate footnotes.

Printed in the United States
By Bookmasters